Screw Theory in Robotics

Screw Theory in Robotics

An Illustrated and Practicable Introduction to Modern Mechanics

Jose M. Pardos-Gotor

CRC Press
Taylor & Francis Group
Boca Raton London New York

CRC Press is an imprint of the
Taylor & Francis Group, an **informa** business

MATLAB® is a trademark of The MathWorks, Inc. and is used with permission. The MathWorks does not warrant the accuracy of the text or exercises in this book. This book's use or discussion of MATLAB® software or related products does not constitute endorsement or sponsorship by The MathWorks of a particular pedagogical approach or particular use of the MATLAB® software.

First edition published 2022
by CRC Press
6000 Broken Sound Parkway NW, Suite 300, Boca Raton, FL 33487-2742

and by CRC Press
2 Park Square, Milton Park, Abingdon, Oxon, OX14 4RN

© 2022 Jose M. Pardos-Gotor

CRC Press is an imprint of Taylor & Francis Group, LLC

Reasonable efforts have been made to publish reliable data and information, but the author and publisher cannot assume responsibility for the validity of all materials or the consequences of their use. The authors and publishers have attempted to trace the copyright holders of all material reproduced in this publication and apologize to copyright holders if permission to publish in this form has not been obtained. If any copyright material has not been acknowledged please write and let us know so we may rectify in any future reprint.

Except as permitted under U.S. Copyright Law, no part of this book may be reprinted, reproduced, transmitted, or utilized in any form by any electronic, mechanical, or other means, now known or hereafter invented, including photocopying, microfilming, and recording, or in any information storage or retrieval system, without written permission from the publishers.

For permission to photocopy or use material electronically from this work, access www.copyright.com or contact the Copyright Clearance Center, Inc. (CCC), 222 Rosewood Drive, Danvers, MA 01923, 978-750-8400. For works that are not available on CCC please contact mpkbookspermissions@tandf.co.uk

Trademark notice: Product or corporate names may be trademarks or registered trademarks and are used only for identification and explanation without intent to infringe.

ISBN: 978-1-032-10736-3 (hbk)
ISBN: 978-1-032-10747-9 (pbk)
ISBN: 978-1-003-21685-8 (ebk)

DOI: 10.1201/9781003216858

Typeset in Times
by SPi Technologies India Pvt Ltd (Straive)

Access the Support Material: https://github.com/DrPardosGotor/Screw-Theory-in-Robotics

To Aitana and Guillermo for making me much better.

With all my heart, I want to dedicate this book to Nuria,
*the firm rock pillar of my internally
to whom I owe the one and only delight
the bright love enlightening eternally.*

Contents

Preface xvii
Acknowledgments xix
List of Abbreviations xxi
Author xxiii
Introduction xxv

Chapter 1 Introduction 1

    1.1 Motivation 1
        1.1.1 A Historical Quest! 1
        1.1.2 A Hundred Years of Menacing Robots! 1
        1.1.3 A Century of Helping Robots! 2
        1.1.4 And Only 50 Years of Commercial Robots! 2
        1.1.5 The Mathematical Complexity of Robotics 3
        1.1.6 Here Comes Screw Theory in Robotics 3
        1.1.7 The Future of Robotics 4
    1.2 About This Book 6
    1.3 Preview 9
        1.3.1 Outline 9
        1.3.2 Chapter 1: Introduction 9
        1.3.3 Chapter 2: Mathematical Tools 10
        1.3.4 Chapter 3: Forward Kinematics 10
        1.3.5 Chapter 4: Inverse Kinematics 11
        1.3.6 Chapter 5: Differential Kinematics 13
        1.3.7 Chapter 6: Inverse Dynamics 13
        1.3.8 Chapter 7: Trajectory Generation 14
        1.3.9 Chapter 8: Robotics Simulation 14
        1.3.10 Chapter 9: Conclusions 15
    1.4 Audience 15
    1.5 Further Reading 16
    Note 16

Chapter 2 Mathematical Tools 17

    2.1 Rigid Body Motion 17
    2.2 Homogeneous Representation 18
        2.2.1 Standard Rigid Body Motion 18
        2.2.2 Homogeneous Basic Transformations 18
        2.2.3 Motion Composition in the SPATIAL "S" Reference System 19

vii

		2.2.4	Motion Composition with STATIONARY and MOBILE Coordinate Systems 20
		2.2.5	Geometrical Interpretation 21
		2.2.6	Exercise: Homogeneous Rotation 22
		2.2.7	Exercise: Homogeneous Rotation Plus Translation 23
	2.3	Exponential Representation 24
		2.3.1	Modern Rigid Body Motion...................................... 24
		2.3.2	Screw Rotation (Orientation)................................... 24
		2.3.3	Rigid Body Motion TWIST 25
		2.3.4	Rigid Body Force WRENCH 26
		2.3.5	Exponential Coordinates for a SCREW Motion 26
		2.3.6	Exercise: Exponential Rotation................................ 29
		2.3.7	Exercise: Exponential Rotation Plus Translation..... 30
	2.4	Summary 31
	Notes 32

Chapter 3 Forward Kinematics.. 33

	3.1	Problem Statement in Robotics... 33
		3.1.1	Kinematics Concept 33
		3.1.2	Kinematics Mathematical Approach........................ 33
		3.1.3	Forward Kinematics (FK) 33
	3.2	Denavit–Hartenberg Convention (DH) 33
		3.2.1	Kinematics Treatment 33
		3.2.2	DH FK Homogeneous Matrix Product 34
		3.2.3	Puma Robots (e.g., ABB IRB120)............................ 35
	3.3	Product of Exponentials Formulation 36
		3.3.1	A New Kinematics Treatment.................................. 36
		3.3.2	General Solution to Forward Kinematics................. 37
		3.3.3	Puma Robots (e.g., ABB IRB120)............................ 38
		3.3.4	Puma Robots (e.g., ABB IRB120) "Tool-Up" 40
		3.3.5	Bending Backwards Robots (e.g., ABB IRB1600) 41
		3.3.6	Gantry Robots (e.g., ABB IRB6620LX)................. 42
		3.3.7	Scara Robots (e.g., ABB IRB910SC) 44
		3.3.8	Collaborative Robots (e.g., UNIVERSAL UR16e)... 45
		3.3.9	Redundant Robots (e.g., KUKA IIWA)................... 46
		3.3.10	Many DoF Robots (e.g., RH0 UC3M Humanoid) 47
	3.4	Summary 48
	Notes 49

Contents

Chapter 4   Inverse Kinematics..51
    4.1   Problem Statement in Robotics and Analytical Difficulty..51
        4.1.1   Kinematics Concept..51
        4.1.2   Inverse Kinematics Mathematical Approach..........51
        4.1.3   Analytical Difficulty to Solve Inverse Kinematics..52
    4.2   Numeric vs. Geometric Solutions..53
        4.2.1   A Numeric Approach to Solve Inverse Kinematics..53
        4.2.2   An Example of a Numeric Algorithm....................54
        4.2.3   A Geometric Approach to Solve Inverse Kinematics..55
        4.2.4   An Example of a Geometric Algorithm.................55
        4.2.5   Puma Robot Inverse Kinematics Algorithms..........56
    4.3   Canonical Subproblems for Inverse Kinematics...................57
        4.3.1   A Key Idea to Solve Inverse Kinematics................57
        4.3.2   Paden–Kahan Subproblem One (PK1) – One Rotation..59
            4.3.2.1   ROTATION around ONE Single AXIS Applied to a POINT....................................59
            4.3.2.2   PK1 Subproblem Simplification................59
        4.3.3   Paden–Kahan Subproblem Two (PK2) – Two Crossing Rotations...60
            4.3.3.1   ROTATION around TWO Subsequent CROSSING AXES Applied to a POINT...60
        4.3.4   Paden–Kahan Subproblem Three (PK3) – Rotation to a Distance...61
            4.3.4.1   ROTATION at a Given DISTANCE Applied to a POINT....................................61
            4.3.4.2   PK3 Subproblem Simplification................63
        4.3.5   Pardos–Gotor Subproblem One (PG1) – One Translation...63
            4.3.5.1   TRANSLATION along a SINGLE AXIS Applied to a POINT..........................63
            4.3.5.2   PG1 Extension - TRANSLATION along a SINGLE AXIS Applied to a PLANE..64
        4.3.6   Pardos-Gotor Subproblem Two (PG2) – Two Crossing Translations...65
            4.3.6.1   TRANSLATION along Two Subsequent CROSSING AXES Applied to a POINT....................................65
        4.3.7   Pardos-Gotor Subproblem Three (PG3) – Translation to a Distance.......................................66

		4.3.7.1	TRANSLATION to a Given DISTANCE Applied to a POINT 66
	4.3.8	\multicolumn{2}{l}{Pardos-Gotor Subproblem Four (PG4) – Two Parallel Rotations .... 67}
		4.3.8.1	ROTATION around TWO Subsequent PARALLEL AXES Applied to a POINT ... 67
		4.3.8.2	PG4 Extension - ROTATION around TWO PARALLEL AXES Applied to a LINE 68
	4.3.9	\multicolumn{2}{l}{Pardos-Gotor Subproblem Five (PG5) – Rotation of a Line or Plane .... 69}
		4.3.9.1	ROTATION around ONE Single AXIS Applied to a Perpendicular LINE or PLANE 69
	4.3.10	\multicolumn{2}{l}{Pardos-Gotor Subproblem Six (PG6) – Two Skewed Rotations .... 70}
		4.3.10.1	ROTATION around TWO Subsequent SKEW AXES Applied to a POINT 70
	4.3.11	\multicolumn{2}{l}{Pardos-Gotor Subproblem Seven (PG7) – Three Rotations to a Point .... 74}
		4.3.11.1	ROTATION around THREE Subsequent AXES (ONE SKEW + TWO PARALLEL) Applied to a POINT 74
	4.3.12	\multicolumn{2}{l}{Pardos-Gotor Subproblem Eight (PG8) – Three Rotations to A Pose .... 77}
		4.3.12.1	ROTATION around THREE Subsequent PARALLEL AXES Applied to a POSE (Position Plus Orientation) or COORDINATE SYSTEM 77
4.4	\multicolumn{3}{l}{Product of Exponentials Approach .... 79}
	4.4.1	\multicolumn{2}{l}{General Solution to Inverse Kinematics .... 80}
	4.4.2	\multicolumn{2}{l}{Puma Robots (e.g., ABB IRB120) .... 82}
		4.4.2.1	Inverse Kinematics Puma Robot ABB IRB120 Problem Definition 83
		4.4.2.2	First Algorithm for ABB IRB120 IK "PK3+PK2+PK2+PK1" 84
		4.4.2.3	Second Algorithm for ABB IRB120 IK "PG7+PK2+PK1" 86
		4.4.2.4	Third Algorithm for ABB IRB120 IK "PG5+PG4+PK2+PK1" 87
		4.4.2.5	Fourth Algorithm for ABB IRB120 IK "PG5+PG4+PG6+PK1" 88
		4.4.2.6	Comparison between the Four Algorithms for ABB IRB120 IK 88

	4.4.2.7	Comment on the Implementation of the Algorithms for ABB IRB120 IK 89
	4.4.2.8	Performance Contrast for Both Numeric and Geometric ABB IRB120 IK Algorithms 89
	4.4.2.9	RST - Robotics System Toolbox™ 90
	4.4.2.10	ST24R - Screw Theory Toolbox for Robotics 90
4.4.3	Puma Robots (e.g., ABB IRB120) "Tool-Up." 91	
	4.4.3.1	Inverse Kinematics PUMA ABB IRB120 "Tool-Up" Problem Definition 91
	4.4.3.2	First Algorithm for ABB IRB120 "Tool-Up" IK "PG7+PG6+PK1" 92
4.4.4	Bending Backwards Robots (e.g., ABB IRB1600) ... 94	
	4.4.4.1	Inverse Kinematics ABB IRB1600 Problem Definition 94
	4.4.4.2	First Algorithm for ABB IRB1600 IK "PG7+PG6+PK1" 95
4.4.5	Gantry Robots (e.g., ABB IRB6620LX) 96	
	4.4.5.1	Inverse Kinematics ABB IRB6620LX Problem Definition 96
	4.4.5.2	First Algorithm for ABB IRB6620LX IK "PG1+PG4+PG6+PK1" 97
4.4.6	Scara Robots (e.g., ABB IRB910SC) 99	
	4.4.6.1	Inverse Kinematics ABB IRB910SC Problem Definition 99
	4.4.6.2	First Algorithm for ABB IRB910SC IK "PG1+PG4+PK1" 100
	4.4.6.3	Second Algorithm for ABB IRB910SC IK "PG1+PK3+PK1+PK1" 101
	4.4.6.4	Comments on the SCARA Robot (ABB IRB910SC) IK Implementation 102
4.4.7	Collaborative Robots (e.g., UNIVERSAL UR16e) ... 102	
	4.4.7.1	Inverse Kinematics UNIVERSAL UR16e Problem Definition 103
	4.4.7.2	First Algorithm for UNIVERSAL UR16e IK "PG5+PG3+PK1+PG8" 104
	4.4.7.3	Comments on the UNIVERSAL UR16e IK Complete Solution Implementation 106
4.4.8	Redundant Robots (e.g., KUKA IIWA) 107	
	4.4.8.1	Inverse Kinematics KUKA IIWA Problem Definition 108
	4.4.8.2	First Algorithm for KUKA IIWA IK "PK1+PK3+PK2+PK2+PK2+PK1" 109
	4.4.8.3	Comments on the KUKA IIWA IK Complete Solution Implementation 112

		4.4.9	Many DoF Robots (e.g., RH0 UC3M Humanoid) 113
	4.5	Summary 114	
	Notes 116		

Chapter 5 Differential Kinematics 119

5.1 Problem Statement in Robotics 119
5.2 The Analytic Jacobian 120
    5.2.1 A Traditional Description 120
    5.2.2 Analytic Jacobian to Forward Differential Kinematics 121
    5.2.3 Analytic Jacobian for Inverse Differential Kinematics 121
    5.2.4 Scara Robot (e.g., ABB IRB910SC) 122
        5.2.4.1 Forward Differential Kinematics with Analytic Jacobian 122
        5.2.4.2 Inverse Differential Kinematics with Analytic Jacobian 123
    5.2.5 Puma Robot (e.g., ABB IRB120) 126
5.3 The Geometric Jacobian 127
    5.3.1 Robot Spatial Geometric Jacobian 128
    5.3.2 The Classical Adjoint Transformation (Ad) 128
    5.3.3 Twist Velocity Concept 129
    5.3.4 Trajectory Generation 129
    5.3.5 Robot Tool Geometric Jacobian 130
    5.3.6 Link Spatial and Link Tool Geometric Jacobian 130
    5.3.7 The New Adjoint Transformation (A_{ij}) 131
    5.3.8 General Solution to Differential Kinematics 131
        5.3.8.1 The Kinematics Mapping 132
        5.3.8.2 The Geometric Forward Differential Kinematics 132
        5.3.8.3 The Geometric Inverse Differential Kinematics 133
    5.3.9 Puma Robots (e.g., ABB IRB120) 135
        5.3.9.1 Geometric Jacobian by Definition 135
        5.3.9.2 Forward Differential Kinematics with Geometric Jacobian 136
        5.3.9.3 Inverse Differential Kinematics with Geometric Jacobian 137
    5.3.10 Puma Robots (e.g., ABB IRB120) "Tool-Up" 138
    5.3.11 Bending Backwards Robots (e.g., ABB IRB1600) 139
    5.3.12 Gantry Robots (e.g., ABB IRB6620LX) 140
    5.3.13 Scara Robots (e.g., ABB IRB910SC) 141
        5.3.13.1 Geometric Jacobian by Inspection 141

Contents   xiii

      5.3.13.2 Geometric Jacobian by Definition..........142
      5.3.13.3 Forward Differential Kinematics with
          Geometric Jacobian..............................143
      5.3.13.4 Inverse Differential Kinematics with
          Geometric Jacobian..............................144
   5.3.14 Collaborative Robots
      (e.g., UNIVERSAL UR16e)...............................146
   5.3.15 Redundant Robots (e.g., KUKA IIWA)................146
 5.4 Summary..148
 Notes..149

Chapter 6 Inverse Dynamics..151

 6.1 Problem Statement in Robotics......................................151
 6.2 The Lagrange Characterization.......................................152
   6.2.1 General Non-Recursive Solution to Inverse
      Dynamics..158
   6.2.2 Puma Robots (e.g., ABB IRB120)....................161
   6.2.3 Puma Robots (e.g., ABB IRB120) "Tool-Up".......165
   6.2.4 Bending Backwards Robots
      (e.g., ABB IRB1600)...166
   6.2.5 Gantry Robots (e.g., ABB IRB6620LX)..............166
   6.2.6 Scara Robots (e.g., ABB IRB910SC)..................168
   6.2.7 Collaborative Robots (e.g., UNIVERSAL
      UR16e)..168
   6.2.8 Redundant Robots (e.g., KUKA IIWA)..............168
 6.3 Robot Dynamics Control..170
   6.3.1 Robotics Control in the Joint Space....................170
   6.3.2 Robotics Control in the Task Space.....................171
 6.4 Spatial Vector Algebra..172
   6.4.1 Coordinate Transforms......................................173
   6.4.2 Mechanics of a Constrained Rigid Body
      System...174
 6.5 The Newton–Euler Equations...174
   6.5.1 General Recursive Solution to Inverse
      Dynamics RNEA with POE..............................174
   6.5.2 Puma Robots (e.g., ABB IRB120)....................180
   6.5.3 Puma Robots (e.g., ABB IRB120) "Tool-Up".......185
   6.5.4 Bending Backwards Robots
      (e.g., ABB IRB1600)...186
   6.5.5 Gantry Robots (e.g., ABB IRB6620LX)..............187
   6.5.6 Scara Robots (e.g., ABB IRB910SC)..................188
   6.5.7 Collaborative Robots (e.g., UNIVERSAL UR16e)...188
   6.5.8 Redundant Robots (e.g., KUKA IIWA)..............189
 6.6 Summary..190
 Notes..191

Chapter 7   Trajectory Generation 193

- 7.1 Concepts and Definitions 193
  - 7.1.1 Point-to-Point Position Straight-line Trajectories ... 194
  - 7.1.2 Trapezoidal Position Trajectory 195
  - 7.1.3 Polynomial Position Trajectory 195
  - 7.1.4 Spline Position Trajectory 196
  - 7.1.5 Rotation Motion Trajectory 197
  - 7.1.6 Trajectory Tracking and Control 197
- 7.2 Trajectory Planning 197
  - 7.2.1 General Solution to Trajectory Generation 198
  - 7.2.2 Puma Robots (e.g., ABB IRB120) 199
  - 7.2.3 Puma Robots (e.g., ABB IRB120) "Tool-Up" 203
  - 7.2.4 Bending Backwards Robots (e.g., ABB IRB1600) ... 204
  - 7.2.5 Gantry Robots (e.g., ABB IRB6620LX) 207
  - 7.2.6 Scara Robots (e.g., ABB IRB910SC) 208
  - 7.2.7 Collaborative Robots (e.g., UNIVERSAL UR16e) ... 210
  - 7.2.8 Redundant Robots (e.g., KUKA IIWA) 211
- 7.3 Summary 212
- Notes 213

Chapter 8   Robotics Simulation 215

- 8.1 Robotics Simulation 215
  - 8.1.1 Why Code in MATLAB®? 216
- 8.2 Screw Theory Toolbox for Robotics (ST24R) 218
- 8.3 Forward Kinematics Simulations 218
  - 8.3.1 General Solution to Forward Kinematics Simulation 219
  - 8.3.2 Puma Robots (e.g., ABB IRB120) 222
  - 8.3.3 Puma Robots (e.g., ABB IRB120) "Tool-Up" 222
  - 8.3.4 Bending Backwards Robots (e.g., ABB IRB1600) ... 223
  - 8.3.5 Gantry Robots (e.g., ABB IRB6620LX) 223
  - 8.3.6 Scara Robots (e.g., ABB IRB910SC) 224
  - 8.3.7 Collaborative Robots (e.g., UNIVERSAL UR16e) ... 224
  - 8.3.8 Redundant Robots (e.g., KUKA IIWA) 225
- 8.4 Inverse Kinematics Simulations 225
  - 8.4.1 General Solution to Inverse Kinematics Simulation 226
  - 8.4.2 Puma Robots (e.g., ABB IRB120) 228
  - 8.4.3 Puma Robots (e.g., ABB IRB120) "Tool-Up" 230
  - 8.4.4 Bending Backwards Robots (e.g., ABB IRB1600) ... 231
  - 8.4.5 Gantry Robots (e.g., ABB IRB6620LX) 232
  - 8.4.6 Scara Robots (e.g., ABB IRB910SC) 233
  - 8.4.7 Collaborative Robots (e.g., UNIVERSAL UR16e) ... 234
  - 8.4.8 Redundant Robots (e.g., KUKA IIWA) 235

	8.5	Differential Kinematics Simulations 236
		8.5.1 General Solution to Differential Kinematics Simulation 238
		8.5.2 Puma Robots (e.g., ABB IRB120) 239
		8.5.3 Puma Robots (e.g., ABB IRB120) "Tool-Up" 240
		8.5.4 Bending Backwards Robots (e.g., ABB IRB1600) ... 241
		8.5.5 Gantry Robots (e.g., ABB IRB6620LX) 242
		8.5.6 Scara Robots (e.g., ABB IRB910SC) 243
		8.5.7 Collaborative Robots (e.g., UNIVERSAL UR16e) ... 245
		8.5.8 Redundant Robots (e.g., KUKA IIWA) 246
	8.6	Inverse Dynamics Simulations 247
		8.6.1 General Solution to ID Simulation 249
		8.6.2 Puma Robots (e.g., ABB IRB120) 250
		8.6.3 Puma Robots (e.g., ABB IRB120) "Tool-Up" 251
		8.6.4 Bending Backwards Robots (e.g., ABB IRB1600) ... 252
		8.6.5 Gantry Robots (e.g., ABB IRB6620LX) 253
		8.6.6 Scara Robots (e.g., ABB IRB910SC) 254
		8.6.7 Collaborative Robots (e.g., UNIVERSAL UR16e) ... 255
		8.6.8 Redundant Robots (e.g., KUKA IIWA) 256
	8.7	Summary 257
	Notes	 258

Chapter 9 Conclusions 263

    9.1   Summary 263
          9.1.1   Introduction 265
          9.1.2   Mathematical Tools 266
          9.1.3   Forward Kinematics 266
          9.1.4   Inverse Kinematics 268
          9.1.5   Differential Kinematics 268
          9.1.6   Inverse Dynamics 270
          9.1.7   Trajectory Generation 271
          9.1.8   Robotics Simulation 272
    9.2   Future Prospects 273

Epigram 275
References 277
Index 281

Preface

In the last three decades, there has been a phenomenal growth of activity in robotics, both in terms of research and in capturing the general public's illusion of endless futuristic possibilities.

Growth in robotics coincides with an increasing number of courses at most major research universities on various aspects of the subject. These courses are available in many departments, such as Systems Engineering and Automation, Computer Science, Mathematics, or Electrical and Mechanical Engineering. Several excellent textbooks support this education, covering different topics in kinematics, dynamics, control, sensing, vision, AI, planning, and navigation for robotic mechanisms. Given the state of knowledge on the robot subject and the vast diversity of students interested in the field, there is an inherent utility for a book that presents the more abstract mathematical formulation given by the screw theory in robotics. This work attempts to provide a visual approach of this formulation to an audience who perhaps did not have enough examples of building up the necessary knowledge.

Although the advantages of screw theory are obvious, few teach it in engineering courses, so there are still proportionally few postgraduates who know how to apply it. Some numerous projects and publications demonstrate the crucial advances achieved when applying the geometric formalisms of Lie's algebra in various robotics specialties. The screw theory has proven superior to other mathematical techniques in robotics. Therefore, it is essential to communicate and disseminate these methodologies among as many students as possible who might be working with robots in the future.

This book will emphasize geometric techniques and provide a modern visual presentation, so concepts are better understood. We best capture the key physical features of a robot with a geometric description. However, these screw theory tools, inaccessible to many students, require a new language (e.g., screws, twists, wrenches, adjoint transformation, geometric Jacobian, spatial vector algebra). The rules for the manipulation of this language can seem kind of obscure. In fact, at the heart of the screw theory, there is a high-level but straightforward geometric interpretation of mechanics. The alternative is the most standard algebraic alternatives, which unfortunately make us often buried in the calculation details.

When Brockett (1983) showed how to use the Lie group structure to describe mathematically the kinematic chains in rigid bodies' motion, there was a breakthrough in making classical screw theory accessible to modern robotics. In presenting it, we must highlight the powerful Product of Exponentials (POE) as a fundamental tool of the screw theory.

Overall, we can realize that a good theory is the fastest way to obtain a better functional performance, and the only thing to do in exchange is to study this mathematics. The selection of topics for this text targets a modern screw theory approach for robot mechanics. The content is an excellent introduction to the subject (e.g., mathematical tools, kinematics, differential kinematics, dynamics, trajectory generation, simulation) and the best way to appreciate this approach's benefits. Other aspects such as control are briefly introduced and used in simulations but left out to a thorough extent for later works.

This book covers the screw theory fundamentals with a very graphical approach. Its contents are very focused, practical, and visual so that any student can quickly grasp the advantages of the theory and the algorithmic definitions. In the future, we will be able to extend those possibilities to other problems and applications. The main goal of this text is to provide valuable tools for roboticists. We present all necessary basic theoretical notions but with great emphasis on applications and exercises with real industrial manipulators. This choice permits that once we get the concepts for these robot mechanics, it will be possible to expand the approach to a great variety of robotics architectures.

The examples and exercises in this work demonstrate that geometric solutions based on the screw theory are more suitable for real-time robotics applications compared to typical numerical iterative algorithms.

In short, this text aims to help students, engineers, scientists, researchers, and practitioners of robotics who want to develop their advanced projects toward the application of screw theory methodologies. We will facilitate the comprehension and access to the topics addressed through convenient examples presented in a clear graphical structure. This book's design intends to build a functional bridge between the mathematical fundamentals, which are extensive, and the practical technological robotic applications.

Another component of this book is the additional material included to reinforce the knowledge-making operative of the formulations. Here, we grant access to code, examples, and simulations for real manipulators, which help many robotics structure applications, even outside the field of industrial manipulators. The MATLAB® environment, including Simulink® and Simscape™, is also a valuable supplement for robotics simulation, permitting students to explore robots' mechanics interactively.

This book presents a personal visual approach on how to explain the fundamental concepts of screw theory. The text is more a companion to essential and excellent textbooks that thoroughly cover the subject's mathematical foundations. Particularly influential are those written by Corke; Davidson & Hunt; Lynch & Park; Mason; Murray, Li, and Sastry; Selig; Siciliano and Khatib. All these have contributed to modern robotics in an inspirational way.

In reading this book, we hope that many will feel the enthusiasm about robotics' technological and social prospects, with the elegance of the underlying screw theory for developing effective and efficient robotics algorithms, solutions, and applications.

MATLAB® is a registered trademark of The Math Works, Inc.
For product information, please contact:
The Math Works, Inc.
3 Apple Hill Drive
Natick, MA 01760-2098
Tel: 508-647-7000
Fax: 508-647-7001
E-mail: info@mathworks.com
Web: http://www.mathworks.com

Acknowledgments

It is my pleasure to acknowledge the many people who have been the sources of inspiration in writing this book. I am deeply grateful to all of them.

- I want to express my appreciation to Professor Dr. Carlos Balaguer Bernaldo de Quirós for his wise guidance as I explored new paths.
- The colleagues from "RoboticsLab" and professors from the Carlos III University of Madrid for his invaluable support.
- The professors of the ETSII of the Polytechnic University of Madrid for introducing me to the fascinating world of robotics.
- Professors R.M. Murray and J.A. Sethian for his extraordinary works were the inspiration for some of my developments.
- Professors P. Corke, K.M. Lynch, J.M. Selig, and B. Siciliano for their contributions to the Tutorial "Screw Theory for Robotics" of the IROS2018 (2018 IEEE/RSJ International Conference on Intelligent Robots and Systems).
- The professors P. Corke, S. Grazioso, A. Jardón, T. Lee, and P. Wensing for their contributions to the Tutorial "Review on Screw Theory & Geometric Robot Dynamics" of the IROS2020 (2020 IEEE/RSJ International Conference on Intelligent Robots and Systems).
- ENDESA, the company of my life for always giving me support in my professional and personal development.

List of Abbreviations

Some of the specific screw theory nomenclature in this book is listed below. The list is not comprehensive. Some symbols have multiple meanings, and a few meanings have more than one symbol, then their context helps to disambiguate them.

θ, θ_i	generalized magnitude of a screw motion
μ_i^*	Inertia transformed matrix for the i-link
υ	the moment that encodes the action line and the pitch of the Twist
υ_i	velocity of a link
υ_{ST}	Spatial Velocity
υ_{TCP}	velocity of TCP in terms of the inertial frame
ξ, ξ_i	Twist as the infinitesimal screw motion
$\xi^\wedge$	Twist represented as a matrix
τ	the moment that encodes the action line and the pitch of the Wrench
τ_i	joint generalized force
ω	axis of the Twist for a screw motion
ω_{ST}	Angular velocity of the TCP
Γ	generalized forces
Γ_{ijk}	Christoffel symbols
Ψ_i	Inertia Tensor of the i-link
$\mathscr{F}, \mathscr{F}_i$	Wrench as the infinitesimal screw force
$\mathscr{F}^\wedge$	Wrench represented as a matrix
3D	3-dimensional
a	spatial acceleration vectors
a_i	acceleration of a link
Ad	Adjoint transformation
A_{ij}	elements of the adjoint transformation matrix
C	Coriolis matrix
C_{ij}	elements of the Coriolis matrix
Canonical	It refers to fundamental & basic subproblems
CM	Center of Mass
DH	Denavit-Hartenberg
DoF	Degrees of Freedom
f	axis of the Wrench for a screw motion
f_i	force on the link
FK	Forward Kinematics
H	homogeneous transformation matrix
$H_{ST}(\theta)$	kinematics formulation for a manipulator
$H_{ST}(0)$	Tool coordinate frame "H" at the reference pose
I	Identity matrix
IK	Inverse Kinematics
J_a	Analytic Jacobian
J_{ST}	Geometric Jacobian of the Tool

J_{SL}	Geometric Jacobian of the Link
Kv	velocity gain matrix
Kp	position gain matrix
L	Lagrangian
m_i	mass of the i-link
M	manipulator Inertia matrix
M_{ij}	elements of the manipulator Inertia matrix
MATLAB®	MAtrix LABoratory of Mathworks®
n	number of DoF for a mechanism
N	manipulator Potential matrix
N_ξ	gravity Twist matrix
$N_\mathscr{F}$	gravity Wrench matrix
N_{SYM}	gravity Symbolic matrix
p	translation vector
PG	Pardos-Gotor canonical subproblems
PID	Proportional Integral Derivative controller
PK	Paden-Kahan canonical subproblems
POE	Product of Exponentials
R	rotation matrix
RH0	robot humanoid model zero (UC3M)
S	Spatial frame system (usually inertial)
S_i	spatial Motion Subspace for a joint
Screw	general movement of a rigid body
SE(3)	3D Lie, special Euclidean group
se(3)	Lie algebra for SE(3)
Simulink®	Simulation environment of Mathworks®
Simscape™	Physics model environment of Mathworks®
T	Tool frame system (usually at end-effector)
TCP	Tool Centre Point
Tool	The robot end-effector of interest
UC3M	University Carlos III of Madrid
X	Plücker coordinate transformation
[ ]^	skew-symmetric matrix
[,]	Lie bracket operation for two Twists

Author

Dr. Jose M. Pardos-Gotor has an extensive international career in the energy industry. He works for the multinational Enel group and has developed projects in Europe and Latin America in Change Management, Sustainability, Innovation, R&D, Commodity Markets & Trading, Energy Management, and Power Generation.

He received his MSc in industrial engineering from Polytechnic University of Madrid (UPM). He got his PhD in systems engineering from the University Carlos III of Madrid (UC3M), where he has been an associate professor in the Systems Engineering & Automation Department. He researches production systems, industrial automation, and ROBOTICS. He has worked in some national programs on robotics and has published about robot mechanics. Besides, he has organized and lectured several tutorials at international conferences (e.g., IEEE/RSJ IROS) in screw theory in robotics.

Introduction

Academia recognizes the importance of screw theory in robotics but only a few postgraduates know how to exploit it because the engineering study programs do not widely include this subject. However, the geometric methods based on the Lie algebras and some extensions of six-dimensional vectors are superior to other mathematical techniques for robotics.

This book's publication arises because there are not many texts specializing in applied screw theory methodologies to treat robot mechanics through visual examples and exercises with real manipulators. We present new canonical subproblems and algorithms to solve inverse kinematics, geometric formulations for differential kinematics, new definitions for the potential matrix in the Lagrange dynamics expression, and the application of the product of exponentials to the inverse dynamics recursive Newton–Euler algorithm. The companion material includes the Screw Theory Toolbox for Robotics (ST24R), the code for all exercises, the simulators for robotics architectures built with the MATLAB® environment, and simulation videos.

We hope this book will be a source of inspiration and enthusiasm for those who aspire to develop robots beneficial to our society, allowing us to create effective and efficient applications. We genuinely believe this good screw theory is the best foundation to get a great robotics future.

DATA AVAILABILITY STATEMENT

The exercises, examples, and code that support this book's findings are openly available on "Github" at https://github.com/DrPardosGotor/Screw-Theory-in-Robotics.

The videos that support the findings of this book are openly available on "Youtube" at https://www.youtube.com/user/DrPardosGotor/videos.

1 Introduction

"In the beginning was Mechanics."

—Max von Laue

1.1 MOTIVATION

1.1.1 A Historical Quest!

For millennia, the study of mechanical systems that interact with each other has aroused interest in the scientific world. The history of robotics began in the ancient world. Greek mythology already presents metallic humanoids that do the work of servants of the Gods. When the Greeks controlled Egypt, several generations of engineers in Alexandria built viable automata powered by hydraulic or steam systems. In Byzantium, this knowledge continued, and then the ability passed to the Arabs, who made automatic clocks and even humanoids. Through medieval Spain, Europeans acquired this knowledge. They devised the famous brazen talking heads like Roger Bacon's, and tradition keeps that Albertus Magnus constructed a complete android that could perform domestic tasks and speak with people.

As a great inspiration, we want to mention the works of Leonardo da Vinci during the 15th and 16th centuries. In several of Leonardo's codices, we can find studies of human anatomy, which are undoubtedly works of great interest for those interested in bioengineering. Also, we find animal anatomy studies, among which we can see some bird studies (see Figure 1.1) in which there was the mechanical analysis of a wing. We can appreciate the draw is a chain of rigid bodies with three joints. It is an image that directly evokes the constitution of an industrial robotic manipulator. Additionally, Leonardo made extensive and remarkable studies on the structure of mechanisms which are an inspirational reference for robotics research.

In the 17th and 18th centuries, the interest in automation continued, especially in France. There were several mechanical animals and androids with truly complex mechanisms.

Nikola Tesla demonstrated a prototype of a remote-controlled boat by the end of the 19th century. This automaton had a shocking and sophisticated level of self-control to maneuver without human intervention.

1.1.2 A Hundred Years of Menacing Robots!

At the beginning of the 20th century, we mark the contemporary inception of the "Robot" idea within the Karel Capek play "R.U.R. - Rossum's Universal Robots." The concept stems from the verb "robota," which means "to work" in Slavic language. Moreover, the oeuvre title could mean "universal reasoning robots" and is the name of a company that manufactures robots.

DOI: 10.1201/9781003216858-1

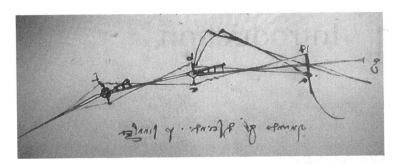

FIGURE 1.1 A bird's wing study from a codex on the flight of birds by Leonardo da Vinci. (Photograph by the author.)

Curiously, the first use of the word robot in the 20th century was to define a machine designed as an artificial human that helps people perform heavy work. Again, the first modern robot concept has the shape of an android. There is a warning message about this technology since the robots opposed society, starting a revolution to destroy humanity.

In 1926, Fritz Lang's classic film "Metropolis" showed a super industrialized society controlled by an android. Once again, a robot becomes a humanoid enemy of people. Furthermore, this foundational concept continues to be maintained throughout the 20th century, as we saw in the series of "Terminator" movies. Even today, we are bombarded with much fake news, insisting on the idea that humanoid robots are dangerous because they are already going to steal our jobs.

1.1.3 A Century of Helping Robots!

From a different perspective, some people saw robotics as a technology with a great potential to help humankind. The collective imaginary also nourished positive stories, such as those of Isaac Asimov. He defined the famous "Three Laws of Robotics." A robot does not harm a human being; a robot obeys a human being except where such orders would conflict with the first law; a robot protects its existence as long as it does not conflict with the first or second laws. This robotics philosophy is still present in many works of literature and film.

As excellent examples, there is also a suitable positive image for some robots fixed in the collective imagination of society worldwide, such as the robots in the movie "Wall-e" or the saga "Star Wars." Additionally, some real robots in medical technology create a world where robotics plays a fantastic service to society.

1.1.4 And Only 50 Years of Commercial Robots!

Let us do a historical review of what were commercial robotics applications. It is a journey of approximately half a century, in which robots with practical application have been developed especially for the industrial world. Furthermore, this is where there are the most critical advances in engineering, science, and technology.

We can recall some milestones of contemporary robotics, such as the first mechanical telemanipulation at the Argonne National Lab (1948), the first robot used in the automotive industry, Unimation (1960), the original programming language for robots developed at Stanford, called Wave (1973) or the first of a kind all-electric robotics manipulator from ASEA (1973).

These first industrial manipulators were machines not flexible enough and dangerous for human work environments. Thus, in recent years, the trend continues to develop collaborative robots (cobots), whose design paradigm is inherently safe and allows interaction with humans. UNIVERSAL UR series and KUKA IIWA are good examples of the current cobots.

Depending on the applications, robots today have a great variety of morphologies (e.g., stationary, wheeled, legged, flying, swimming, modular, hybrid, soft). In any case, we must take advantage of everything learned from industrial manipulators, especially in terms of mechanics, to extend this knowledge to other models and robotic structures. We must work so that when the future arrives, people perceive robots as a beneficial system. They must be excellent products and versatile machines that serve competently and with productivity to complement humans' capabilities.

1.1.5 THE MATHEMATICAL COMPLEXITY OF ROBOTICS

Anyone who has experience working with the robot's motion equations has realized the enormous complexity involved in solving them. Apparently, according to classical mechanics, getting the motion equations of a chain of coupled rigid bodies may seem relatively straightforward. The only need is to have a reference coordinate system and apply the equations Newton–Euler's or Lagrange's to obtain the corresponding differential equations. The Lagrange approach provides an exciting high-level formulation to represent important parameters and give closed-form implementations, while the Newton–Euler formulation is used more for recursive algorithms.

When the system is humanoid, the mechanical analysis's complexity grows even more. The first reason is the enormous number of Degrees of Freedom (DoF). The second cause is because stable biped locomotion problems appear. Besides, we need to solve complex collision-free trajectory planning problems.

It is crucial to have simpler motion equations formulations when the robots are increasingly complex. It is desirable to have an explicit representation of these equations, which we can manipulate at a high level. The parameters of the mechanical system's kinematics and dynamics must be capable of being represented transparently. Furthermore, the algorithms used must be independent of the chosen coordinate systems (i.e., algorithms not linked to any local reference system) to be flexible in analyzing kinematics and dynamics.

1.1.6 HERE COMES SCREW THEORY IN ROBOTICS

Many researchers use differential geometry, Lie algebras, and screw theory mathematical tools to study sets of linked rigid bodies.

The very cornerstone of the modern revival of these theories came with connecting the Lie groups theory to robot kinematics by introducing the Product of Exponentials (POE) (Brockett, 1983). There were excellent introductions to Lie groups theory and the special Euclidean group SE(3) and its algebra se(3) (Murray et al., 2017). They showed the geometric meaning of these theories related to the Theory of Screws (Ball, 1900). The new developments introduced in this book follow that way to a great extent. Therefore, we use the same terminology for the formulations of this text.

Some researchers studied the Lie theory applied to the properties of manipulators (Paden & Sastry, 1988). Many used the mathematics of Lie to formulate robot dynamics (Park et al., 1995). Some derived the motion equations for open chains of rigid bodies, using screw theory and the Lagrange equations (Brockett et al., 1993). Some presented iterative versions of the Newton–Euler and Lagrange formulations to solve robot dynamics with Lie theory (Selig, 2005). There was a valuable functional space formulation for applications on force/position control of the robot tool (Khatib, 1987). There were simulations of robotic mechanisms with dynamics algorithms (Lilly & Orin, 1994). Some researchers develop iterative versions for the motion equations of linked rigid bodies, resulting in algorithms independent of the coordinate system (Ploen, 1997). There were also geometric versions for the solution of those equations (Featherstone, 2016). Moreover, we must highlight the pioneering developments of Featherstone on robot dynamics by the introduction of the Spatial Vector.

For formulations of the dynamics of a chain of rigid bodies, we have several alternatives. On the one hand, there are geometric formulations based on the Lagrange equations, and on the other, the recursive developments of the Newton–Euler equations. The two are equivalent in terms of effectiveness, but the second is more efficient in computational cost. All avant-garde techniques employ the representation of the screw theory with six-dimensional vectors or some extension of this concept.

Robots need a system that allows solving the problem of Inverse Dynamics (ID) and control of this chain of rigid bodies, even in the presence of disturbances and errors. We use two control paradigms for robotics, control in the workspace and control in the joint space. They benefit from Lie groups theory to develop control algorithms (Murray et al., 2017). We will see examples of ID robust control for some industrial robots with these methodologies. Of course, there are many more developments, but control is a subject that is mostly beyond the scope of this text, and we hope to address it thoroughly in a possible future edition of this text.

We also have to highlight the works in the "RoboticsLab" at the University Carlos III of Madrid (UC3M). The whole team has carried out exciting works for robots' mechanical analysis. The new screw theory methods and algorithms presented in this book have got tests with several industrial manipulators.

1.1.7 THE FUTURE OF ROBOTICS

Automation improves the quality of production by offering precise repeatability. These tasks would be impossible to achieve by only using humans. However, today's robots do not look or act like the beings portrayed in science fiction movies. Instead, these machines are carrying out basic tasks inside factories to boost productivity. At a minimum, we are still far away from some foreseeable future with robots carrying out more significant tasks.

Engineering and computer scientists are devising ways to make robots wiser, dexterous, and more human-like in cognitive abilities. In warehouses and factories, they are already working alongside humans. They are even starting to perform functions that have typically been the domain of humans. However, no matter which sector they serve, robots are far less advanced than many thought they would be nowadays.

Some say it is advantageous for robots outside of a factory to look more like humans, and the humanoids come in. However, their utility in real life is still to prove. Even it is the case for some awe-inspiring models that can run, jump and flip (e.g., several from Boston Dynamics). Recently, the robot dog "Spot," also from this company, was made available in the real world. The jury regarding the commercial success of these robots is still out, even though it is a promising start like the invention "Digit" from Agility Robotics, designed for the delivery of packages vehicle-to-door. We must wait and see for the possibility of armies of these machines in the years to come.

Ultimately, we wish to have humanoid robots working in commercial activities. Anyhow, we do not have to worry about all the hype of artificial intelligence (AI). No machine is going to chase us up the stairs anytime soon. We do not come anywhere near to what a human can do. Perhaps it will take one or several centuries to see such a level of cleverness in robotics.

One of the problems we have is the inexistence of something as good as human muscle. Besides, developing robots with high-performing intellectual capacity is a challenging dare. If we want to make robots react as people do physically, the challenge is even more formidable. Of course, the human body has soft materials such as muscle and skin. There is a new research area of "soft robotics" for robot bodies inspired by the efficiency of the soft materials found in nature.

Right now, the trend is to work on what some robotics specialists describe as "multiplicity." It is related to the already operational idea that humans and machines work in collaboration. We must ensure that society sees robots as systems that can cooperate with people and not perceive as threats. The world of cobots has already reached the industry, but we must take it even further in terms of skills.

Machine learning and AI are technologies broadly used in robotics. Coupled with AI comes legislative regulation that will inevitably slow down the pace of robotic automation progress.

Advanced automation will change the jobs that people do. Still, it will not be the case where a robot fully replaces a human being because it is too complex a set of mind and body, which we do not yet know enough to emulate well from the technology. However, of course, automation creates opportunities that may mean the replacement of a job. Investing in education is the best way to take advantage of the impact robots will have on our societies without significant adverse effects. But unfortunately, it seems that society is ill prepared not only for what lies ahead but also for what is happening right now. This book then comes to make a humble contribution to the necessary education of the fundamental subject for robotics of advanced mechanics.

Human beings create environments adapted to people's habitability. For that reason, a humanoid robot could be one of the best tools to serve society's needs well. However, the massive industrial production of reliable humanoids is still far from our capacity. One of the main reasons for this is that robotics presents a formidable challenge from a computational perspective. Only for the mechanical aspects, the complexity given by the high number of restrictions and DoF is enormous for these systems.

When dealing with highly complex systems, elegant mathematical formulations are necessary because they allow us to develop solutions with optimal performance.

For this, research in robotics proposes using techniques based on the mathematics of the Lie groups and screw theory. This approach leads to geometric solutions, numerically stable and with an unequivocal mechanical interpretation.

The dream of all of us who work in robotics, in one way or another, is to be able to build an android that does daily tasks in the real world, such as perhaps doing the shopping, cleaning our house, or helping people in any need. However, in this book, we get started with the classical industrial manipulators, working from the very basics and hoping many can extend the acquired knowledge to more sophisticated mechanisms, such as a humanoid robot.

"Man is the measure of all things…"[1]

—**Protagoras (Vth century B.C.)**

1.2 ABOUT THIS BOOK

This book aims to be a helpful tool for those scientists, technologists, engineers, or mathematicians involved in robotics research. It is a discipline with a significant expansion in recent years, not only because of the increase in business activity but also because it has managed to capture the general public's imagination. There is a shared and widespread vision, both by specialists and by the media, which presents the robot as one of the technologies with the most significant potential for growth and with the most remarkable capacity to transform society crucially in a future full of exciting possibilities and almost unlimited services for people.

Today's robotics still has many challenges to solve before it can get closer to that future of science fiction that many foresee. Automatic robot systems are still expensive and inflexible mechanisms. As an example, it is enough to remember that the slow programming required for adapting robots to a new product makes this type of solution not productive enough in many factories. Another general and highly relevant problem in robotics is the lack of progress in object manipulation applications. We all hope that with the importation into the world of robotics of AI tools, together with the creative effort of those who believe in the potential of this technology, truly advanced robots can be developed to provide high-value services to people. Moreover, this service is what matters!

This publication's premise is that in the long run, abstraction saves a lot of time in exchange for an investment of effort and patience to learn somewhat different and less widespread mathematical tools. Critical theories remain as fundamental foundations for many years in scientific developments. This fact happens too with the differential geometry used for the algorithm developments presented here. We are confident that this book's contents can be exciting and inspiring examples for using mathematical resources such as the POE, geometric Jacobian, or Spatial Vector Algebra. They are most certainly helpful in improving the performance of many robotics applications.

Different approaches to robotics problems are worth exploring to find innovative solutions. Furthermore, many times, the truly novel thing comes if we respect and study the classics, as we will see with screw theory, Lie algebras, and its associated tools. It is a precious paradox that a mathematical theory developed in the 19th century can help solve problems that do not seem solvable with more modern engineering ideas of the 20th century. After all this, we hope that the result of reading this text is that geometry still has a lot to offer for a future of more intelligent, flexible, and practical robotics.

AI applications are succeeding justly, both in research areas and in commercial solutions. Their achievements are spectacular, and we must apply them with equal success in robotics. However, we cannot trust all commercial advances in robotic systems to only these technologies. There is nothing that fails more than the success of an overly practical approach. On the contrary, the grand theories have brought tremendous technical advances and the best suitable and truly practical solutions in almost all the arts and sciences. We can rely on the magnificent ideas of geometry developed for decades and even centuries to advance with robotics. They are concepts full of beauty and practical potential.

Curious paradoxes emerge in this book when different approaches are under comparison in solving some exercises. For instance, this happens with screw theory and Denavit–Hartenberg (DH) convention, Geometry and Calculus, or even Engineering and Mathematics. One more is the defense of the importance of mechanics as a fundamental discipline for the future of advanced robotics. We are making a tremendous and necessary effort in AI, but we should not make less effort with the developments of robot mechanics.

We must remember the fascinating clairvoyance of Norbert Wiener when he anticipated the profound relationship between the brain and the body of a human being. We must deeply emulate from the mechanical point of view the person's body before reaching an advanced imitation of a person's intelligence. This relationship between robotics and the unresolved duality of mind–body is something that perhaps has not yet been sufficiently explored and investigated. Moreover, this is a research field of total actuality.

> "Theoretically, if we could build a machine whose mechanical structure duplicated human physiology, then we could have a machine whose intellectual capacities duplicated those of human beings."
>
> —N. Wiener

This book will show how the exciting technological expectations raised by robotics technology are more achievable if we apply elegant mathematical theories, which are more productive. These methods are what we use as a basis for the exercises and examples presented in this text. We hope that this work will be a beneficial inspiration in creating more robust and innovative algorithms that will help bring the future robotics of our dreams.

This text will present the basic and canonical applications of the screw theory in robotics, focusing on explaining robot mechanics in an illustrated way. The intent is not to provide a thorough explanation of all mathematics involved but rather to give the reader a *vade mecum* of screw theory mechanics. The rewarding models and examples can be the cornerstone for further very performing applications.

This work aims to facilitate robotics stakeholders (e.g., students, engineers, scientists, and practitioners) to go in-depth into these methodologies through representative practical graphical instances and exercises. The idea is to present the necessary mathematical background, theory, and examples in an integrated and visual fashion.

Code is at the core of this work, as the programs illuminate the topics and examples discussed. All this additional material is available at an internet repository linked to this book. This site helps to reinforce the concepts and make the formulas and algorithms operational.

This book's scope focuses on a modern approach for robot mechanics based on screw theory and its methods. Other critical aspects for robotics, such as path planning, are left out for later works because they go beyond this text's scope.

The intention to present this book's contents in such a graphical way has been to create an innovative bridge between great books with thorough mathematical foundations and those intended primarily for technicians. This volume somehow lies in the middle, linking grand theories of the screw theory mathematics with real robots' practical applications. The inception of this work has been the belief that there is nothing more valuable than a good theory. This spirit and its fruitful outcomes stand for the field of robot mechanics, with a new style and approach for both theory notions and presentations.

This book will surely encourage more deep research of screw theory in robotics. We can discover that it is possible to develop better algorithms and solutions for new robotics challenges and applications.

Emphasizing the main goals and targets for this text:

- It offers a screw theory mechanics visual approach.
- It presents new ideas to solve some prototypical robot mechanics.
- It provides a solid base of understanding and learning screw theory basics.
- It makes the abstract screw theory for robotics concepts tangible and practical.
- It demonstrates that we tackle better complex robot mechanical problems with the powerful tools of the screw theory.
- It gives gratification by solving complex kinematics problems with an elegant approach and minimal code.
- It limits the number of equations to those necessary for screw theory fundamentals.
- It fosters the research for good theories and not so much for implementations as the best way to speed up great results.
- It sparks inspiration for developing new and better solutions using these screw theory methodologies for further applications.

Working in robotics, we must pay attention to the great benefits and advantages of this approach for projects and applications:

- It provides a global and geometric representation of the mechanics, which dramatically simplifies the analysis for robotics.
- It makes the equations of motion very easily treatable because the matrix exponential is the basic mathematical primitive.

Introduction

- It gives closed geometric solutions for the inverse kinematics and dynamics of manipulators with many DoF, better than other numerical solutions.
- It can develop more efficient and practical algorithms for many robots (e.g., industrial manipulators, humanoids, mobile, drones).

Learning the mechanics illustrated in this book is worthwhile. We can grasp the full potential of screw theory to complete amazement. It brings real-time robot applications to solve inverse kinematics with closed-form geometric algorithms or work with velocities without differentiation. This book is for robotics enthusiasts striving to improve robot applications' efficiency and performance. And this journey starts here!

1.3 PREVIEW

1.3.1 Outline

Robotics is a relatively young field with the ultimate goal of creating machines that can behave like humans. This attempt leads us to examine ourselves to understand how we learn and perform complex tasks. The fundamental questions in robotics are, in the end, questions about humans, and that is what makes robotics such a fascinating challenge.

The disciplines involved in robotics are numerous (e.g., mechanics, electronics, control, mathematics, software, AI), which is one of the difficulties in achieving substantial advances in this science. This book focuses on some specific mathematical foundations for mechanics. The core of this text is the screw theory in robotics.

Working with this book will give immediate gratification, particularly for anyone who has approached robot kinematics and dynamics with numeric iterative algorithms since their solutions do not lead to suitable online implementations. Conversely, with the screw theory tools, it is easy to rapidly end up with elegant geometric solutions for great real-time applications.

The structure of the book has a design for the best convenience of the reader. It can work going from beginning to end or go directly to the topic or chapter of interest. This setup makes possible the flexible approach to the text because the chapters, sections, and examples have, as much as possible, maximum completeness and independence.

We find out what is yet to come on the following pages.

1.3.2 Chapter 1: Introduction

We get started with the motivation to write and read this work devoted to robotics. There is a brief journey through historical references to robotics. The artworks of the last century (i.e., literature and cinema) have inserted the robot's image in the collective unconsciousness of the general public. That is society's thoughts, fears, and expectations regarding robots.

The preview of this book presents the goals and **the benefits of the screw theory in robotics**. It will emerge from them that the learning of the mechanics theory and practice illustrated here is worthwhile. There is a review of the book's contents to quickly grasp what to expect from each chapter allowing a rapid focus of the study.

We describe this book's intended audience and the basic concepts needed to understand and follow the text. The prominent concerned people are probably engineers interested in screw theory in robotics, looking for clear basic examples, exercises, and illustrated applications to complement other theory-focused texts.

We point to further essential reading for those who want to expand their knowledge in the mathematics underlying the screw theory or those who need to learn more about applying these methodologies in different specialized robotics fields. The books included in this list are well-known classics, indispensable for anyone interested in advanced robotics.

1.3.3 Chapter 2: Mathematical Tools

This chapter addresses how to describe a rigid body motion in a three-dimensional (3D) physical space with better mathematical tools.

The classic approach uses the homogeneous representation to quantitatively describe a rigid body's position and orientation as it moves. Here, we will find a concise explanation of how to understand and skillfully use this standard tool. There are some exercises to practice.

The more fundamental and newest approach focuses on **the exponential representation, the modern geometric interpretation of the screw theory**, to describe the rigid body motions. The description of rotation and translation for a rigid body pose operates in terms of the screw theory. New theory terminology comes out with the key terms "Twist" and "Wrench" introduced as the infinitesimal version of a "Screw" motion and force, respectively.

More importantly, we show how to apply the "Screw," "Twist," and "Wrench" concepts to the rotation and translation of the typical revolute and prismatic robot joints.

1.3.4 Chapter 3: Forward Kinematics

The movements of the mechanism (e.g., robot manipulator) joints define the pose (i.e., position and orientation) of an open rigid body chain end-effector. Given the set of joint magnitudes, the Forward Kinematics (FK) problem solves the position and orientation of a frame attached to the robot Tool Center Point (TCP). Analytically, the FK problem is relatively easy because there is always a unique end-effector configuration for any given set of joint magnitudes.

We include the DH convention for the FK, as this is yet the standard approach for many robot mechanical problems. We use a prototypical exercise for a Puma-type robot to contrast the DH approach with the most modern screw theory.

We will go over the POE formula describing the FK of any rigid body's open chain. It provides an intuitive and visual interpretation of the links and robot motion, with the twists of the joints. The POE offers other advantages, like eliminating the need to have frames associated with all links. With the POE, the frames can be chosen arbitrarily, for instance, only two (e.g., for the base and the tool or between two links), which might be the most interesting for a specific application.

Introduction

There are some examples of the FK solved for some typical and commercial robot mechanisms to consolidate the knowledge: Puma type with six DoF (e.g., ABB IRB120, ABB IRB1600), Gantry type with six DoF (e.g., ABB IRB6620LX), Scara with four DoF (e.g., ABB IRB910SC), a Cobot with six DoF (e.g., UNIVERSAL UR16e), and a Redundant manipulator with seven DoF (i.e., KUKA IIWA). Of course, the POE is helpful for the more complex FK; such is the case presented for the humanoid robot RH0 of the UC3M with 21 DoF. These exercises aim to illustrate the use of POE while becoming a precedent for other robot endeavors.

1.3.5 Chapter 4: Inverse Kinematics

The inverse kinematics problem must obtain the joint magnitudes, which, once applied, make the robot tool achieve the desired configuration (i.e., position and orientation). The analytical difficulty of Inverse Kinematics (IK) is quite significant. For a particular tool pose, multiple solutions may exist for the joints' values or no solution at all, out of a system of nonlinear coupled equations.

We first examine the typical approach to solve the robot with several DoF inverse kinematics, which usually is a numerical iterative algorithm. The result is neither very efficient nor very effective, but it has become standard because of the IK analytical difficulty. Besides, the numeric approach only gives one solution out of all possible.

Then we present **the idea of solving the inverse kinematics problem employing the techniques and tools from the screw theory and, in particular, the POE. This concept has paramount importance in this book. It provides tools and methodologies to build geometric closed-form efficient and effective algorithms to solve many mechanical problems**, particularly robotics inverse kinematics.

We will see an example of the classic Puma robot, with both numeric and geometric algorithms. The performance is overwhelmingly better in favor of the screw theory approach.

It is possible to develop geometric algorithms using the joint screw exponentials to solve inverse kinematics for really complex mechanisms with many DoF. This method's cornerstone is the existence of several subproblems (i.e., canonical), which frequently occur in the inverse kinematics analysis for standard robot mechanism designs. These canonical IK subproblems are numerically stable and geometrically meaningful. To solve the inverse kinematics complexity, we must reduce the entire IK problem into appropriate canonical subproblems whose solutions are known.

One of the more original and valuable sections of this book is presenting a new set of canonical IK subproblems. Paden initially introduced this method (Paden & Sastry 1988), building on an unpublished work (Kahan, 1983). That is why the first subproblems are the three already classical **Paden–Kahan (PK) canonical subproblems (i.e., PK1, PK2, and PK3)**.

- **PK1**: ONE ROTATION about one single axis, applied to a POINT.
- **PK2**: TWO consecutive ROTATIONS about CROSSING axes, applied to a POINT.
- **PK3**: ONE ROTATION to a given DISTANCE, applied to a POINT.

The set of canonical subproblems is by no means exhaustive. It is possible to develop additional problems to solve the inverse kinematics of different robots. For example, we show how to solve the IK of manipulators with prismatic joints and mechanisms with parallel or skewed rotations. For doing that, we present some innovative Pardos-Gotor (PG) canonical subproblems (i.e., PG1 to PG8). These algorithms take their inspiration from the classic PK examples. The new subproblems are helpful for many mechanisms and encourage creating others for different robotics architectures. The PG subproblems cover:

- **PG1**: ONE TRANSLATION along a single axis, applied to a POINT.
- **PG2**: TWO consecutive TRANSLATIONS along CROSSING axes, applied to a POINT.
- **PG3**: ONE TRANSLATION to a given DISTANCE, applied to a POINT.
- **PG4**: TWO consecutive ROTATIONS about PARALLEL axes, applied to a POINT.
- **PG5**: ONE ROTATION about a single axis, applied to a LINE or PLANE.
- **PG6**: TWO consecutive ROTATIONS about SKEWED axes, applied to a POINT.
- **PG7**: THREE consecutive ROTATIONS about one SKEWED and two PARALLEL axes, applied to a POINT.
- **PG8**: THREE consecutive ROTATIONS about PARALLEL axes, applied to a POSE or COORDINATE SYSTEM.

Another critical section of this text is in this chapter, where **we introduce some systematic, elegant, and geometrically meaningful solutions based on the POE for the inverse kinematics of some well-known robotics architectures**: Puma type with six DoF (e.g., ABB IRB120, ABB IRB1600), Gantry type with six DoF (e.g., ABB 6620LX), Scara with four DoF (e.g., ABB IRB910SC), a Cobot with six DoF (e.g., UNIVERSAL UR16e), and a Redundant manipulator with seven DoF (i.e., KUKA IIWA). The ideas stemmed from the previous exercises are helpful to develop solutions for more complex IK, such as the case presented for the humanoid robot RH0 of the UC3M with 21 DoF. The exercises' goal is to illustrate the techniques for training on manipulating the POE to solve IK.

- **The POE allows for EFFICIENT and EFFECTIVE algorithms**. There are no iterations, and the calculation has convergence guaranteed. On top of all this, **the geometric closed-form formulation provides the complete Set of Solutions, if they exist, for the inverse kinematics problem**. What a great advantage this is! We can choose the better one for the robot's next move or the optimal one to follow a trajectory from the set of solutions. We remark the POE power in contrast with the typical numerical solutions, which usually provide only one approximate solution.
- We can practice this approach's computational performance with the "**ST24R**" (**Screw Theory Toolbox for Robotics**) and all the examples of the book, which are in an internet repository (Pardos-Gotor, 2021a).

Introduction 13

1.3.6 CHAPTER 5: DIFFERENTIAL KINEMATICS

Differential Kinematics (DK) refers to the robot velocity. It defines the relationship between the end-effector's velocities (linear and angular) and joint velocities.

Central to differential kinematics is the Jacobian concept, as this operator relates the velocities for both the forward and inverse differential kinematics problems. Besides, the Jacobian can solve by integrating the joint velocities the inverse kinematics problem when we do not have IK closed-form solution. This feature will be handy also for trajectory generation. **The critical idea illustrated in this chapter is the difference between the concepts of analytic and geometric Jacobian.**

The geometric screw theory Jacobian can be obtained directly by definition without any differentiation, using the joints' twist. We introduce more screw theory terminology, as the concept of "Velocity Twist" inherently linked to the geometric Jacobian and robot velocity.

A Scara robot example serves to introduce the concept of robot singularity, which corresponds to those configurations at which the Jacobian matrix drops rank. Singular configurations can demand unacceptable velocities for some joints, eventually dangerous for the mechanism. Therefore, the robot motion design must avoid any singular configuration.

The advantages of using the screw theory geometric Jacobian to work with robot velocities are showed with examples: Puma type with six DoF (e.g., ABB IRB120, ABB IRB1600), Gantry type with six DoF (e.g., ABB 6620LX), Scara with four DoF (e.g., ABB IRB910SC), a Cobot with six DoF (e.g., UNIVERSAL UR16e) and a Redundant manipulator with seven DoF (i.e., KUKA IIWA). Moreover, this geometric Jacobian concept and the Spatial Velocity approach can be used and extended to more complex mechanics.

1.3.7 CHAPTER 6: INVERSE DYNAMICS

The subject of robot dynamics is still an open area of research because they can have very complex structures with a lot of DoF. The nonlinear equations of motion with the existence of constraints make the formulation of robot dynamics tricky.

Our goal will be limited to provide robot dynamics algorithms and formulations that rely on methods from the screw theory and some constructed extensions from it. This chapter's dynamics approach is geometric with many advantages to capture the physical features of a robot.

These tools have been primarily inaccessible to engineers because the screw theory dynamics requires a new language and notations (e.g., Twist, Wrench, Robot Inertia Matrix, Link Inertia Matrix, Link Jacobian, Robot Coriolis Matrix, Robot Potential Matrix). For this reason, the first target of this chapter is to address these concepts correctly. Then, **we introduce a geometric expression to solve the ID problem, based on the classical Lagrange screw theory tools.**

This chapter presents **three novel expressions for the Potential or Gravity matrix. Two of which are genuinely new complete geometric formulas (i.e., Gravity Twist and Gravity Wrench matrices)**. They allow to complete the screw theory forward and ID approach with a geometric formulation, and their convergence is assured.

Featherstone introduced the Spatial Vector Algebra and formulated the dynamics of rigid multibody systems by extending constructs from classical screw theory. This chapter presents the well-recognized Recursive Newton–Euler Algorithm (RNEA) Featherstone's machinery to solve the ID problem. **We introduce an innovation in this book to use the screw theory POE for the kinematics analysis of the RNEA instead of the standard DH.** This idea provides flexibility and clarity to the recursive algorithm. The efficiency of this algorithm is much better. The ID problem results obtained with the RNEA give the joint forces and torques with two or three faster orders of magnitude. Therefore, this approach is more convenient for real-time applications.

A set of exercises with the typical robots endorse all these screw theory dynamics benefits for both approaches, with the classical of Lagrange and the recursive of Newton–Euler: Puma type with six DoF (e.g., ABB IRB120, ABB IRB1600), Gantry type with six DoF (e.g., ABB 6620LX), Scara with four DoF (e.g., ABB IRB910SC), a Cobot with six DoF (e.g., UNIVERSAL UR16e) and a Redundant manipulator with seven DoF (i.e., KUKA IIWA).

1.3.8 Chapter 7: Trajectory Generation

A trajectory specifies as a function of time the robot end-effector path (i.e., set of pose configurations). We can see the trajectory as a combination of a path, a purely geometric description of the sequence of poses, and the timing between configurations. Then, the trajectory generation consists of constructing the path plus time scaling so that the robot reaches the desired configurations each time. There are some examples of trajectory generation solved for some commercial robots with typical architectures with two different approaches. We first use the inverse kinematics geometric solutions and second with the integration of joint velocities supported by the differential kinematics with the geometric Jacobian.

The trajectory generation has a clear relationship with the solution for the inverse and differential kinematics problem. An inverse kinematics geometric solution for any robot tool target can generate a joint trajectory with the path formed by a series of points and the time scaling to develop the necessary position, velocities, and accelerations. Another way to create the trajectory uses the velocity of the robot joints at a specific configuration and the inverse DK solution, which, once integrated, can give us some joint path increments.

There are examples of both techniques presented for the architectures reviewed in previous chapters: Puma type with six DoF (e.g., ABB IRB120, ABB IRB1600), Gantry type with six DoF (e.g., ABB 6620LX), Scara with four DoF (e.g., ABB IRB910SC), a Cobot with six DoF (e.g., UNIVERSAL UR16e), and a Redundant manipulator with seven DoF (i.e., KUKA IIWA).

1.3.9 Chapter 8: Robotics Simulation

We know that there are hundreds of simulation tools for robotics. For instance, very popular nowadays is the open-source Gazebo with its integration with ROS (Robot

Introduction 15

Operating System). For this book, the robot simulator used is not so relevant, as we emphasize and stress the importance and value of the theory. However, practicing the exercises with some simulation software is critical to check out the performance of the algorithms and learn.

We have found that the MATLAB® package of tools, with its integrated robot simulation software Simulink® and Simscape™, suits our purpose very well. This software environment is precious because it provides very diffused and recognized instruments to support any robotics developments.

The simulators included in this chapter allow students to interactively explore all the robot mechanics exercises presented throughout this book. Therefore, there are simulations for FK, IK, DK, Trajectory Generation, and ID. These examples implement commercial robot architectures, making them more valuable simulations: Puma type with six DoF (e.g., ABB IRB120, ABB IRB1600), Gantry type with six DoF (e.g., ABB 6620LX), Scara with four DoF (e.g., ABB IRB910SC), a Cobot with six DoF (e.g., UNIVERSAL UR16e), and a Redundant manipulator with seven DoF (i.e., KUKA IIWA). In addition, there are clarifying simulation videos available in an internet repository.

1.3.10 Chapter 9: Conclusions

The closing chapter emphasizes the significant advantages of this screw theory in the robotics approach for mechanics. Hopefully, these arguments will entice anyone involved in this field to continue the pathway to dominate screw theory in robotics. Good luck!

1.4 AUDIENCE

We expect this book will be handy primarily for engineering undergraduate, postgraduate, and PhD students. Nonetheless, the text is also helpful for mathematicians, scientists, technicians, and practitioners involved in robotics. The material collected is also suitable for advanced university courses in screw theory robot mechanics.

This book will also serve as a reference companion to crucial topics in robotics mechanics addressed with screw theory for no longer students. We believe the contents will be helpful for start-ups and research departments involved in robotics.

This volume assumes some familiarity with using and programming MATLAB® because the code examples reinforce the main concepts and ideas. Knowledge of Simulink® and Simscape™ is recommended but not essential.

The primary prerequisite is a proper course in linear algebra. A study on control is also helpful for the robot simulations but not strictly necessary for following the core screw theory material. Some mathematical maturity is also desirable, although anyone who can master the concepts in Chapter 2 will have no difficulty with the rest of this book.

This text treats the screw theory topics in a very visual and unified way with plenty of exercises and figures. It is appropriate for students with only freshman-level physics, ordinary differential equations, linear algebra, and a little computing background.

For practitioners interested in exploring the potential of screw theory for robotics but who do not have time to go through the approach entirely, this book directly introduces the fundamental theorems and conclusions necessary for practical applications in robotics engineering. All the required mathematical background is in this book, but only with the essential expressions. Thorough mathematical developments and demonstrations are available in other texts. Our book's goal is to help speed up the learning process for these topics.

For those looking for more examples, exercises, and simulations about applying the screw theory for robotics, who do not find them in other reference books with much more focus on theoretical aspects of the subject, this volume and the additional repositories offer many practical resources. Besides, the many figures and software code are a crucial vehicle to very quickly convey the concepts.

When looking for a complete but practical and illustrated introduction to screw theory in robotics, this text is the response! This book is a possession to improve the efficiency and effectiveness of robot applications through elegant geometric solutions.

1.5 FURTHER READING

We recommend some fundamental readings on screw theory in robotics and beyond:
- Brockett, R.W. (1983). *Robotic manipulators and the product of exponentials formula.* In Proc. Int. Symp. Math. Theory of Networks and Systems, Beer Sheba, Israel.
- Corke, P. (2017). *Robotics, Vision & Control: Fundamental Algorithms in MATLAB.* Springer.
- Davidson, J. K., & Hunt, K.H. (2004). *Robots and Screw Theory.* Oxford University Press.
- Featherstone, R. (2016). *Robot Dynamics Algorithms.* Springer.
- Lynch, K.M., & Park, F.C. (2017). *Modern Robotics – Mechanics, Planning & Control.* Cambridge University Press.
- Murray, R.M., Li, Z., & Sastry, S.S. (2017). *A Mathematical Introduction to Robotic Manipulation.* C.R.C. Press. (Original work published 1994).
- Park, F.C., Bobrow, J.E., & Ploen, S.R. (1995). *A Lie Group Formulation of Robot Dynamics. The International Journal of Robotics Research*, 14(6):609-618.
- Selig, J.M. (2005). *Geometric Fundamentals of Robotics.* Springer-Verlag.
- Siciliano, B., & Khatib, O. (2016). *Book of Robotics.* Springer.

Our book aspires to be a worthy companion and complement to these fantastic texts.

NOTE

1 This significant sentence of Protagoras concerns the geometric and mechanical configuration of the civilized world. We do not understand it in the modern relativistic or anthropocentric sense. Quite the contrary, well thought out, it can be luminous anticipation consistent with the Christian doctrine of Man and the scholastic logic of the world.

2 Mathematical Tools

"Geometry is knowledge of the eternally existent."

—**Pythagoras**

2.1 RIGID BODY MOTION

A rigid body is a solid object in which deformation is zero or negligible. The distance between any two given points on a rigid body remains constant regardless of external forces exerted. In that sense, the typical robotics manipulator is a rigid multibody, and we must have some mathematical language to deal with such a mechanism. For the chain of linked rigid bodies that constitute a robot, we are looking for practical mechanics solutions to our robotics challenges to permit advanced projects and applications. The problem is that the mathematical difficulty for solving mechanics increases with the number of Degrees of Freedom (DoF) with complex mechanisms.

For a mechanism with few DoF, such as the historical case of the study of a bird's wing of Leonardo Da Vinci (see Figure 1.1), we can solve the kinematics quite well with geometric or algebraic solutions. If we want to solve the inverse kinematics for a typical industrial arm with six DoF, the task is more complicated. The trouble increases exponentially for a humanoid robot like the "RH0" of the UC3M with 21 DoF.

Given the mathematical difficulties, the most common approach is to use numerical procedures. They are iterative algorithms, and their convergence is not guaranteed, or the resolution speed is not sure, which is not the optimal way for real-time robot applications.

In many projects, we need to solve the kinematics problem in real time (e.g., tracking a trajectory), and therefore we need to get closed-form solutions. It would be a lot easier to have a mathematical approach to get this kind of geometric solution for different robotics structures and mechanisms with many DoF. Therefore, we propose and defend this elegant approach of the screw theory in robotics (Lynch and Park, 2017). Nonetheless, we get started with some mathematical basics for the representation of any rigid body motion.

There are two basic rigid body exercises in this chapter solved with both homogeneous and screw theory representations. These illustrate the use of the two mathematical tools introduced in this chapter with their code in MATLAB® (MathWorks, 2021a).

2.2 HOMOGENEOUS REPRESENTATION

2.2.1 Standard Rigid Body Motion

Rigid motions are sets of rotations and displacements. We can use a simple representation in terms of a matrix (4×4), called homogeneous representation (Strang, 2009). Using a homogeneous matrix "H" (see Equation 2.1) is very convenient as the significance of its elements is evident. The rotation matrix "R" represents the rigid body orientation. The displacement vector "D" represents the position of the same body. The perspective transformation and the scale factor are not used in robotics, leaving the practical homogeneous transformation between two configurations (e.g., Spatial and Tool) as given by Equation 2.2.

$$Homogeneous_{4\times4} = \begin{bmatrix} Rotation_{3\times3} & Displacement_{3\times1} \\ Perspective_{1\times3} & Scale_{1\times1} \end{bmatrix} \quad (2.1)$$

$$H_{ST} = \begin{bmatrix} R_{3\times3} & D_{3\times1} \\ 0 & 1 \end{bmatrix} = \begin{bmatrix} U_X & V_X & W_X & p_X \\ U_Y & V_Y & W_Y & p_Y \\ U_Z & V_Z & W_Z & p_Z \\ 0 & 0 & 0 & 1 \end{bmatrix} \quad (2.2)$$

2.2.2 Homogeneous Basic Transformations

There are four fundamental homogeneous transformations (Figure 2.1). Three are for unique rotation about the reference axis X, Y, and Z. One is for displacement along the axes XYZ. More complex motions can be developed by composing these fundamental transformations, as we will show now.

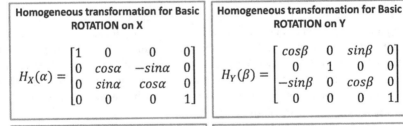

FIGURE 2.1 Four basic homogeneous transformations.

Mathematical Tools

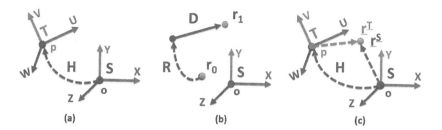

FIGURE 2.2 (a) A rotated and translated system. (b) A rotated and translated vector. (c) A trasformed vector from mobile to reference system.

There are three typical uses of the homogeneous matrix representation for robotics:

1. Represent the orientation and position of a rotated and translated system "*UVW*" concerning a fixed reference system, "*XYZ*," which is the same as representing a rotation and translation performed on the reference system (Figure 2.2a).
2. Rotate and translate a vector "r_0" concerning a fixed reference system "*XYZ*" with the Equation 2.3, so we get the new vector "r_1" referred to the same system (Figure 2.2b).

$$r_1^S = H_{ST} \cdot r_0^S \rightarrow \begin{bmatrix} r_1^X \\ r_1^Y \\ r_1^Z \\ 1 \end{bmatrix} = H_{ST} \begin{bmatrix} r_0^X \\ r_0^Y \\ r_0^Z \\ 1 \end{bmatrix} \qquad (2.3)$$

3. Transform a vector "r^T" expressed in coordinates of "*UVW*" (e.g., mobile system) with the Equation 2.4, to its expression in the reference system "*XYZ*" "r^S" (Figure 2.2c).

$$r^S = H_{ST} \cdot r^T \rightarrow \begin{bmatrix} r^X \\ r^Y \\ r^Z \\ 1 \end{bmatrix} = H_{ST} \begin{bmatrix} r^U \\ r^V \\ r^W \\ 1 \end{bmatrix} \qquad (2.4)$$

2.2.3 Motion Composition in the Spatial "S" Reference System

It is possible to combine different transformations by the multiplication of the corresponding homogeneous "*H*" matrices. Be aware that **the PRODUCT of homogeneous matrices is NOT COMMUTATIVE!** For example, let us compare the completely different results between:

- The rotation plus translation (Figure 2.3a) referred to the Spatial frame with Equation 2.5.

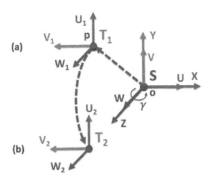

FIGURE 2.3 (a) The rotation plus translation referred to the Spatial frame. (b) The translation plus rotation referred to the Spatial frame.

- The translation plus rotation (Figure 2.3b) referred to the Spatial frame with Equation 2.6.

$$H_{ST1}^{S} = H_{XYZ}(p)H_{Z}(\gamma) = \begin{bmatrix} \cos\gamma & -\sin\gamma & 0 & p_X \\ \sin\gamma & \cos\gamma & 0 & p_Y \\ 0 & 0 & 1 & p_Z \\ 0 & 0 & 0 & 1 \end{bmatrix} \quad (2.5)$$

$$H_{ST2}^{S} = H_{Z}(\gamma)H_{XYZ}(p) = \begin{bmatrix} \cos\gamma & -\sin\gamma & 0 & p_X\cos\gamma - p_Y\sin\gamma \\ \sin\gamma & \cos\gamma & 0 & p_X\sin\gamma + p_Y\cos\gamma \\ 0 & 0 & 1 & p_Z \\ 0 & 0 & 0 & 1 \end{bmatrix} \quad (2.6)$$

2.2.4 Motion Composition with Stationary and Mobile Coordinate Systems

If the MOTION definition with rotations and translations is on the SPATIAL (stationary) system "OXYZ," the composed homogeneous matrices must be PRE-Multiplied. As an example, the transformation "H^{S}_{ST1}" of the Equation 2.7 represents (consistently over the spatial system "OXYZ") a rotation "α" on "OX," followed by a rotation "β" on "OY" and a rotation "γ" on "OZ" (Figure 2.4a).

$$H_{ST1}^{S} = H_{Z}(\gamma)H_{Y}(\beta)H_{X}(\alpha) = \begin{bmatrix} C_\beta C_\gamma & -C_\alpha S_\gamma + S_\beta C_\gamma S_\alpha & S_\alpha S_\gamma + S_\beta C_\gamma C_\alpha & 0 \\ C_\beta S_\gamma & C_\alpha C_\gamma + S_\beta S_\gamma S_\alpha & -S_\alpha C_\gamma + S_\beta S_\gamma C_\alpha & 0 \\ -S_\beta & C_\beta S_\alpha & C_\beta C_\alpha & 0 \\ 0 & 0 & 0 & 1 \end{bmatrix}$$
(2.7)

If the MOTION definition with rotations and translations is on the TOOL (mobile) system "OUVW," the composed homogeneous matrices must be POST-Multiplied. As an example, the transformation "H^{T}_{ST2}" of the Equation 2.8 represents

Mathematical Tools

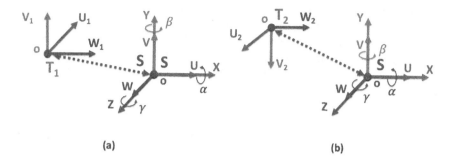

FIGURE 2.4 (a) A motion composition on the Spatial system. (b) A motion composition on the Mobile system.

(consistently over the mobile system "OUVW") a rotation "α" on "OU," followed by a rotation "β" on "OV" and a rotation "γ" on "OW" (Figure 2.4b).

$$H_{ST2}^{T} = H_U(\alpha)H_V(\beta)H_W(\gamma) = \begin{bmatrix} C_\beta C_\gamma & -C_\beta S_\gamma & S_\beta & 0 \\ C_\alpha S_\gamma + S_\beta C_\gamma S_\alpha & C_\alpha C_\gamma - S_\beta S_\gamma S_\alpha & -C_\beta S_\alpha & 0 \\ S_\alpha S_\gamma - S_\beta C_\gamma C_\alpha & S_\alpha C_\gamma + S_\beta S_\gamma C_\alpha & C_\beta C_\alpha & 0 \\ 0 & 0 & 0 & 1 \end{bmatrix}$$

(2.8)

You can check how the results of Equations 2.7 and 2.8 are different!

2.2.5 Geometrical Interpretation

This representation of the rigid body motion with homogeneous matrices constitutes a language to relate two frames of the robot: the stationary at the base (**"S" spatial system**) and the mobile (**"T" tool system**) linked to the robot end-effector.

We can describe the pose of the robot end-effector concerning its base by associating the base of the manipulator to a spatial coordinate system "S" (OXYZ) and the tool to a mobile system "T" (noap) (Figure 2.5).

Expressed in the spatial system "S," which is typically the base or stationary system in the case for many manipulators, the components of the mobile tool system "T" are:

- **p - position**: it is the point origin of the tool "T" reference system, which frequently is the "Tool center Point" (TCP) associated with the robot's end-effector.

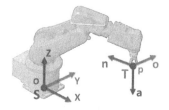

FIGURE 2.5 Tool reference system geometrical interpretation.

- **n - normal**: vector that forms the orthogonal trio with the "o" and "a." It is equivalent to the "X" axis of the tool "T" reference system.
- **o - orientation**: vector perpendicular to "a" in the plane defined by the robot's body or tool. It is equivalent to the "Y" axis of the tool "T" reference system.
- **a - approach**: vector approaching the robot end-effector towards the task direction. It is equivalent to the "Z" axis of the tool "T" reference system.

These components (noap) of the coordinate system attached to the tool "T" (Figure 2.5) can be defined almost by inspection on many occasions, as they are directly their expression in the spatial system "S," as in Equation 2.9.

$$H_{ST} = \begin{bmatrix} n & o & a & p \\ 0 & 0 & 0 & 1 \end{bmatrix} = \begin{bmatrix} n_X & o_X & a_X & p_X \\ n_Y & o_Y & a_Y & p_Y \\ n_Z & o_Z & a_Z & p_Z \\ 0 & 0 & 0 & 1 \end{bmatrix} \quad (2.9)$$

2.2.6 Exercise: Homogeneous Rotation

Transform a vector $r^T(3,2,1)$ expressed in coordinates of the $T(UVW)$ system to its expression r^S in coordinates of the reference system $S(XYZ)$. The system $T(UVW)$ is rotated by an angle ($\gamma = \pi/4$) about the axis "Z" (Figure 2.6). The result is straightforward with the simple product by the fundamental homogeneous transformation for rotation on "Z."

We can solve the exercise with the classical rotation matrix (3×3) on "Z."

$$r^S = \begin{bmatrix} r^X \\ r^Y \\ r^Z \end{bmatrix} = R_Z(\gamma) \cdot r^T = \begin{bmatrix} \cos\gamma & -\sin\gamma & 0 \\ \sin\gamma & \cos\gamma & 0 \\ 0 & 0 & 1 \end{bmatrix} \begin{bmatrix} r^U \\ r^V \\ r^W \end{bmatrix}$$

$$\begin{bmatrix} 0.7 \\ 3.5 \\ 1 \end{bmatrix} = \begin{bmatrix} 0.7071 & -0.7071 & 0 \\ 0.7071 & 0.7071 & 0 \\ 0 & 0 & 1 \end{bmatrix} \begin{bmatrix} 3 \\ 2 \\ 1 \end{bmatrix}$$

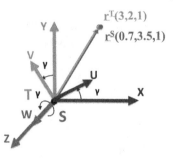

FIGURE 2.6 Exercise 2.2.6. Homogeneous rotation.

Mathematical Tools

The same solution with homogeneous representation (4×4) is:

$$r^S = \begin{bmatrix} r^X \\ r^Y \\ r^Z \\ 1 \end{bmatrix} = H_Z(\gamma) \cdot r^T = \begin{bmatrix} \cos\gamma & -\sin\gamma & 0 & 0 \\ \sin\gamma & \cos\gamma & 0 & 0 \\ 0 & 0 & 1 & 0 \\ 0 & 0 & 0 & 1 \end{bmatrix} \begin{bmatrix} r^U \\ r^V \\ r^W \\ 1 \end{bmatrix}$$

$$\begin{bmatrix} 0.7 \\ 3.5 \\ 1 \\ 1 \end{bmatrix} = \begin{bmatrix} 0.7071 & -0.7071 & 0 & 0 \\ 0.7071 & 0.7071 & 0 & 0 \\ 0 & 0 & 1 & 0 \\ 0 & 0 & 0 & 1 \end{bmatrix} \begin{bmatrix} 3 \\ 2 \\ 1 \\ 1 \end{bmatrix}$$

The code for this Exercise 2.2.6 is in the internet hosting for the software of this book[1].

2.2.7 Exercise: Homogeneous Rotation Plus Translation

Transform a vector $r(-3,4,-11)$ expressed in coordinates of the $T(UVW)$ system to its expression in coordinates of the reference system $S(XYZ)$ (Figure 2.7). The system $T(UVW)$ is rotated "$\pi/2$" on the axis OX and then translated by a vector $p(8,-4,12)$, concerning $S(XYZ)$. As we refer to the motion definition in the spatial coordinate systems, matrices' composition goes with pre multiplication.

$$r^S = H_{ST} \cdot r^T = H_{XYZ}(p) H_X(\alpha) \cdot r^T$$

$$\begin{bmatrix} r^X \\ r^Y \\ r^Z \\ 1 \end{bmatrix} = \begin{bmatrix} 1 & 0 & 0 & p_X \\ 0 & 1 & 0 & p_Y \\ 0 & 0 & 1 & p_Z \\ 0 & 0 & 0 & 1 \end{bmatrix} \begin{bmatrix} 1 & 0 & 0 & 0 \\ 0 & \cos\alpha & -\sin\alpha & 0 \\ 0 & \sin\alpha & \cos\alpha & 0 \\ 0 & 0 & 0 & 1 \end{bmatrix} \begin{bmatrix} r^U \\ r^V \\ r^W \\ 1 \end{bmatrix}$$

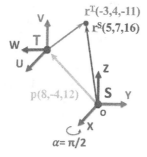

FIGURE 2.7 Exercise 2.2.7. Homogeneous rotation plus translation.

$$\begin{bmatrix} 5 \\ 7 \\ 16 \\ 1 \end{bmatrix} = \begin{bmatrix} 1 & 0 & 0 & 8 \\ 0 & 1 & 0 & -4 \\ 0 & 0 & 1 & 12 \\ 0 & 0 & 0 & 1 \end{bmatrix} \begin{bmatrix} 1 & 0 & 0 & 0 \\ 0 & 0 & -1 & 0 \\ 0 & 1 & 0 & 0 \\ 0 & 0 & 0 & 1 \end{bmatrix} \begin{bmatrix} -3 \\ 4 \\ -11 \\ 1 \end{bmatrix}$$

The code for this Exercise 2.2.7 is in the internet hosting for the software of this book[2].

2.3 EXPONENTIAL REPRESENTATION

2.3.1 Modern Rigid Body Motion

The screw theory allows defining the generalized motion of a rigid body with only one rotation about a straight line followed by a single translation parallel to that line. This movement is what we call the "**SCREW**" (Ceccarelli, 2000). The infinitesimal version of this screw movement is what we know as a "**TWIST**," which describes the instantaneous velocity of a body in terms of its linear and angular components (Millman and Parker, 1997). The fundamental tools that we will use in this book to represent the kinematics of mechanisms and robots will be the twists and screws of their joints, using the map between the group and Lie algebra **EXPONENTIAL of matrices**.

Using twists and the exponential of the screws to describe a rigid body's kinematics provides many advantages. The first is that it facilitates the global description of the movement, which does not suffer from singularities because of the local coordinates (Selig, 2005). The second advantage is that **screw theory always gives an entirely geometric description of the rigid body motion, which significantly simplifies the analysis of mechanisms in robotics**.

Hereafter, we propose a concise review of the fundamental mathematical foundations of the screw theory (Ohwovoriole and Roth, 1981). We will begin by taking a short historical tour of its development, which helps get the basic concepts. In addition to rigid body kinematics, the complementary concepts for dynamics will be presented, including the wrench as the screw theory expression for generalized forces on a rigid body.

2.3.2 Screw Rotation (Orientation)

Benjamin Olinde-Rodrigues (1795–1851) was a French banker and mathematician (Figure 2.8). His doctorate dissertation contains the result now called **Rodrigues' formula**, an efficient algorithm for rotating a vector in space, given an axis "ω" and angle of rotation "θ." It provides an algorithm to compute the exponential map from its Lie algebra so(3) to the three-dimensional (3D) special orthogonal space SO(3) without computing the full matrix exponentially. **The exponential coordinates are the canonical representation of the rotation group** as Equation 2.10. Rodrigues' formula uses the skew-symmetric expression of Equation 2.11 for "ω," which is helpful to transform a cross product of vectors to a matrix product.

$$R_\omega(\theta) = I_3 + \hat{\omega} \sin\theta + \hat{\omega}^2 (1 - \cos\theta) = e^{\hat{\omega}\theta} \qquad (2.10)$$

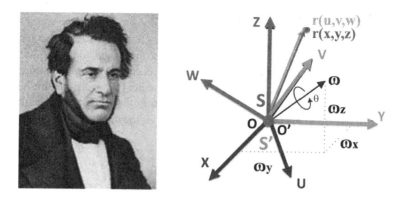

FIGURE 2.8 Benjamin Olinde-Rodrigues and his general coordinate frame rotation formula.

$$\hat{\omega} = \begin{bmatrix} 0 & -\omega_Z & \omega_Y \\ \omega_Z & 0 & -\omega_X \\ -\omega_Y & \omega_X & 0 \end{bmatrix} \quad (2.11)$$

2.3.3 Rigid Body Motion Twist

Michel Chasles (1793–1880) (Figure 2.9a) was a French scientist and studied at the "École Polytechnique" in Paris under Poisson. Chasles' theorem says that any general rigid body motion is equivalent to a translation along a line (screw axis) followed (or preceded) by a rotation about that line. That rigid body motion can be represented by a screw motion called "**TWIST**" (ξ), vector (6×1) of Equation 2.12, where "υ" is referred to as the moment and encodes the action line in space, and the pitch of the screw, and "ω" is the direction of the screw axis.

$$\xi_{6\times 1} = \begin{bmatrix} \upsilon \\ \omega \end{bmatrix} \quad (2.12)$$

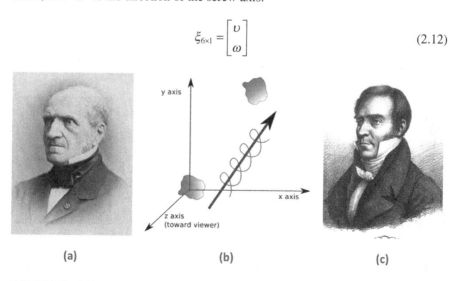

FIGURE 2.9 (a) Michel Chasles. (b) Screw motion (Twist) and force (Wrench). (c) Louis Poinsot.

2.3.4 Rigid Body Force WRENCH

Louis Poinsot (1777–1859) (Figure 2.9c) was a French mathematician and physicist. Poinsot's theorem says that any system of forces affecting a rigid body is equivalent to a single force along a line (screw axis) followed (or preceded) by a unique torque about that line. The screw force called "**WRENCH**" ($\mathcal{F}$), vector (6×1) of Equation 2.13, can represent a rigid body force, where "τ" is referred to as the moment and encodes the action line in space and the pitch of the screw, and "f" is the direction of the screw axis.

$$\mathcal{F}_{6\times 1} = \begin{bmatrix} f \\ \tau \end{bmatrix} \tag{2.13}$$

Robert Ball (1840–1913) (Figure 2.10a) was an Irish astronomer and is best known for his contributions to kinematics described in his treatise "The Theory of Screws" (Ball, 1900).

Marius Sophus Lie (1842–1899) (Figure 2.10c) was a Norwegian mathematician. He mainly created the theory of continuous transformation groups (**Lie groups**) symmetry and applied it to the study of geometry and differential equations. The special Euclidean Lie group SE(3) represents the motion of a rigid body in 3D space. The SE(3) has associated a Lie algebra se(3) that describes the group's local structure, equivalent to matrices called "twists." The primary connection between the Lie group and its corresponding algebra is the exponential transformation, equivalent to a homogeneous transformation such as Equation 2.14.

$$H_{4x4} = e^{\hat{\xi}\theta} H$$
$$\hat{\xi} = \begin{bmatrix} \hat{\omega} & v \\ 0 & 0 \end{bmatrix} \tag{2.14}$$

2.3.5 Exponential Coordinates for a SCREW Motion

Screws are a geometric description of twists. This representation allows an elegant and geometric treatment of rigid body motions, presented as screws

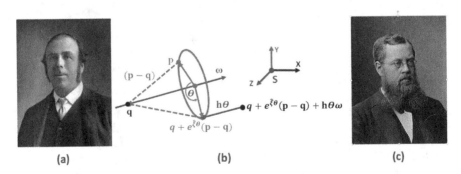

FIGURE 2.10 (a) Robert Ball. (b) Rotation & Translation Screw components. (c) Marius Sophus Lie.

Mathematical Tools

(Figure 2.10b). Then, the exponential coordinates for a screw motion are given by Equation 2.15.

$$e^{\hat{\xi}\theta} = \begin{bmatrix} e^{\hat{\omega}\theta} & (I_3 - e^{\hat{\omega}\theta})(\omega \times \upsilon) + \omega\omega^T \upsilon\theta \\ 0 & 1 \end{bmatrix} \quad (2.15)$$

The exponential map for a "twist" (ξ) gives a rigid body's relative motion. This transformation maps elements from their initial coordinates to their final ones after applying the movement (Brockett, 1983). Be careful because the exponential does not pass a point directly from one coordinate frame to another. Therefore, the initial map $H_{ST}(0)$ must be identified for getting the complete transformation as Equation 2.16.

$$H_{ST}(\theta) = e^{\hat{\xi}\theta} H_{ST}(0) \quad (2.16)$$

Screw theory simplifies the analysis of mechanisms with its geometric description of any general rigid body motion.

Besides, screws permit a description of the motion which does not suffer from singularities due to local coordinates. Any three-angle representation for orientation has singularities (e.g., Euler angles), which is a severe problem, also known as "Gimbal Lock."

Two simplifications are very interesting for practical robotics applications from the screw expression in terms of homogeneous transformation (Davidson and Hunt, 2004). These are the cases for pure rotation (e.g., revolute joints) and pure translation (e.g., prismatic joints).

In the case of **Revolute Joints** (see Figure 2.11), considering "ω" as the direction of the screw axis, "q" any point on that axis and "θ" the magnitude (i.e., angle) of the motion (i.e., rotation), the fundamental formulations are:

- The expression for the **TWIST** of a pure **ROTATION** is given by Equation 2.17. The moment vector of the twist is the cross product of the axis of the screw and any point on that axis. The negative sign comes from the definition

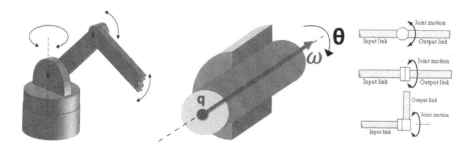

FIGURE 2.11 Pure ROTATION of a revolute joint.

of the exponential coordinates for a screw (see Equation 2.15). For further understanding, review Figure 2.9b and the recommended bibliography (Murray et al., 2017).

$$\xi = \begin{bmatrix} \upsilon \\ \omega \end{bmatrix} = \begin{bmatrix} -\omega \times q \\ \omega \end{bmatrix} \qquad (2.17)$$

- The expression for the **WRENCH** of a pure **ROTATION** is given by Equation 2.18.

$$\mathcal{F} = \begin{bmatrix} 0 \\ \omega \end{bmatrix} \qquad (2.18)$$

- The expression for the **SCREW** exponential of a pure **ROTATION** is given by Equation 2.19.

$$e^{\hat{\xi}\theta} = \begin{bmatrix} e^{\hat{\omega}\theta} & (I_3 - e^{\hat{\omega}\theta})(\omega \times \upsilon) \\ 0 & 1 \end{bmatrix} \qquad (2.19)$$

In the case of **Prismatic Joints** (see Figure 2.12), considering "υ" as the direction of the screw axis, "q" any point on that axis, and "θ" the magnitude (i.e., distance) of the motion (i.e., displacement), the fundamental formulations are:

- The expression for the **TWIST** of a pure **TRANSLATION** is given by Equation 2.20.

$$\xi = \begin{bmatrix} \upsilon \\ 0 \end{bmatrix} \qquad (2.20)$$

- The expression for the **WRENCH** of a pure **TRANSLATION** is given by Equation 2.21. The moment vector of the wrench is the cross product of the

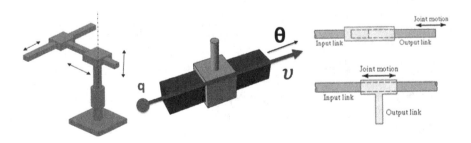

FIGURE 2.12 Pure TRANSLATION of a prismatic joint.

Mathematical Tools

axis of the screw and any point on that axis. The negative sign comes from the definition of the exponential coordinates for a screw (see Equation 2.15). For further understanding, review Figure 2.9b and the recommended bibliography (Murray et al., 2017).

$$\mathcal{F} = \begin{bmatrix} \upsilon \\ -\upsilon \times q \end{bmatrix} \quad (2.21)$$

- The expression for the **SCREW** exponential of a pure **TRANSLATION** is given by Equation 2.22.

$$e^{\hat{\xi}\theta} = \begin{bmatrix} I_3 & \upsilon\theta \\ 0 & 1 \end{bmatrix} \quad (2.22)$$

2.3.6 EXERCISE: EXPONENTIAL ROTATION

Transform a vector $r^T(3,2,1)$ expressed in coordinates of the mobile T(UVW) system to its expression r^S in coordinates of the reference system S(XYZ). The system T(UVW) is rotated by an angle $\gamma(\pi/4)$ to the axis Z. Figure 2.13 represents this motion, and you can check it is the same as the homogeneous approach of Figure 2.6. In this case, the result is obtained by Rodrigues' formula (see Equation 2.23).

$$r^S = \begin{bmatrix} r^X \\ r^Y \\ r^Z \end{bmatrix} = e^{\hat{\omega}\gamma} \cdot r^T = \left[I_3 + \hat{\omega}\sin\gamma + \hat{\omega}^2(1-\cos\gamma) \right] \begin{bmatrix} r^U \\ r^V \\ r^W \end{bmatrix} \quad (2.23)$$

$$\begin{bmatrix} 0.7 \\ 3.5 \\ 1 \end{bmatrix} = \begin{bmatrix} \begin{bmatrix} 1 & 0 & 0 \\ 0 & 1 & 0 \\ 0 & 0 & 1 \end{bmatrix} + \begin{bmatrix} 0 & -1 & 0 \\ 1 & 0 & 0 \\ 0 & 0 & 0 \end{bmatrix}\sin\frac{\pi}{4} + \begin{bmatrix} 0 & -1 & 0 \\ 1 & 0 & 0 \\ 0 & 0 & 0 \end{bmatrix}^2 \left(1-\cos\frac{\pi}{4}\right) \end{bmatrix} \begin{bmatrix} 3 \\ 2 \\ 1 \end{bmatrix}$$

The code for this Exercise 2.3.6 is in the internet hosting for the software of this book[3].

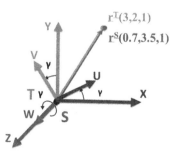

FIGURE 2.13 Exercise 2.3.6. Exponential rotation.

2.3.7 EXERCISE: EXPONENTIAL ROTATION PLUS TRANSLATION

Transform a vector $r(-3,4,-11)$ expressed in coordinates of the mobile T(UVW)" system to its expression in coordinates of the reference system S(XYZ). The system T(UVW) is rotated "$\pi/2$" on the axis OX and then translated by a vector $p(8,-4,12)$, concerning S(XYZ).

We can compare this exercise with Exercise 2.2.7 and check that the result is the same. Therefore, Figure 2.14 represents this motion, and you can see it is the same as the homogeneous approach of Figure 2.7. It might seem that the exponential (Equation 2.24) approach is more complicated than the homogeneous one, especially if this is your first time with this learning. We will see how the benefits of the POE emerge forward when solving robot inverse kinematics and dynamics problems.

$$r^S = H_{ST} \cdot r^T = e^{\hat{\xi}_1 \theta_1} e^{\hat{\xi}_2 \theta_2} H_{ST}(0) \cdot r^T \qquad (2.24)$$

$$v_1 = \frac{p}{\|p\|} = \frac{p}{\sqrt{8^2 + (-4^2) + 12^2}} = \begin{bmatrix} 8/\sqrt{224} \\ -4/\sqrt{224} \\ 12/\sqrt{224} \end{bmatrix} \; ; \; \theta_1 = \|p\|$$

$$\xi_1 = \begin{bmatrix} v_1 \\ 0 \end{bmatrix}$$

$$\omega_2 = \begin{bmatrix} 1 \\ 0 \\ 0 \end{bmatrix} \; ; \; \theta_2 = \frac{\pi}{2}$$

$$\xi_2 = \begin{bmatrix} -\omega_2 \times o \\ \omega_2 \end{bmatrix} = \begin{bmatrix} 0 \\ \omega_2 \end{bmatrix}$$

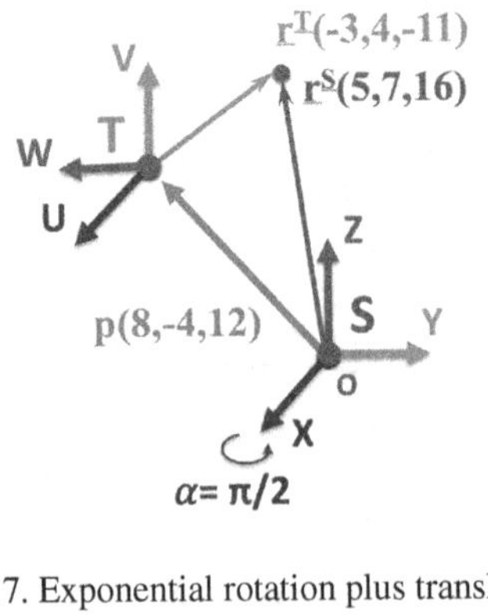

FIGURE 2.14 Exercise 2.3.7. Exponential rotation plus translation.

Mathematical Tools

$$H_{ST}(0) = \begin{bmatrix} I & 0 \\ 0 & 1 \end{bmatrix}$$

$$\begin{bmatrix} 5 \\ 7 \\ 16 \\ 1 \end{bmatrix} = \begin{bmatrix} 1 & 0 & 0 & 8 \\ 0 & 1 & 0 & -4 \\ 0 & 0 & 1 & 12 \\ 0 & 0 & 0 & 1 \end{bmatrix} \begin{bmatrix} 1 & 0 & 0 & 0 \\ 0 & 0 & -1 & 0 \\ 0 & 1 & 0 & 0 \\ 0 & 0 & 0 & 1 \end{bmatrix} \begin{bmatrix} 1 & 0 & 0 & 0 \\ 0 & 1 & 0 & 0 \\ 0 & 0 & 1 & 0 \\ 0 & 0 & 0 & 1 \end{bmatrix} \begin{bmatrix} -3 \\ 4 \\ -11 \\ 1 \end{bmatrix}$$

The code for this Exercise 2.3.7 is in the internet hosting for the software of this book[4] (Pardos-Gotor, 2021a). It is advisable to review the details of the Screw Theory Toolbox for Robotics "ST24R" (Pardos-Gotor, 2018) to understand better the concepts that worked in this example.

2.4 SUMMARY

This chapter presents the necessary mathematical tools to follow this book, including all algorithms, exercises, and simulations to come. Of course, to thoroughly study the screw theory and Lie algebras, it is advisable to refer to other specialized texts. On the other hand, if what we need is to have the main conclusions and formulations practical to robotics, there is everything indispensable to complete this text's developments in this chapter.

The robot's mathematical description was the rigid multibody mechanism for the typical industrial manipulators and similar systems and excluding elastic or soft robots. Besides, we have reviewed the standard mathematical homogeneous representation.

We make a very brief historical itinerary for the development of the screw theory. There is the introduction of some new concepts: "**TWIST**," "**SCREW**," and "**WRENCH**." As a critical tool, we present **the screw EXPONENTIAL**, which has an equivalent homogeneous representation. The extension for the robotic multibody systems mechanics of the **Product of Exponentials (POE)** fundamental formulation stems from there.

The most characteristic joints used in robotics (i.e., revolute and prismatic) have their representation in terms of screw theory (i.e., twist, screw, and wrench). Check the expressions from Equations 2.17 to 2.22 because of their great importance.

There were two examples in this chapter to solve a simple rotation and a composition of rotation plus translation. They have solutions with both the homogeneous representation (i.e., Exercises 2.2.6 and 2.2.7) and screw theory (i.e., Exercises 2.3.6 and 2.3.7).

In the next chapter, we will learn how to apply the POE to the Forward Kinematics of any robotic mechanism. There will be several examples of real industrial manipulators.

"The abstraction saves time in the long run, in return for an initial investment of effort and patience in learning some mathematics."

—**(Murray et al., 2017, p. xiv)**

NOTES

1 Pardos-Gotor, J.M. (2021). *Screw theory in robotics*. Github. https://github.com/ DrPardosGotor/ Screw-Theory-in-Robotics/blob/master/Exercises/Exercise_2_2_6.m
2 Pardos-Gotor, J.M. (2021). *Screw theory in robotics*. Github. https://github.com/ DrPardosGotor/ Screw-Theory-in-Robotics/blob/master/Exercises/Exercise_2_2_7.m
3 Pardos-Gotor, J.M. (2021). *Screw theory in robotics*. Github. https://github.com/ DrPardosGotor/ Screw-Theory-in-Robotics/blob/master/Exercises/Exercise_2_3_6.m
4 Pardos-Gotor, J.M. (2021). *Screw theory in robotics*. Github. https://github.com/ DrPardosGotor/ Screw-Theory-in-Robotics/blob/master/Exercises/Exercise_2_3_7.m

3 Forward Kinematics

"Never confuse motion with action."

—Benjamin Franklin

3.1 PROBLEM STATEMENT IN ROBOTICS

3.1.1 Kinematics Concept

Kinematics is the study of the relationship between a robot joints magnitudes and the robot links and tool pose (i.e., position and orientation) in 3D (Craig, 2004).

3.1.2 Kinematics Mathematical Approach

The standard mathematical approach for robot kinematics is the Denavit–Hartenberg (DH) representation. Nevertheless, another mathematical formulation of the kinematics using the Product of Exponentials (POE) formalism stemmed from screw theory, which provides numerous advantages, even though this approach is more abstract (Siciliano and Khatib, 2016). We attempt to provide the foundations to apply the screw theory POE kinematics formulation to robots with different configurations and many DoF (e.g., humanoids, drones, legged robots, multi-fingered hands), based on extensions from the examples solved here for robot manipulators.

The DH and POE methods require a similar linear algebra treatment (Strang, 2009), but the latter offers a higher-level approach with a clearer geometrical meaning.

3.1.3 Forward Kinematics (FK)

Classically, we construct the FK mapping composing all the individual joints motion for the whole open-chain manipulator (Abraham and Marsden, 1999). This concept computes the tool (end-effector) motion in terms of the given magnitudes of the joints. Applying a movement (i.e., angles for rotations and displacements for translations) to the robot joints (i.e., θ_1–θ_n), we get the configuration "$H_{ST}(\theta)$" (i.e., pose) of the manipulator tool (i.e., end-effector).

We present the use of the POE to get the tool's pose for any robot with an open-chain configuration. We can compute the individual joints' motion with the twist associated with the joint axis, and therefore getting a clear geometric description of the FK.

We recall that **FK is an easy problem since it always has a solution, which is unique.**

3.2 DENAVIT–HARTENBERG CONVENTION (DH)

3.2.1 Kinematics Treatment

Jacques Denavit (1930–2012) and **Richard Hartenberg** (1907–1997) introduced many of the critical concepts of kinematics for serial-link manipulators (Denavit & Hartenberg, 1955).

The DH algorithm needs to go from the "Spatial Coordinate System" (S) to the "Tool Coordinate System" (T) through one coordinate system on each robot link. It may seem surprising that DH only uses four parameters to define the relative link motions since each joint's pose has, in general, six independent parameters. The method simplifies by cleverly choosing the link frames so that individual cancellations occur. Most textbooks still prefer the kinematics formulation of DH, so it has become through the years the quasi-standard approach used in robotics. That is why we take for granted the knowledge of this convention applied to a robot manipulator FK. Nonetheless, hereafter we summarize the classical DH method for completing this chapter with the standard convention.

3.2.2 DH FK Homogeneous Matrix Product

For a typical manipulator, the expression for the DH FK has the form of a homogeneous matrix product. Any homogeneous matrix (H) represents the transformation of the coordinate system associated with each robot link. Transformations in any coordinate system are possible for whatever series of joints and links. It systematizes coordinate systems selection, guaranteeing that the system passes through a concrete sequence of four simple movements. It allows using a standardized agreement, which serves as a common language and develops kinematics calculation tools.

Defining the DH parameters is still the most frequent approach in robotics. They are available for industrial manipulators and used by almost any simulation and programming robotic system. DH representation's fundamental problem is that it cannot represent any motion on the Y axis because all actions are about the X axis and Z axis. Besides, there are several ways to implement the algorithm (e.g., Classical DH and Modified DH parameters). The classical DH parameters are:

- "d_i": it is **the length along the previous $Z(i\text{-}1)$ to the common normal. It is a variable parameter in prismatic joints. It is the translation on the $Z_{(i\text{-}1)}$ axis**.
- "θ_i": It is **the angle about the previous $Z(i\text{-}1)$ from the old $X_{(i\text{-}1)}$ to the new X_i axis. It is a variable parameter in revolute joints. It is the rotation about the $Z_{(i\text{-}1)}$ axis**.
- "r_i" or "a_i": It is **the length along with $X\underline{i}$ (the common normal)**. For a revolute joint, this is the radius about previous $Z_{(i\text{-}1)}$. The distance on the X_i axis.
- "α_i": It is **the angle about Xi (the common normal), from old $Z_{(i\text{-}1)}$ to new Z_i**. The X_i axis angle between successive $Z_{(i\text{-}1)}$ & Z_i axes.

We must realize that only "di" and "θi" are the parameters that can have actual joint variables. From the parameters, we get the **basic link step DH matrix transformation** by Equation 3.1.

$$H_{i-1,i} = H_{Z_{i-1}}(d_i) H_{Z_{i-1}}(\theta_i) H_{X_i}(r_i) H_{X_i}(\alpha_i) \tag{3.1}$$

Forward Kinematics

$$H_{i-1,i} = \begin{bmatrix} 1 & 0 & 0 & 0 \\ 0 & 1 & 0 & 0 \\ 0 & 0 & 1 & d_i \\ 0 & 0 & 0 & 1 \end{bmatrix} \begin{bmatrix} \cos\theta_i & -\sin\theta_i & 0 & 0 \\ \sin\theta_i & \cos\theta_i & 0 & 0 \\ 0 & 0 & 1 & 0 \\ 0 & 0 & 0 & 1 \end{bmatrix} \begin{bmatrix} 1 & 0 & 0 & r_i \\ 0 & 1 & 0 & 0 \\ 0 & 0 & 1 & 0 \\ 0 & 0 & 0 & 1 \end{bmatrix} \begin{bmatrix} 1 & 0 & 0 & 0 \\ 0 & \cos\alpha_i & -\sin\alpha_i & 0 \\ 0 & \sin\alpha_i & \cos\alpha_i & 0 \\ 0 & 0 & 0 & 1 \end{bmatrix}$$

$$H_{i-1,i} = \begin{bmatrix} C\theta_i & -C\alpha_i S\theta_i & S\alpha_i S\theta_i & r_i C\theta_i \\ S\theta_i & C\alpha_i C\theta_i & -S\alpha_i C\theta_i & r_i S\theta_i \\ 0 & S\alpha_i & C\alpha_i & d_i \\ 0 & 0 & 0 & 1 \end{bmatrix}$$

Repeating the process for all the robot links, we obtain **the expression for the manipulator FK by the product of the different links DH matrices transformations**, following Equation 3.2. This definition establishes the relationship between the spatial and tool coordinates.

$$H_{ST}(\theta) = \prod_{i=1}^{n} H_{i-1,i} = \begin{bmatrix} n & o & a & p \\ 0 & 0 & 0 & 1 \end{bmatrix} \tag{3.2}$$

3.2.3 Puma Robots (e.g., ABB IRB120)

This exercise solves the FK for this Puma-type robot (i.e., ABB IRB120) with the DH approach. The expression for the FK based on DH has the form of a product of homogeneous matrices. Any homogeneous matrix (H) represents the movement of the coordinate system associated with each link (Figure 3.1). The definition of this link coordinate matrix is not evident, and it is necessary to carefully follow the steps of the DH algorithm.

We can obtain the parameters for the robot following the rules of the DH classic algorithm. With these values (Figure 3.2), the FK expression is the one of Equation 3.3.

$$H_{ST}(\theta) = \prod_{i=1}^{6} H_{i-1,i} = H_{0,1} H_{1,2} H_{2,3} H_{3,4} H_{4,5} H_{5,6} = \begin{bmatrix} n & o & a & p \\ 0 & 0 & 0 & 1 \end{bmatrix} \tag{3.3}$$

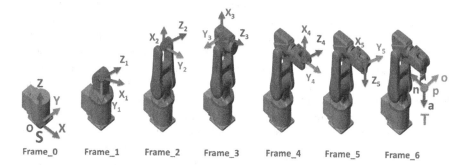

FIGURE 3.1 Puma robot DH kinematics scheme and links coordinate frames.

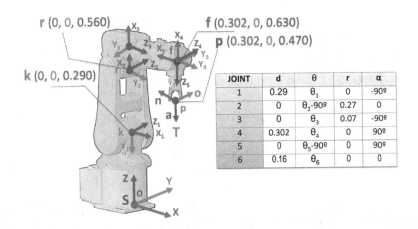

FIGURE 3.2 Puma robot ABB IRB120 DH parameters.

The code for this Exercise 3.2.3 is in the internet hosting for the software of this book[1].

3.3 PRODUCT OF EXPONENTIALS FORMULATION

3.3.1 A New Kinematics Treatment

With the screw theory (Ball, 1900), it is possible to work with a description of the kinematics more geometric for any open-chain manipulator with multiple DoF, generating the general motion of a joint with its twist. We must choose a reference configuration for the robot, as the initial pose corresponds to the joints' zero motion. This reference or home configuration is necessary as the POE is a relative map from one pose to another (Selig, 2005).

Roger Ware Brockett presented the formalism of POE to represent the kinematics of an open-link manipulator (Brockett, 1983). This POE is an elegant treatment that combines with it a great deal of analytical sophistication. The payoff is multiple advantages for solving both kinematics and dynamics problems. Besides, it allows an optimal kinematic design of mechanisms (Park, 1991).

This treatment is somewhat a different path from many textbooks, which prefer a DH formulation (Millman and Parker, 1997). Nevertheless, among other advantages, the twists' geometric significance makes the POE a superior alternative to using the DH parameters, as will be demonstrated along with more complex problems of inverse kinematics, differential kinematics, and inverse dynamics (Davidson and Hunt, 2004).

The POE kinematics analysis gets started by defining the two related coordinate frames of interest, for example, the base frame (Spatial "S"), which is stationary, and the end-effector frame (Tool "T") which is mobile. An advantage compared with the DH approach is that it is unnecessary to fix a coordinate frame to all robot links, but only to those of study. Another advantage of POE is the flexibility to select the robot reference configuration. For example, the spatial frame does not need to be the base frame by obligation, or the operational frame of interest does not have to be compulsory the tool frame. In general, it is much more practical to use these flexibilities, so

Forward Kinematics

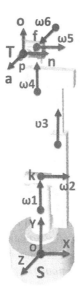

FIGURE 3.3 Stanford robot kinematics analysis with the screw theory.

the kinematics problem remains as simple as possible for practical application with these geometrical considerations (Li, 1990).

As a summary, we can say that **the POE is a straightforward formalization of the FK problem for all joints, which can be easily defined knowing only the axis of the twists ("ω" for rotation and "v" for translation) and any point on the rotational axis.** Beware that the POE represents a relative movement, and therefore we must apply it to the initial pose of the tool to obtain the complete FK solution (Lynch and Park, 2017).

It is easy to analyze a manipulator to get the necessary information to define the FK with the POE. We can see an example with a Sanford-type manipulator scheme in Figure 3.3. **It is easy and quick to obtain each joints' axis and a point on those axes even by inspection. With this simple information, it is effortless to define all the joints' twists and get the POE FK map.**

3.3.2 General Solution to Forward Kinematics

Under the screw theory formalism and using the POE, the FK problem must obtain the configuration for **the tool pose (i.e., orientation plus position)**, knowing the joints' motion (i.e., **magnitudes $\theta_1 ... \theta_n$**).

The general solution to any manipulator FK problem follows this algorithm:

- Select the **Spatial coordinate system "S"** (usually stationary and the robot's base) and the **Mobile coordinate system "T"** (typically the tool). There is no rule and complete flexibility to make this definition, and we can choose the most convenient frames according to the application.
- Define the **Axis of each joint**, which are the axes "ω_i" for revolute joints and "v_i" for the prismatic joints, and a point "q_i" on those axes.

- Obtain the **Twists** "ξ_i" for the joints, knowing for each revolute joint its axis and a point on that axis (see Equation 3.4), and for each prismatic joint its axis (see Equation 3.5).

$$\xi_i = \begin{bmatrix} v_i \\ \omega_i \end{bmatrix} = \begin{bmatrix} -\omega_i \times q_i \\ \omega_i \end{bmatrix} \quad (3.4)$$

$$\xi_i = \begin{bmatrix} v_i \\ 0 \end{bmatrix} \quad (3.5)$$

- Get **the pose of the tool at the reference** "$H_{ST}(0)$" (home) robot position. This configuration of the end-effect happens when all the joint magnitudes are zero. This tool pose comes as a homogeneous matrix.
- Solve the **FK mapping** "$H_{ST}(\theta)$," with the product of all joint screw exponentials (POE) and "$H_{ST}(0)$" (see Equation 3.6). The joints' magnitudes (θ_i) are inputs, and the output is the FK map, which represents the tool pose given by a homogeneous matrix (i.e., ***noap***).

$$H_{ST}(\theta) = \prod_{i=1}^{n} e^{\hat{\xi}_i \theta_i} H_{ST}(0) = \begin{bmatrix} n\ o\ a\ p \\ 0\ 0\ 0\ 1 \end{bmatrix} \quad (3.6)$$

The FK expression has a POE given by the twist (ξ) and Magnitude (θ) associated with each joint. Each exponential represents the relative motion of a joint with a very geometrical meaning. We must do not forget that the exponential mapping is relative to some element. Therefore, we must always multiply the POE by the home configuration of the tool "$H_{ST}(0)$" if we want to get the tool's absolute final pose (Murray et al., 2017).

It is very illustrative to solve the FK problem for typical robot architectures (Pardos-Gotor, 2018). These exercises will permit us to consolidate the skills very quickly to manage the POE. These examples have different joints, configurations, and a diverse selection of base and tool frames. The solved robot architectures are Puma (e.g., ABB IRB120), Bending-Backwards (ABB IRB1600), Gantry (e.g., ABB 6620LX), Scara (e.g., ABB IRB910SC), Collaborative (e.g., UNIVERSAL UR16e), and a Redundant manipulator (e.g., KUKA IIWA). The POE is a handy tool for much more complex FK too. We briefly introduce a humanoid robot example with 21 DoF (i.e., UC3M RH0). All these exercises aim to encourage the POE to formulate the FK of more challenging robotics structures.

There are various tools to put these exercises into practice (Corke, 2017). In our case, it is advisable to review the Screw Theory Toolbox for Robotics "ST24R" (Pardos-Gotor, 2021a) to understand better the concepts working in the examples of the following sections.

3.3.3 Puma Robots (e.g., ABB IRB120)

This exercise solves the FK for this Puma-type robot (i.e., ABB IRB120) with the screw theory approach. We recall that the expression for the FK has the form of POE, and any exponential represents the relative movement of each robot link. The

Forward Kinematics

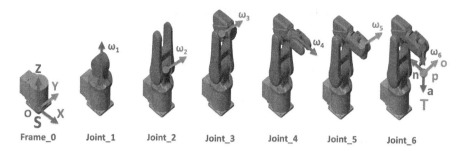

FIGURE 3.4 Puma robot with its screw theory kinematics analysis.

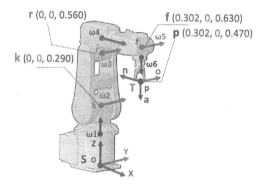

FIGURE 3.5 Puma robot ABB IRB120 POE parameters.

kinematics analysis of mechanisms is much easier to define with this POE approach than with the DH convention. You can compare the analysis made for this manipulator with the DH algorithm (see Figure 3.1) with the screw theory clearer schematic (Figure 3.4).

The kinematics parameters are evident from Figure 3.5.

- It is straightforward to define the axis of each joint "$\omega_1...\omega_6$".

$$\omega_1 = \begin{bmatrix} 0 \\ 0 \\ 1 \end{bmatrix} \omega_2 = \begin{bmatrix} 0 \\ 1 \\ 0 \end{bmatrix} \omega_3 = \begin{bmatrix} 0 \\ 1 \\ 0 \end{bmatrix} \omega_4 = \begin{bmatrix} 1 \\ 0 \\ 0 \end{bmatrix} \omega_5 = \begin{bmatrix} 0 \\ 1 \\ 0 \end{bmatrix} \omega_6 = \begin{bmatrix} 0 \\ 0 \\ -1 \end{bmatrix}$$

- It is easy to obtain the Twists ($\xi_1...\xi_6$), knowing each joint's axis and any point on those axes.

$$\xi_1 = \begin{bmatrix} -\omega_1 \times o \\ \omega_1 \end{bmatrix} \xi_2 = \begin{bmatrix} -\omega_2 \times k \\ \omega_2 \end{bmatrix} \xi_3 = \begin{bmatrix} -\omega_3 \times r \\ \omega_3 \end{bmatrix}$$

$$\xi_4 = \begin{bmatrix} -\omega_4 \times f \\ \omega_4 \end{bmatrix} \xi_5 = \begin{bmatrix} -\omega_5 \times f \\ \omega_5 \end{bmatrix} \xi_6 = \begin{bmatrix} -\omega_6 \times f \\ \omega_6 \end{bmatrix}$$

- We get "$H_{ST}(0)$" as the tool's pose at the reference (home) robot position.

$$H_{ST}(0) = T_{xyz}\begin{bmatrix} p_x \\ p_y \\ p_z \end{bmatrix} R_Y(\pi) = \begin{bmatrix} -1 & 0 & 0 & p_x \\ 0 & 1 & 0 & 0 \\ 0 & 0 & -1 & p_z \\ 0 & 0 & 0 & 1 \end{bmatrix}$$

- With the previous values, we define the FK by the POE by Equation 3.7.

$$H_{ST}(\theta) = \prod_{i=1}^{6} e^{\hat{\xi}_i \theta_i} H_{ST}(0) = e^{\hat{\xi}_1 \theta_1} e^{\hat{\xi}_2 \theta_2} e^{\hat{\xi}_3 \theta_3} e^{\hat{\xi}_4 \theta_4} e^{\hat{\xi}_5 \theta_5} e^{\hat{\xi}_6 \theta_6} H_{ST}(0) = \begin{bmatrix} n & o & a & p \\ 0 & 0 & 0 & 1 \end{bmatrix} \quad (3.7)$$

It is interesting to compare the two approaches for FK applied to the robot ABB IRB120: the solution through the DH convention of Exercise 3.2.3 and this POE solution with this Exercise 3.3.3. Perhaps the advantages of screw theory and the POE formulation are not evident for FK exercises, but they are revealed later for more complex mechanical problems.

The code for this Exercise 3.3.3 is in the internet hosting for the software of this book[2].

3.3.4 Puma Robots (e.g., ABB IRB120) "Tool-Up"

This FK exercise for the same Puma robot, but with a different home configuration, illustrates the freedom that screw theory and POE provide to the kinematics analysis. Here, we define the spatial coordinate system with a different orientation. We choose the robot reference pose with a tool-up direction (Figure 3.6). Besides, we do not assume knowing the actual sign of rotation for the joints of the commercial robot ABB IRB120. The solution follows the same steps as the general FK algorithm with POE. To obtain the necessary parameters is trivial, seeing Figure 3.6.

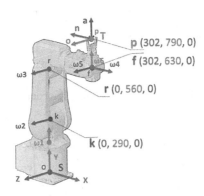

FIGURE 3.6 Puma robot ABB IRB120 "Tool-Up" POE parameters.

Forward Kinematics

- To define the axis of each joint, "$\omega_1...\omega_6$."

$$\omega_1 = \begin{bmatrix} 0 \\ 1 \\ 0 \end{bmatrix} \omega_2 = \begin{bmatrix} 0 \\ 0 \\ 1 \end{bmatrix} \omega_3 = \begin{bmatrix} 0 \\ 0 \\ 1 \end{bmatrix} \omega_4 = \begin{bmatrix} 1 \\ 0 \\ 0 \end{bmatrix} \omega_5 = \begin{bmatrix} 0 \\ 0 \\ 1 \end{bmatrix} \omega_6 = \begin{bmatrix} 0 \\ 1 \\ 0 \end{bmatrix}$$

- To obtain the Twists ($\xi_1... \xi_6$), knowing each joint's axis and a point on those axes.

$$\xi_1 = \begin{bmatrix} -\omega_1 \times o \\ \omega_1 \end{bmatrix} \quad \xi_2 = \begin{bmatrix} -\omega_2 \times k \\ \omega_2 \end{bmatrix} \quad \xi_3 = \begin{bmatrix} -\omega_3 \times r \\ \omega_3 \end{bmatrix}$$

$$\xi_4 = \begin{bmatrix} -\omega_4 \times f \\ \omega_4 \end{bmatrix} \quad \xi_5 = \begin{bmatrix} -\omega_5 \times f \\ \omega_5 \end{bmatrix} \quad \xi_6 = \begin{bmatrix} -\omega_6 \times f \\ \omega_6 \end{bmatrix}$$

- To get $H_{ST}(0)$ as the tool's pose at the reference (home) robot position.

$$H_{ST}(0) = T_{xyz}\begin{bmatrix} p_x \\ p_y \\ p_z \end{bmatrix} R_X\left(\frac{-\pi}{2}\right) R_Z(\pi) = \begin{bmatrix} -1 & 0 & 0 & p_x \\ 0 & 0 & 1 & p_y \\ 0 & 1 & 0 & 0 \\ 0 & 0 & 0 & 1 \end{bmatrix}$$

- Then, solving the problem to get $H_{ST}(\theta)$, applying the POE.

$$H_{ST}(\theta) = e^{\hat{\xi}_1\theta_1} e^{\hat{\xi}_2\theta_2} e^{\hat{\xi}_3\theta_3} e^{\hat{\xi}_4\theta_4} e^{\hat{\xi}_5\theta_5} e^{\hat{\xi}_6\theta_6} H_{ST}(0) = \begin{bmatrix} n & o & a & p \\ 0 & 0 & 0 & 1 \end{bmatrix}$$

The code for this Exercise 3.3.4 is in the internet hosting for the software of this book[3].

3.3.5 Bending Backwards Robots (e.g., ABB IRB1600)

We face a slightly different manipulator architecture for this FK problem, the one of a Bending Backwards robot (Figure 3.7). This design does not need to rotate the robot to reach for things behind it. Instead, we simply swing the arm backwards, which significantly extends the working range of the robot. The POE algorithm is the same. Knowing the joint's motion (i.e., magnitudes $\theta_1...\theta_6$), we get the pose for the tool, with the kinematics parameters recognizable in Figure 3.7.

- To define the axis of each joint, "$\omega_1...\omega_6$."

$$\omega_1 = \begin{bmatrix} 0 \\ 0 \\ 1 \end{bmatrix} \omega_2 = \begin{bmatrix} 0 \\ 1 \\ 0 \end{bmatrix} \omega_3 = \begin{bmatrix} 0 \\ 1 \\ 0 \end{bmatrix} \omega_4 = \begin{bmatrix} 1 \\ 0 \\ 0 \end{bmatrix} \omega_5 = \begin{bmatrix} 0 \\ 1 \\ 0 \end{bmatrix} \omega_6 = \begin{bmatrix} 1 \\ 0 \\ 0 \end{bmatrix}$$

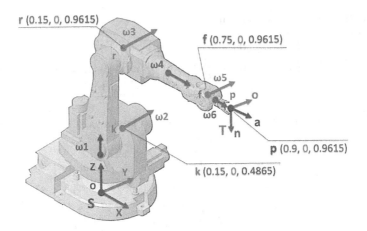

FIGURE 3.7 Bending Backwards robot ABB IRB1600 POE parameters.

- To obtain the Twists ($\xi_1 \ldots \xi_6$), knowing each joint's axis and a point on those axes.

$$\xi_1 = \begin{bmatrix} -\omega_1 \times o \\ \omega_1 \end{bmatrix} \quad \xi_2 = \begin{bmatrix} -\omega_2 \times k \\ \omega_2 \end{bmatrix} \quad \xi_3 = \begin{bmatrix} -\omega_3 \times r \\ \omega_3 \end{bmatrix}$$

$$\xi_4 = \begin{bmatrix} -\omega_4 \times f \\ \omega_4 \end{bmatrix} \quad \xi_5 = \begin{bmatrix} -\omega_5 \times f \\ \omega_5 \end{bmatrix} \quad \xi_6 = \begin{bmatrix} -\omega_6 \times f \\ \omega_6 \end{bmatrix}$$

- To get $H_{ST}(0)$ as the tool's pose at the reference (home) robot position.

$$H_{ST}(0) = T_{xyz}\begin{bmatrix} p_x \\ p_y \\ p_z \end{bmatrix} R_Y\left(\frac{\pi}{2}\right) = \begin{bmatrix} 0 & 0 & 1 & p_x \\ 0 & 1 & 0 & 0 \\ -1 & 0 & 0 & p_z \\ 0 & 0 & 0 & 1 \end{bmatrix}$$

- Then, solving the problem to get $H_{ST}(\theta)$, applying the POE.

$$H_{ST}(\theta) = e^{\hat{\xi}_1\theta_1}e^{\hat{\xi}_2\theta_2}e^{\hat{\xi}_3\theta_3}e^{\hat{\xi}_4\theta_4}e^{\hat{\xi}_5\theta_5}e^{\hat{\xi}_6\theta_6}H_{ST}(0) = \begin{bmatrix} n & o & a & p \\ 0 & 0 & 0 & 1 \end{bmatrix}$$

The code for this Exercise 3.3.5 is in the internet hosting for the software of this book[4].

3.3.6 Gantry Robots (e.g., ABB IRB6620LX)

The FK problem for this Gantry robotic architecture combines the advantages of both a linear axis (first joint) and rotation robot axes (five joints). From the mathematical point of view, this is the first example including a prismatic joint. The treatment of the

Forward Kinematics

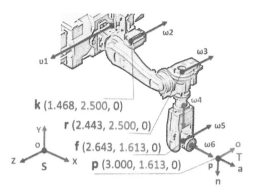

FIGURE 3.8 Gantry robot ABB IRB6620LX POE parameters.

POE is the same. The exercise is for the ABB IRB6620LX, which shows its robot structure in Figure 3.8, to illustrate the application. We get the tool's pose from the joints' motion magnitudes ($\theta_1...\theta_6$), with the kinematics parameters recognizable in Figure 3.8.

- To define the axis of each joint "$v_1, \omega_2...\omega_6$."

$$v_1 = \begin{bmatrix} 0 \\ 0 \\ 1 \end{bmatrix} \quad \omega_2 = \begin{bmatrix} 0 \\ 0 \\ -1 \end{bmatrix} \quad \omega_3 = \begin{bmatrix} 0 \\ 0 \\ -1 \end{bmatrix} \quad \omega_4 = \begin{bmatrix} 0 \\ -1 \\ 0 \end{bmatrix} \quad \omega_5 = \begin{bmatrix} 0 \\ 0 \\ -1 \end{bmatrix} \quad \omega_6 = \begin{bmatrix} 1 \\ 0 \\ 0 \end{bmatrix}$$

- To obtain the Twists ($\xi_1... \xi_6$), knowing each joint's axis and a point on the axes of the revolute joints.

$$\xi_1 = \begin{bmatrix} v_1 \\ 0 \end{bmatrix} \quad \xi_2 = \begin{bmatrix} -\omega_2 \times k \\ \omega_2 \end{bmatrix} \quad \xi_3 = \begin{bmatrix} -\omega_3 \times r \\ \omega_3 \end{bmatrix}$$

$$\xi_4 = \begin{bmatrix} -\omega_4 \times f \\ \omega_4 \end{bmatrix} \quad \xi_5 = \begin{bmatrix} -\omega_5 \times f \\ \omega_5 \end{bmatrix} \quad \xi_6 = \begin{bmatrix} -\omega_6 \times f \\ \omega_6 \end{bmatrix}$$

- To get $H_{ST}(0)$, the tool's pose at the reference (home) robot position.

$$H_{ST}(0) = T_{xyz}\begin{bmatrix} p_x \\ p_y \\ p_z \end{bmatrix} R_Y\left(\frac{\pi}{2}\right) R_Z\left(\frac{-\pi}{2}\right) = \begin{bmatrix} 0 & 0 & 1 & p_X \\ -1 & 0 & 0 & p_Y \\ 0 & -1 & 0 & 0 \\ 0 & 0 & 0 & 1 \end{bmatrix}$$

- Then, solving the problem to get $H_{ST}(\theta)$, applying the POE.

$$H_{ST}(\theta) = e^{\hat{\xi}_1\theta_1} e^{\hat{\xi}_2\theta_2} e^{\hat{\xi}_3\theta_3} e^{\hat{\xi}_4\theta_4} e^{\hat{\xi}_5\theta_5} e^{\hat{\xi}_6\theta_6} H_{ST}(0) = \begin{bmatrix} n & o & a & p \\ 0 & 0 & 0 & 1 \end{bmatrix}$$

The code for this Exercise 3.3.6 is in the internet hosting for the software of this book[5].

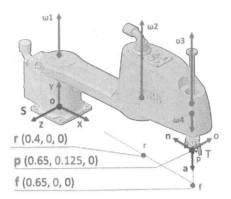

FIGURE 3.9 Scara robot ABB IRB910SC POE parameters.

3.3.7 SCARA ROBOTS (E.G., ABB IRB910SC)

The Scara robot has a parallel joint axes layout, which is advantageous for many assembly operations. For the sake of showing the freedom given by the screw theory in the selection of the coordinate systems, we have chosen the typical virtual reality spatial frame orientation (see Figure 3.9). This example has only four joints, but the general FK algorithm applies the same to get the tool's pose from the joint magnitudes $(\theta_1...\theta_4)$, with the kinematics parameters recognizable in Figure 3.9.

- To define the axis of each joint "$\omega_1, \omega_2, v_3, \omega_4$."

$$\omega_1 = \begin{bmatrix} 0 \\ 1 \\ 0 \end{bmatrix} \omega_2 = \begin{bmatrix} 0 \\ 1 \\ 0 \end{bmatrix} v_3 = \begin{bmatrix} 0 \\ 1 \\ 0 \end{bmatrix} \omega_4 = \begin{bmatrix} 0 \\ -1 \\ 0 \end{bmatrix}$$

- To obtain the Twists ($\xi_1... \xi_4$), knowing each joint's axis and a point on the axes of the revolute joints.

$$\xi_1 = \begin{bmatrix} -\omega_1 \times o \\ \omega_1 \end{bmatrix} \xi_2 = \begin{bmatrix} -\omega_2 \times r \\ \omega_2 \end{bmatrix} \xi_3 = \begin{bmatrix} v_3 \\ 0 \end{bmatrix} \xi_4 = \begin{bmatrix} -\omega_4 \times f \\ \omega_4 \end{bmatrix}$$

- To get $H_{ST}(0)$, the tool's pose at the reference (home) robot position.

$$H_{ST}(0) = T_{xyz} \begin{bmatrix} p_x \\ p_y \\ p_z \end{bmatrix} R_X\left(\frac{\pi}{2}\right) R_Z(-\pi) = \begin{bmatrix} -1 & 0 & 0 & p_X \\ 0 & 0 & -1 & p_Y \\ 0 & -1 & 0 & 0 \\ 0 & 0 & 0 & 1 \end{bmatrix}$$

Forward Kinematics

- Then, solving the problem to get $H_{ST}(\theta)$, applying the POE.

$$H_{ST}(\theta) = e^{\hat{\xi}_1\theta_1}e^{\hat{\xi}_2\theta_2}e^{\hat{\xi}_3\theta_3}e^{\hat{\xi}_4\theta_4}H_{ST}(0) = \begin{bmatrix} n\, o\, a\, p \\ 0\, 0\, 0\, 1 \end{bmatrix}$$

The code for this Exercise 3.3.7 is in the internet hosting for the software of this book[6].

3.3.8 Collaborative Robots (e.g., UNIVERSAL UR16e)

This FK exercise deals with a typical collaborative manipulator, such as the robot UR16e of UNIVERSAL (Figure 3.10), made for human-robot cooperation in the workspace. We use the same general approach to FK based on the POE, with the joint magnitudes ($\theta_1...\theta_6$) as inputs and getting the tool's pose as output. The kinematics parameters are recognizable in Figure 3.10.

- To define the axis of each joint, "$\omega_1...\omega_6$."

$$\omega_1 = \begin{bmatrix} 0 \\ 0 \\ 1 \end{bmatrix} \omega_2 = \begin{bmatrix} 0 \\ 1 \\ 0 \end{bmatrix} \omega_3 = \begin{bmatrix} 0 \\ 1 \\ 0 \end{bmatrix} \omega_4 = \begin{bmatrix} 0 \\ 1 \\ 0 \end{bmatrix} \omega_5 = \begin{bmatrix} 0 \\ 0 \\ -1 \end{bmatrix} \omega_6 = \begin{bmatrix} 0 \\ 1 \\ 0 \end{bmatrix}$$

- To obtain the Twists ($\xi_1... \xi_6$), knowing each joint's axis and a point on those axes.

$$\xi_1 = \begin{bmatrix} -\omega_1 \times o \\ \omega_1 \end{bmatrix} \xi_2 = \begin{bmatrix} -\omega_2 \times k \\ \omega_2 \end{bmatrix} \xi_3 = \begin{bmatrix} -\omega_3 \times r \\ \omega_3 \end{bmatrix}$$

$$\xi_4 = \begin{bmatrix} -\omega_4 \times f \\ \omega_4 \end{bmatrix} \xi_5 = \begin{bmatrix} -\omega_5 \times f \\ \omega_5 \end{bmatrix} \xi_6 = \begin{bmatrix} -\omega_6 \times p \\ \omega_6 \end{bmatrix}$$

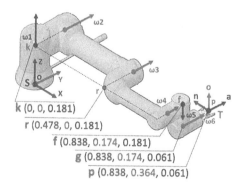

FIGURE 3.10 Collaborative robot UNIVERSAL UR16e POE parameters.

- To get $H_{ST}(0)$, the tool's pose at the reference (home) robot position.

$$H_{ST}(0) = T_{xyz}\begin{bmatrix}p_x\\p_y\\p_z\end{bmatrix}R_X\left(\frac{-\pi}{2}\right)R_Z(\pi) = \begin{bmatrix}-1 & 0 & 0 & p_X\\0 & 0 & 1 & p_Y\\0 & 1 & 0 & p_Z\\0 & 0 & 0 & 1\end{bmatrix}$$

- Then, solve the problem to get $H_{ST}(\theta)$ applying the POE.

$$H_{ST}(\theta) = e^{\hat{\xi}_1\theta_1}e^{\hat{\xi}_2\theta_2}e^{\hat{\xi}_3\theta_3}e^{\hat{\xi}_4\theta_4}e^{\hat{\xi}_5\theta_5}e^{\hat{\xi}_6\theta_6}H_{ST}(0) = \begin{bmatrix}n\,o\,a\,p\\0\,0\,0\,1\end{bmatrix}$$

The code for this Exercise 3.3.8 is in the internet hosting for the software of this book[7].

3.3.9 Redundant Robots (e.g., KUKA IIWA)

Extending POE formulation to robots with more DoF is straightforward. For instance, this exercise applies FK to the lightweight robot KUKA IIWA, made for human-robot collaboration in the workspace (Figure 3.11). This manipulator is redundant as it counts with seven joints. Nonetheless, the general approach of screw theory to solve FK with the POE works without problem. Knowing the seven joint magnitudes ($\theta_1...\theta_7$) inputs, we get the tool's pose as output, with the kinematics parameters recognizable in Figure 3.11.

- To define the axis of each joint, "$\omega_1...\omega_7$."

$$\omega_1 = \begin{bmatrix}0\\0\\1\end{bmatrix} \omega_2 = \begin{bmatrix}0\\1\\0\end{bmatrix} \omega_3 = \begin{bmatrix}0\\0\\1\end{bmatrix} \omega_4 = \begin{bmatrix}0\\-1\\0\end{bmatrix} \omega_5 = \begin{bmatrix}0\\0\\1\end{bmatrix} \omega_6 = \begin{bmatrix}0\\1\\0\end{bmatrix} \omega_7 = \begin{bmatrix}0\\0\\1\end{bmatrix}$$

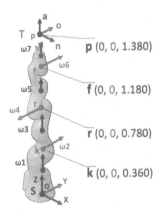

FIGURE 3.11 Redundant robot KUKA IIWA POE parameters.

Forward Kinematics

- To obtain the Twists ($\xi_1 \ldots \xi_7$), knowing each joint's axis and a point on those axes.

$$\xi_1 = \begin{bmatrix} -\omega_1 \times o \\ \omega_1 \end{bmatrix} \xi_2 = \begin{bmatrix} -\omega_2 \times k \\ \omega_2 \end{bmatrix} \xi_3 = \begin{bmatrix} -\omega_3 \times o \\ \omega_3 \end{bmatrix} \xi_4 = \begin{bmatrix} -\omega_4 \times r \\ \omega_4 \end{bmatrix}$$

$$\xi_5 = \begin{bmatrix} -\omega_5 \times o \\ \omega_5 \end{bmatrix} \xi_6 = \begin{bmatrix} -\omega_6 \times f \\ \omega_6 \end{bmatrix} \xi_7 = \begin{bmatrix} -\omega_7 \times o \\ \omega_7 \end{bmatrix}$$

- To get HST(0), the tool's pose at the reference (home) robot position.

$$H_{ST}(0) = T_{xyz} \begin{bmatrix} p_x \\ p_y \\ p_z \end{bmatrix} = \begin{bmatrix} 1 & 0 & 0 & 0 \\ 0 & 1 & 0 & 0 \\ 0 & 0 & 1 & p_z \\ 0 & 0 & 0 & 1 \end{bmatrix}$$

- Then, solving the problem to get HST(θ) applying the POE.

$$H_{ST}(\theta) = e^{\hat{\xi}_1 \theta_1} e^{\hat{\xi}_2 \theta_2} e^{\hat{\xi}_3 \theta_3} e^{\hat{\xi}_4 \theta_4} e^{\hat{\xi}_5 \theta_5} e^{\hat{\xi}_6 \theta_6} e^{\hat{\xi}_7 \theta_7} H_{ST}(0) = \begin{bmatrix} n\ o\ a\ p \\ 0\ 0\ 0\ 1 \end{bmatrix}$$

The code for this Exercise 3.3.9 is in the internet hosting for the software of this book[8].

3.3.10 Many DoF Robots (e.g., RH0 UC3M Humanoid)

The advantages of formulating the FK with the POE are beneficial to robotics mechanisms with many DoF. For example, we will introduce a reference to a Robot Humanoid (RH0) of the University Carlos III of Madrid (UC3M).

Humanoid kinematics presents a tremendous computational difficulty for real-time applications because of the high number of DoF (e.g., 21 for the RH0). Nonetheless, the humanoid FK problem for any link of the robot body and particularly for the hands is expressed quickly with the POE.

This book's scope does not include presenting all the details for this formulation because it involves another approach called Sagittal Kinematics Division (SKD). The idea behind this SKD model is to analyze and control the humanoid considering its two halves separately (left and right of his body), similar to the control that the cerebral hemispheres perform for locomotion of the human body (Figure 3.12). The SKD model considers the mechanics of the humanoid theoretically divided by the sagittal plane. In this way, it is possible to analyze a humanoid using two manipulator robots (left and right), coupled with the common parts (i.e., pelvis and trunk).

Applying SKD to RH0, we appreciate two manipulators. Each one has six virtual DoF to define the pose of the feet referred to the spatial coordinate system (i.e., θ_{VZ1} to θ_{VZ6} of the left foot and θ_{VD1} to θ_{VD6} of the right), plus eleven actual DoF (i.e., θ_{12} to $\theta_7 + \theta_{13} + \theta_{18}$ to θ_{21} in left side and θ_1 to $\theta_6 + \theta_{13}$ to θ_{17} in the right side) of the

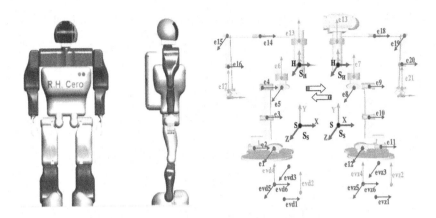

FIGURE 3.12 Humanoid robot UC3M RH0 POE parameters.

humanoid mechanism. The DoF corresponding to the movement of the column "θ_{13}" is common to the two manipulators. Then, we define the kinematics of this RH0 according to the corresponding POE of the screw theory. The FK is defined for the right and left manipulators that form the RH0 humanoid by Equations 3.8 and 3.9, assuming the kinematics chain goes up to the hand, we would consider locating the hypothetical tool. All added, it is enough to demonstrate how the formulation showed with the POE FK can solve complicated mechanisms.

$$G_{SH}(\theta) = e^{\hat{\xi}_{VZ1}\theta_{VZ1}} \ldots e^{\hat{\xi}_{VZ6}\theta_{VZ6}} \cdot e^{\hat{\xi}_{12}\theta_{12}} \ldots e^{\hat{\xi}_{7}\theta_{7}} \cdot e^{\hat{\xi}_{13}\theta_{13}} \cdot e^{\hat{\xi}_{18}\theta_{18}} \ldots e^{\hat{\xi}_{21}\theta_{21}} G_{SH}(0) \quad (3.8)$$

$$G_{SH}(\theta) = e^{\hat{\xi}_{VD1}\theta_{VD1}} \ldots e^{\hat{\xi}_{VD6}\theta_{VD6}} \cdot e^{\hat{\xi}_{1}\theta_{1}} \ldots e^{\hat{\xi}_{6}\theta_{6}} \cdot e^{\hat{\xi}_{13}\theta_{13}} \ldots e^{\hat{\xi}_{17}\theta_{17}} G_{SH}(0) \quad (3.9)$$

This example illustrates how the POE quickly formulates the FK of different mechanisms, manipulators, and robots, even with complex architectures.

3.4 SUMMARY

We have introduced the concept of **FK** in connection to a manipulator. It describes the relationship between the joints' motion and the resulting movement of the open-chain rigid bodies that form the robot. More precisely, usually, the FK problem's interest is to obtain the configuration or pose (i.e., orientation plus position) for the tool or end-effector, given the motion of the robot joints (i.e., magnitudes $\theta_1 \ldots \theta_n$). **The FK problem always has a unique solution.**

We get started presenting the standard mathematical approach for robot kinematics of the DH convention. Of course, the DH formulation is helpful to define FK problems, and we compare it with the screw theory approach. We use the commercial robot ABB IRB120 to contrast DH and POE FK methods.

The POE formalizes the screw theory approach for FK. It has many advantages, even though they are not so evident in juxtaposition with DH when we speak of FK. We can mention some excellent characteristics of working with POE:

Forward Kinematics

- There is no fixed rule to define the stationary or spatial reference frame and the tool coordinate system. There is total flexibility to choose the position and orientation.
- There is no need to define a coordinate system for each robot link. We only need to specify the coordinate systems of interest (e.g., base and tool).
- We only need to identify each joint's axis and any point on the rotational axis to define the FK map.
- The twists of the joints have an exact geometrical meaning, and we can define them by inspection. In any case, with the previous identification of the joint axis and point, the definition of the twists is immediate.
- Because the exponential is an excellent mathematical artifact, it is very convenient to solve the FK mapping "$H_{ST}(\theta)$" with the product of all joint screw exponentials and the tool configuration at home position "$H_{ST}(0)$".

$$H_{ST}(\theta) = \prod_{i=1}^{n} e^{\hat{\xi}_i \theta_i} H_{ST}(0)$$

We have presented some FK examples with typical robot architectures: ABB IRB120, ABB IRB1600, ABB IRB6620LX, ABB IRB910SC, UNIVERSAL UR16e, and KUKA IIWA.

In the next chapter, we will realize how strong are the potential and benefits of using the POE to solve Inverse Kinematics. It is possible to learn how to obtain closed-form geometric and complete solutions for robots with many DoF.

NOTES

1 Pardos-Gotor, J.M. (2021). *Screw Theory in Robotics.* Github. https://github.com/DrPardosGotor/Screw-Theory-in-Robotics/blob/master/Exercises/Exercise_3_2_3.m
2 Pardos-Gotor, J.M. (2021). *Screw Theory in Robotics.* Github. https://github.com/DrPardosGotor/Screw-Theory-in-Robotics/blob/master/Exercises/Exercise_3_3_3.m
3 Pardos-Gotor, J.M. (2021). *Screw Theory in Robotics.* Github. https://github.com/DrPardosGotor/Screw-Theory-in-Robotics/blob/master/Exercises/Exercise_3_3_4.m
4 Pardos-Gotor, J.M. (2021). *Screw Theory in Robotics.* Github. https://github.com/DrPardosGotor/Screw-Theory-in-Robotics/blob/master/Exercises/Exercise_3_3_5.m
5 Pardos-Gotor, J.M. (2021). *Screw Theory in Robotics.* Github. https://github.com/DrPardosGotor/Screw-Theory-in-Robotics/blob/master/Exercises/Exercise_3_3_6.m
6 Pardos-Gotor, J.M. (2021). *Screw Theory in Robotics.* Github. https://github.com/DrPardosGotor/Screw-Theory-in-Robotics/blob/master/Exercises/Exercise_3_3_7.m
7 Pardos-Gotor, J.M. (2021). *Screw Theory in Robotics.* Github. https://github.com/DrPardosGotor/Screw-Theory-in-Robotics/blob/master/Exercises/Exercise_3_3_8.m
8 Pardos-Gotor, J.M. (2021). *Screw Theory in Robotics.* Github. https://github.com/DrPardosGotor/Screw-Theory-in-Robotics/blob/master/Exercises/Exercise_3_3_9.m

4 Inverse Kinematics

"Take to kinematics. It will repay you. It is more fecund than geometry; it adds a fourth dimension to space."

—Pafnuti Chebyshev

4.1 PROBLEM STATEMENT IN ROBOTICS AND ANALYTICAL DIFFICULTY

4.1.1 KINEMATICS CONCEPT

Kinematics is the study of the relationship between the robot tool pose (i.e., position and orientation) in 3D and the robot joint coordinates (Abraham and Marsden, 1999).

4.1.2 INVERSE KINEMATICS MATHEMATICAL APPROACH

The solution of the IK problem is fundamental to generate advanced manipulation applications in robotics (Mason, 2001). Besides, the performance of IK solutions is critical for real-time robot motion applications (Choset and Lynch, 2005).

For more than 50 years, the standard mathematical approach for robot kinematics is the Denavit–Hartenberg (DH) convention (Denavit and Hartenberg 1955). Nevertheless, the se(3) Lie algebra (Selig, 2005) of screw theory gives the mathematical Product of Exponentials (POE) formalism as an alternative that provides multiple benefits, especially to solve the inverse kinematics (IK) problem with geometric closed-form solutions. This book attempts to provide the foundations to apply this kinematics formulation to robotics architectures with many Degrees of Freedom (DoF). Besides, this chapter presents examples solved for some manipulators, which can play cornerstones to succeed in daring to solve other IK endeavors with excellent performance (Siciliano and Khatib, 2016).

The DH and POE methods require a similar linear algebra treatment (Strang, 2009), but the latter offers a higher-level approach with a clearer geometrical meaning.

In connection to a robot manipulator, the IK concept (Marsden and Ratiu, 1999) computes the joints magnitudes in terms of the given end-effector pose or configuration (i.e., position and orientation).

The input to the IK problem is the desired configuration for the tool (i.e., end-effector pose) irrespective of the kinematics approach, which usually comes defined by a homogeneous matrix (see Equation 4.1 right) with the position vector (i.e., "p"),

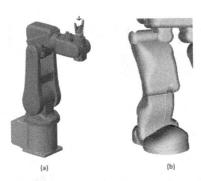

FIGURE 4.1 (a) ABB IRB120 robot with six DoF. (b) A leg of the RH0 Humanoid with six DoF.

and the orientation matrix (i.e., "noa"). The output of IK must give the robot joints' motion (i.e., magnitudes $\theta_1 \ldots \theta_n$), which once applied to the mechanism make the tool reach the expected pose.

4.1.3 Analytical Difficulty to Solve Inverse Kinematics

The first great trouble is that **the IK problem might have NO solution, ONE solution, or SEVERAL solutions**, then solving it is quite a challenge (Tsai, 1999). The complexity of the robot IK problem increases dramatically with the number of joints. For a manipulator with six DoF, we need to find the multiple possible answers out of a set up to 16 solutions, which is the theoretical maximum for this mechanism.

For a real example of a Puma-type robot IK (Figure 4.1a), the problem has a maximum of eight solutions because of their crossing axes. Mathematically this is a nontrivial issue, as we must solve **a system of 12 nonlinear coupled algebraic equations, with six unknowns and from zero to eight solutions** (see Equation 4.1). We can reckon the severe difficulty of the challenge doing deep down in this mathematics system of equations for the Puma robot (see Equation 4.2). For other similar mechanisms, such as the case of the RH0 humanoid leg (see Figure 4.1b), the challenge is quite the same. Besides, the difficulty is even more severe with the increase of the number of DoF of more complex robots.

When the analytical challenge is so big, the path to follow tends to be a numeric approach and optimization algorithms, as the closed-form or geometric solutions seem inaccessible. This numeric way has become the widespread solution for most the IK problem. However, we will see across this chapter how with the POE it is possible to generate a geometric, closed-form, and complete set of solutions for almost any robotics IK problem (Park, 1994).

$$H_{ST}(\theta) = H_{0,1}H_{1,2}H_{2,3}H_{3,4}H_{4,5}H_{5,6} = \begin{bmatrix} n_X & o_X & a_X & p_X \\ n_Y & o_Y & a_Y & p_Y \\ n_Z & o_Z & a_Z & p_Z \\ 0 & 0 & 0 & 1 \end{bmatrix} \quad (4.1)$$

Inverse Kinematics

$$n_x = C_6\left(S_5\left(S_1S_4 + C_4\left(C_1C_2S_3 + C_1C_3S_2\right)\right) + C_5\left(C_1S_2S_3 - C_1C_2C_3\right)\right) - S_6\left(C_4S_1 - S_4\left(C_1C_2S_3 + C_1C_3S_2\right)\right)$$

$$o_x = -C_6\left(C_4S_1 - S_4\left(C_1C_2S_3 + C_1C_3S_2\right)\right) - S_6\left(S_5\left(S_1S_4 + C_4\left(C_1C_2S_3 + C_1C_3S_2\right)\right) + C_5\left(C_1S_2S_3 - C_1C_2C_3\right)\right)$$

$$a_x = -C_5\left(S_1S_4 + C_4\left(C_1C_2S_3 + C_1C_3S_2\right)\right) + S_5\left(C_1S_2S_3 - C_1C_2C_3\right)$$

$$p_x = \frac{27}{100}C_1C_2 - \frac{151}{500}\left(C_1S_2S_3 - C_1C_2C_3\right) + \frac{7}{100}\left(C_1C_2S_3 - C_1C_3S_2\right)$$
$$- \frac{4}{25}\left(C_5S_1S_4 + C_1C_2C_3S_5 + C_1S_2S_3S_5 + C_1C_2C_4C_5S_3 + C_1C_3C_4C_5S_2\right)$$

$$n_y = -C_6\left(S_5\left(C_1S_4 - C_4\left(C_sS_1S_3 + C_3S_1S_2\right)\right) - C_5\left(S_1S_2S_3 - C_2C_3S_1\right)\right) + S_6\left(C_1C_4 + S_4\left(C_2S_1S_3 + C_3S_1S_2\right)\right)$$

$$o_y = C_6\left(C_1C_4 + S_4\left(C_2S_1S_3 + C_3S_1S_2\right)\right) + S_6\left(S_5\left(C_1S_4 - C_4\left(C_2S_1S_3 + C_3S_1S_2\right)\right) - C_5\left(S_1S_2S_3 - C_2C_3S_1\right)\right)$$

$$a_y = C_5\left(C_1S_4 - C_4\left(C_2S_1S_3 + C_3S_1S_2\right)\right) + S_5\left(S_1S_2S_3 - C_2C_3S_1\right)$$

$$p_y = \frac{27}{100}S_1S_2 - \frac{151}{500}\left(S_1S_2S_3 - C_2C_3S_1\right) + \frac{7}{100}\left(C_2S_1S_3 - C_3S_1S_2\right)$$
$$+ \frac{4}{25}\left(C_1C_5S_4 - C_2C_3S_1S_5 + S_1S_2S_3S_5 - C_2C_4C_5S_1S_3 - C_3C_4C_5S_1S_2\right)$$

$$n_z = C_6\left(S_{23}C_5 + C_{23}C_4S_5\right) + C_{23}S_4S_6$$

$$o_z = -S_6\left(S_{23}C_5 + C_{23}C_4S_5\right) + C_{23}C_6S_4$$

$$a_z = S_{23}S_5 - C_{23}C_4C_5$$

$$p_z = \frac{27}{100}C_2 - \frac{151}{500}\left(C_2S_3 + C_3S_2\right) + \frac{7}{100}\left(C_2C_3 - S_2S_3\right)$$
$$+ \frac{4}{25}\left(C_2S_3S_5 + C_3S_2S_5 - C_2C_3C_4C_5 + C_4C_5S_2S_3\right) + \frac{29}{100}$$

(4.2)

4.2 NUMERIC VS. GEOMETRIC SOLUTIONS

4.2.1 A Numeric Approach to Solve Inverse Kinematics

Numeric algorithms have become the preferred method to solve IK. Some texts even propose this as the only approach to solve certain cases, such as under-actuated robots, redundant manipulators, or robots without spherical wrists. Nevertheless, this is not entirely true. Screw theory solves IK for many of these robot architectures with many DoF, using a geometric approach, as we will demonstrate along with this chapter.

The numerical solution considers the IK problem as an adjustment of the joint magnitudes until the Forward Kinematics (FK) matches the desired pose. This approach is an optimization problem used to minimize the error between the forward and inverse formulations.

Once we have chosen the numeric route, there are multiple ways to work out the problem (e.g., Heuristics, AI, Neural Networks, Genetic Algorithms). Nonetheless, in the end, all of them are some sort of optimization with the underlying idea that a direct closed-form geometric solution is not feasible. The resulting implementation is by force an iterative algorithm. Therefore, the convergence or the computation speed is not guaranteed (Sipser, 2021). For some combinations of initial guesses and

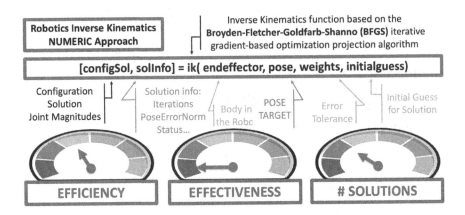

FIGURE 4.2 An example of Inverse Kinematics NUMERIC algorithm performance.

desired end-effector poses, the algorithm may even exit without any robot configuration solution.

4.2.2 AN EXAMPLE OF A NUMERIC ALGORITHM

For instance, an archetypal algorithm is the Broyden-Fletcher-Goldfarb-Shanno (BFGS) iterative gradient-based optimization projection algorithm (see Figure 4.2). A quasi-Newton method uses the gradients of a cost function from previous iterations to generate approximate second derivative information.

Following the characteristic numeric algorithm BFGS, let us take a more detailed look at its practical application (see Figure 4.2). The function to solve the IK expects to work with a specific rigid body tree model. The arguments of the function are the rigid body for which the IK problem stands (e.g., end-effector), the target configuration (e.g., pose), values for the error tolerance in the iterations (e.g., weights), and an initial guess for the solution (e.g., initial guess). In a certain way, we see the general numeric approach as kind of funny, in the sense that we want to get a solution to a complex IK problem, and we need to introduce an excellent initial guess of that solution that we do not know. It works fine for many engineering applications because when a robot develops a task trajectory, the difference between the actual tool configuration and its next target is not significant. After all, usually, the trajectory discretization is small. This fact allows us to use the actual robot pose as the initial guess for the numerical algorithm. Anyhow, the function gives back a configuration as a solution (e.g., configSol) and adds some information regarding the iterative process (e.g., solinfo). We must pay attention to the fact that the output configuration is generally only an approximation to the exact solution. The quality of BFGS is unquestionable, but its implementation suffers many drawbacks because of its nature.

It is essential to consider the fundamental features and typical **Characteristics for the IK Numeric Algorithms:**

- **The computation is NOT EFFICIENT** because the necessary iterations make the convergence speed is not guaranteed.

Inverse Kinematics 55

- **The calculation is NOT EFFECTIVE** because even if a solution exists, it usually is not exact and only an approximation. Even worse, sometimes the algorithm does not answer due to convergence difficulties (e.g., local minimum).
- **There is ONLY ONE solution** because of the optimization nature approach. Therefore, there is no possibility to choose the better solution out of the set of all which are possible.

All considered numeric algorithms are non-suitable for real-time robotics applications because of the lack of certainty for their convergence!

4.2.3 A Geometric Approach to Solve Inverse Kinematics

A closed-form solution can be determined using geometric or algebraic methods. However, this becomes enormously challenging and very difficult to apply with the increase in the number of robot joints and DoF. However, we can prove that this is a solvable problem if we use the screw theory POE for robotics.

Many texts establish that the necessary condition for a closed-form solution of many DoF robots is to have a spherical wrist (i.e., robots with the last three rotation joints axes orthogonal and intersecting in a common point). Besides, the Tool Center Point (TCP) must be at that wrist point. This method simplifies the IK because the tool's position is a function only of the first three joints, and the orientation exclusively depends on the last three joints. Then, we apply the algorithm known as "Kinematic Decoupling" to divide the problem with six DoF into two problems of three DoF. This idea is an innovative approach to keeping up with the geometric solution, but not very realistic. Most commercial manipulators and robots do not have the TCP position in the last three DoF intersection. In other words, usually, some of the last three DoF affect the translation of the TCP in 3D and not only its rotation. With the screw theory, we will learn that this spherical wrist restriction is no longer necessary to get complete geometric solutions.

This chapter will show that the screw theory allows systematic, elegant, and geometrically meaningful solutions for the IK of many mechanical architectures, including those challenging with many DoF.

4.2.4 An Example of a Geometric Algorithm

The Screw Theory Toolbox for Robotics (ST24R) (Pardos-Gotor, 2021a) allows developing this kind of geometric algorithm for many robot architectures, as we will see in the following sections. Here, as an illustrative example (see Figure 4.3 in contrast with the numeric solution of Figure 4.2), we present a function to solve the IK. The function uses the screw theory canonical subproblems (explained in the next section). It is possible to appreciate how the implementation is elegant and straightforward. The function's arguments are the reference configuration of the tool (e.g., TcpInit) and the tool target pose (e.g., TcpGoal). There is no need to define levels of error tolerance or initial guess for the solution. Do not confuse the initial tool configuration with the initial guess of the numeric algorithms. The first is the unique information regarding the home tool pose for the geometric algorithm. The latter means a different guess configuration for each point of the trajectory and demands continuous reevaluation.

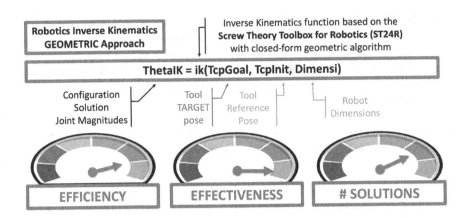

FIGURE 4.3 An example of Inverse Kinematics GEOMETRIC algorithm performance.

It is essential to consider the fundamental features and typical **Characteristics for the IK Geometric Algorithms:**

- **The computation is EFFICIENT** since there are no iterations for the geometric closed-form solution formulation.
- **The calculation is EFFECTIVE** because it gives exact solutions since the direct geometric formulation guarantees the convergence of the resolution.
- **There are MULTIPLE solutions**, as the geometry gives a set of all possible suitable configurations, which facilitates choosing the better solution for each application.

All considered, geometric algorithms are suitable for real-time applications, and screw theory provides a practical and elegant approach for it. Of course, these algorithms are extensible to other architectures beyond manipulator robots (Husty, 1996).

4.2.5 Puma Robot Inverse Kinematics Algorithms

We finish this section with an example of the comparison between the geometric and numeric IK algorithms. Afterward, throughout this chapter, we will explain how to build screw theory IK algorithms. For this example, it suffices to realize the performance difference between both approaches in a practical way. There is no need to worry now about how the algorithm works. We will learn it step by step along the following sections for this robot architecture and many others.

The goal is to compare the performance between the two methods; the first is a numeric "Robotics System Toolbox" (RST) algorithm (MathWorks, 2021a), and the second a "Screw Theory Toolbox for Robotics" (ST24R)[1] algorithm. The focus here is not to enter in detail for any of those two procedures but rather to compare only the results.

The exercise solves the IK for the typical Puma robot architecture (Figure 4.4) with six DoF (Paden and Sastry, 1988). This example's input is a random pose for the tool (i.e., position and orientation) inside the robot's dexterous workspace, as a

Inverse Kinematics

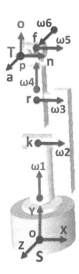

FIGURE 4.4 Puma robot kinematics analysis with the screw theory.

configuration expressed with a homogeneous matrix. We can check out the results and contrast the two different approaches' performance. The numeric "RST" algorithm gives an approximate (i.e., not exact) solution, whereas **the screw theory geometric algorithm provides a set of eight exact solutions. Moreover, in terms of computational costs, the "ST24R" exercise is two or three orders of magnitude faster**, as expected according to its geometric nature.

The code for this Exercise 4.2.5 is in the internet hosting for the software of this book[2].

4.3 CANONICAL SUBPROBLEMS FOR INVERSE KINEMATICS

4.3.1 A Key Idea to Solve Inverse Kinematics

The screw theory approach provides a significant benefit with the chance given by the POE to develop geometric algorithms for complex IK. This method targets to reduce the IK complexity by dividing the problem into simpler canonical subproblems, which have exact geometric solutions with a clear understanding. Besides, these geometric algorithmic implementations are numerically stable. To get a closed-form geometric solution for a robot's IK with many Degrees of Freedom (DoF), we first solve several subproblems. It is the idea of divide and conquer applied to robot mechanics.

The original approach developed the very well-known Paden–Kahan (PK) subproblems to solve many manipulators' IK with revolute joints. This set of subproblems is not exhaustive, and clearly, some manipulators cannot be solved using these canonical subproblems. The canonical subproblems developed by PK serve to solve the IK problems for a single rotation (PK1), two consecutive rotations about crossing axes (PK2), and a rotation to a given distance (PK3). The expression of PK subproblems in terms of screw theory is a POE applied to a point. Built on an unpublished work of Kahan (Kahan, 1983), Paden presented this method for the first time (Paden, 1986).

PK subproblems are a set of archetypical solved geometric problems that frequently occur in IK of common robotic mechanisms. These problems are not comprehensive, but they are helpful to simplify inverse kinematic analysis for many industrial manipulators and other mechanisms (Murray et al., 2017).

For any mechanism with revolute joints, analyzed and defined by a POE equation, PK subproblems are helpful to simplify the problem. Generally, subproblems are applied to solve points in the IK problem (e.g., the intersection of joint axes) to solve the joint movement magnitudes (i.e., angles). The PK tools can apply simplifications to the POE expression.

We will see the three famous PK subproblems (PK1 to PK3), the general solution, and some examples to help to practice the screw theory concepts.

The beauty of the canonical IK methodology is that the set of subproblems is by no means limited. Once we get the basics, it will be possible to develop new subproblems geometrically meaningful and adapted to solve various robotics configurations (Pardos-Gotor, 2018).

Afterward, we introduce a set of new Pardos-Gotor (PG) subproblems (PG1 to PG8). We created them to solve some other common architectures presented in robotics.

The Screw Theory Toolbox for Robotics (ST24R) is the library that includes the software code for all canonical subproblems and many other screw theory functions. This toolbox facilitates the analysis of the mechanisms with formulas of screw theory and Lie algebra. All the details of the "ST24R" software are in an internet repository (Pardos-Gotor, 2021a).

The "ST24R" is free software. It is possible to redistribute or modify it under the GNU Lesser General Public License terms published by the Free Software Foundation, either version 3.7 of the License or any other version. Anyone is more than welcome to improve and extend any function of this toolbox.

It is essential to understand that we develop the "ST24R" software for teaching and didactical purposes. Therefore, this tool is not designed nor optimized for industrial or commercial use. The code's priority is to understand better the screw theory developments and explanations rather than optimize the software execution performance. Nevertheless, the evident advantages of the mathematical approach will lead to efficient applications.

It is crucial to know how we implement the "ST24R" IK functions to give exact solutions if they exist or approximations otherwise. It is essential to be aware of this fact. This coding philosophy is handy for simulations and applications. The outcome is a natural motion for many robots, even for targets lying out of the workspace. In Chapter 8 of this book it is possible to check how for several simulations, giving targets outside the workspace, the robot tool responds with a good approximation. Nonetheless, if there is a need for a different behavior of the manipulator, we can modify all the canonical subproblems programming for other requirements. For instance, it is easy to change the canonical IK functions to get no result unless the exact solution exists, to freeze the robot in such a case.

Inverse Kinematics

It is advisable to review the Screw Theory Toolbox for Robotics "ST24R" (Pardos-Gotor, 2021a) to understand better the concepts working in the examples of the following sections.

4.3.2 PADEN–KAHAN SUBPROBLEM ONE (PK1) – ONE ROTATION

4.3.2.1 ROTATION around ONE Single AXIS Applied to a POINT

The movement is defined by the rotation of a point "p" around an axis "ω" so that the point passes to occupy the position of another point "k" (see Figure 4.5a). The expression for this subproblem is given by Equation 4.3, as a single exponential of rotation twist "ξ" and magnitude "θ," which applied to a point "p" gives result "k."

$$e^{\hat{\xi}\theta} p = k \tag{4.3}$$

Given the points "p" and "k" and the rotation axis "ω," the IK problem must determine the rotation angle "θ" to accomplish the movement. There is a single geometric solution, given by Equation 4.4, for any 3D problem (the complementary to 2π not considered):

$$u = p - r \; ; \; u_p = u - \omega\omega^T u \; ; \; v = k - r \; ; \; v_p = v - \omega\omega^T v$$
$$\theta = atan2\left(\omega^T\left(u_p \times v_p\right), u_p^T \cdot v_p\right) \tag{4.4}$$

4.3.2.2 PK1 Subproblem Simplification

Case for Simplification PK1 (see Figure 4.5b), given by Equation 4.5. A Screw Rotation does not affect any point on the axis "ω" of its twist so that we can cancel the exponential for any "θ" (Equation 4.6).

$$e^{\hat{\xi}\theta} p = p \; \forall p \in \omega \tag{4.5}$$

$$\forall \theta \tag{4.6}$$

We can check the PK1 algorithm with Exercise 4.3.2, the code of which is in the internet hosting for the software of this book[3].

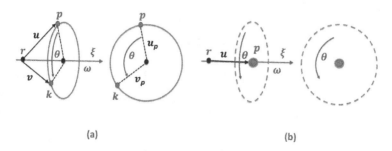

(a)   (b)

FIGURE 4.5 (a) Paden-Kahan IK subproblem ONE (PK1). (b) PK1 simplification when the point is on the axis of rotation.

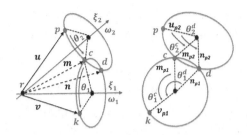

FIGURE 4.6 Paden–Kahan IK subproblem TWO (PK2).

4.3.3 Paden–Kahan Subproblem Two (PK2) – Two Crossing Rotations

4.3.3.1 ROTATION around TWO Subsequent CROSSING AXES Applied to a POINT

This subproblem movement has a first rotation of a point "p" around an axis "ω_2" and then followed by a second rotation around an axis "ω_1" so that the point passes to occupy the position given by another point "k" (Figure 4.6). The axes "ω_2" and "ω_1" cross at some point "r." The expression for this subproblem in terms of the screw theory is the product of two exponentials (see Equation 4.7), defined by the twists "ξ_1" and "ξ_2" and the magnitudes "θ_1" and "θ_2," which applied to a point "p" make two successive rotations to give the point "k."

Given the points "p" and "k" and the rotation axes "ω_2" and "ω_1," the IK problem must determine the necessary rotation angles "θ_2" and "θ_1" to accomplish that motion. The subproblem can have none, one, or two solutions, as we can infer from the graphic representation. In existence, each solution has a couple of angles, "θ_2-θ_1" (the complementary solutions to 2π not considered).

To get the possible solutions, we must be able to determine the intermediate points "c" and "d" where point "p" moves after the first rotation. Subsequently, we solve the subproblem by a division into two PK1 subproblems, one for each screw.

$$e^{\hat{\xi}_1\theta_1}e^{\hat{\xi}_2\theta_2}p = k \qquad (4.7)$$

Therefore, the IK solution needs to find points "c" and "d" in any 3D case, and we obtain them by geometry with the Equation 4.8 and Equation 4.9.

$$u = p - r$$

$$v = k - r$$

$$\alpha = \frac{\left(\omega_1^T \omega_2\right)\omega_2^T u - \omega_1^T v}{\left(\omega_1^T \omega_2\right)^2 - 1}$$

$$\beta = \frac{\left(\omega_1^T \omega_2\right)\omega_1^T v - \omega_2^T u}{\left(\omega_1^T \omega_2\right)^2 - 1}$$

Inverse Kinematics

$$\gamma^2 = \frac{\|u\|^2 - \alpha^2 - \beta^2 - 2\alpha\beta\omega_1^T\omega_2}{\|\omega_1 \times \omega_2\|^2}$$

$$c = r + \alpha\omega_1 + \beta\omega_2 + \gamma\left(\omega_1 \times \omega_2\right) \tag{4.8}$$

$$d = r + \alpha\omega_1 + \beta\omega_2 - \gamma\left(\omega_1 \times \omega_2\right) \tag{4.9}$$

Given "p" and "k" and the rotation axes "ω_2" and "ω_1," we get two possible geometric solutions with couples "θ_2-θ_1" with two PK1 subproblems (the complementary solutions to 2π not considered). The first solution develops the path "p-c-k" with the pair of magnitudes "θ^c_2-θ^c_1" as Equation 4.10. The second solution creates the course "p-d-k" with the couple of magnitudes "θ^d_2-θ^d_1" as Equation 4.11.

$$m = c - r \;\; ; \;\; n = d - r \;\; ; \;\; u_{p2} = u - \omega_2\omega_2^T u \;\; ; \;\; v_{p1} = v - \omega_1\omega_1^T v$$

$$m_{p2} = m - \omega_2\omega_2^T m \;\; ; \;\; m_{p1} = m - \omega_1\omega_1^T m \;\; ; \;\; n_{p2} = n - \omega_2\omega_2^T n \;\; ; \;\; n_{p1} = n - \omega_1\omega_1^T n$$

$$\begin{aligned}\theta^c_2 &= atan2\left(\omega_2^T\left(u_{p2} \times m_{p2}\right), u_{p2}^T \cdot m_{p2}\right) \\ \theta^c_1 &= atan2\left(\omega_1^T\left(m_{p1} \times v_{p1}\right), m_{p1}^T \cdot v_{p1}\right)\end{aligned} \tag{4.10}$$

$$\begin{aligned}\theta^d_2 &= atan2\left(\omega_2^T\left(u_{p2} \times n_{p2}\right), u_{p2}^T \cdot n_{p2}\right) \\ \theta^d_1 &= atan2\left(\omega_1^T\left(n_{p1} \times v_{p1}\right), n_{p1}^T \cdot v_{p1}\right)\end{aligned} \tag{4.11}$$

As we have mentioned, canonical subproblems can be modified to make them useful for solving other IK problems. For example, in the literature there is an interesting extension of this PK2 (Yue-sheng and Ai-ping, 2008).

We can check the PK2 algorithm with Exercise 4.3.3, the code of which is in the internet hosting for the software of this book[4].

4.3.4 PADEN–KAHAN SUBPROBLEM THREE (PK3) – ROTATION TO A DISTANCE

4.3.4.1 ROTATION at a Given DISTANCE Applied to a POINT

This subproblem represents the rotation of a point "p" around an axis "ω" so that the point passes to occupy the position given by the point "c" or "d." This motion must comply with the condition that the distance from either of those two points (i.e., "c" or "d") to a certain point "k" in the 3D space is given by the magnitude "δ" (see Figure 4.7).

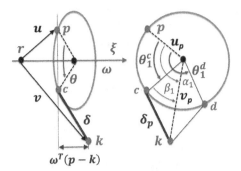

FIGURE 4.7 Paden–Kahan IK subproblem THREE (PK3).

The subproblem definition in terms of the screw theory is the distance "δ" (norm or module) of the exponential defined by the twist "ξ" and the magnitude "θ" applied to a point "p" and then subtracted from the point "k," as defined by Equation 4.12.

Given the points "p" and "k," the rotation axis "ω" and the distance "δ," the IK problem must determine the necessary rotation angles "θ." The IK problem can have none, one, or two solutions, as we can infer from the graphic representation. Each solution of the IK is the angle "θ" (the complementary solutions to 2π not considered).

To get the possible solutions, we must determine the angles "α_1" and "β_1" as a function where point "p" is after the rotation. Afterward, we solve the subproblem simply by geometry.

$$\left\| e^{\hat{\xi}\theta} p - k \right\| = \delta \qquad (4.12)$$

Therefore, first, we need to find angles "α_1" by Equation 4.13 and "β_1" by Equation 4.14 as pure geometry, in which case the solution is trivial for the two possible values as Equation 4.15.

$$u = p - r \ ; \ u_p = u - \omega\omega^T u \ ; \ v = k - r \ ; \ v_p = v - \omega\omega^T v$$
$$\alpha_1 = atan2\left(\omega^T \left(u_p \times v_p\right), u_p^T \cdot v_p\right) \qquad (4.13)$$
$$\delta_p^2 = \delta^2 - \omega^T (p-k)^2$$

$$\beta_1 = \cos^{-1}\left(\frac{\|u_p\|^2 + \|v_p\|^2 - \delta_p^2}{2\|u_p\|\|v_p\|} \right) \qquad (4.14)$$

$$\theta_1^d = \alpha_1 + \beta_1$$
$$\theta_1^c = \alpha_1 - \beta_1 \qquad (4.15)$$

Inverse Kinematics

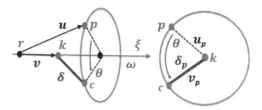

FIGURE 4.8 Paden–Kahan IK subproblem THREE (PK3) case for simplification.

4.3.4.2 PK3 Subproblem Simplification

A screw rotation motion applied to any point "p" does not affect the distance to any point "k" on the axis of the screw twist "ω" (Figure 4.8). The mathematical expression of this fact is given by Equation 4.16. Consequently, we can take out the exponential of the PK3 because it is valid for any "θ" (Equation 4.17).

$$\left\| e^{\hat{\xi}\theta} p - k \right\| = \| p - k \| = \delta \quad \forall k \in \omega \tag{4.16}$$

$$\forall \theta \tag{4.17}$$

We can check the PK3 algorithm with Exercise 4.3.4, the code of which is in the internet hosting for the software of this book[5].

4.3.5 Pardos–Gotor Subproblem One (PG1) – One Translation

Screw theory allows us to develop new canonical subproblems to solve other complex IK problems. Following PK's example, we propose new subproblems, the first three extending the PK to translation and the other five for different geometries.

We dare to give these new canonical subproblems the surname of this book's author, not because of the superior quality of the ideas but for taking responsibility for these implementations. Furthermore, the naming makes it possible to distinguish these algorithms from other equivalent or even more efficient sources in the literature.

Hereafter there are eight canonical Pardos-Gotor subproblems designed with mainly a didactical purpose. All these algorithms give geometric closed-form solutions. The geometries are for the translation along a single axis (PG1), two translations along two subsequent crossing axes (PG2), translation to a given distance (PG3), rotation around two parallel axes (PG4), rotation of a line or plane (PG5), two consecutive rotations around skewed axes (PG6), three consecutive rotations around a skew and two parallel axes (PG7) and three consecutive parallel rotations applied to a coordinate system (PG8).

It is always possible to design other new subproblems for different robot applications!

4.3.5.1 TRANSLATION along a SINGLE AXIS Applied to a POINT

This PG1 subproblem is for the translation along a single axis (see Figure 4.9a). The movement makes the translation of a point "p" along the axis "$\hat{v}$" so that the point

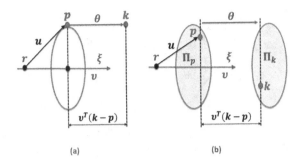

FIGURE 4.9 (a) Pardos-Gotor IK subproblem ONE (PG1). (b) PG1 applied to a plane.

passes to occupy the position given by another point, "k." The expression for this subproblem in terms of the screw theory is provided by a single exponential as Equation 4.18, with the translation twist "ξ" and magnitude "θ," which applied to a point "p" gives point "k."

Given the points "p" and "k" and the translation axis "$\hat{v}$," the IK problem must determine the necessary translation magnitude "θ" to accomplish the movement. **There is a single geometric solution obtained for any 3D problem** by Equation 4.19. This formulation works nicely for any relative position between points "p" and "k." The result "θ" of the IK problem will be positive or negative as a function of the points' position. The good sign of the magnitude will naturally come out correctly with this algorithm's implementation.

In applying this canonical subproblem to actual robotics translation joints, it is essential to pay attention to the characteristics of the mechanisms to adjust the direction of the twist. This way, the result gives the correct sign for the screw magnitude.

$$e^{\hat{\xi}\theta} p = k \qquad (4.18)$$

$$\theta = v^T (k - p) \qquad (4.19)$$

4.3.5.2 PG1 Extension - TRANSLATION along a SINGLE AXIS Applied to a PLANE

The formulation for this PG1 subproblem also permits solving another IK subproblem in the same way. That is the translation of a plane in 3D (see Figure 4.9b).

A screw translation applied to a plane "Πp" to move it on top of another plane, "Πk," can be obtained knowing any point (e.g., "p" and "k") on those planes, Equation 4.20.

This simplification is very interesting to reduce the difficulty of some mechanisms IK problems. Perhaps we do not know precisely if the two points "p" and "k" are on a line parallel to the axis of the screw. However, if the points are part of the two planes, we can define the movement's magnitude. The PG1 computes "θ" to move a plane where "p" is to another plane where "k" is included by Equation 4.21. Both are parallel planes, and the distance between them "θ" is measured on the axis of the

Inverse Kinematics

screw, even if the planes are not perpendicular to the twist axis. The points "p" and "k" might not coincide after the movement, but the expression works, providing the motion's magnitude.

$$e^{\hat{\xi}\theta}\Pi_p = \Pi_k \qquad (4.20)$$

$$\theta = v^T(k-p) \ \forall p \in \Pi_p \wedge \forall k \in \Pi_k \qquad (4.21)$$

We can check the PG1 algorithm with Exercise 4.3.5, the code of which is in the internet hosting for the software of this book[6].

4.3.6 Pardos-Gotor Subproblem Two (PG2) – Two Crossing Translations

4.3.6.1 TRANSLATION along Two Subsequent CROSSING AXES Applied to a POINT

We define this geometry by the first translation of a point "p" along an axis "$\mathring{v}_2$" and then followed by a second translation along an axis "$\mathring{v}_1$" so that the point passes to occupy the position given by "k" (Figure 4.10). The expression (Equation 4.22) is with the product of two exponentials defined by the twists "ξ_1" and "ξ_2" and the magnitudes "θ_1" and "θ_2," which applied to a point "p" give as a result "k." Given points "p" and "k" and the translation axes "v_2" and "$\mathring{v}_1$," the IK solves the translations "θ_2" and "θ_1" to accomplish that movement.

The subproblem can have one solution or none. To get the possible solution, we must determine the intermediate point "c" where point "p" is after the first translation motion, by Equation 4.23. Afterward, we can solve the subproblem by a division into two PG1 subproblems, one for each screw. In existence, the solution is formed by a couple of two magnitudes, "θ_2-θ_1," by Equation 4.24.

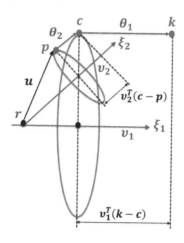

FIGURE 4.10 Pardos-Gotor IK subproblem TWO (PG2).

$$e^{\hat{\xi}_1\theta_1}e^{\hat{\xi}_2\theta_2}p = k \tag{4.22}$$

$$c = k + \frac{\|v_2 \times (p-k)\|}{\|v_2 \times v_1\|} v_1 \cdot sign\left((v_2 \times (p-k))^T \cdot (v_2 \times v_1)\right) \tag{4.23}$$

$$\begin{aligned}\theta_2 &= v_2^T(c-p) \\ \theta_1 &= v_1^T(k-c)\end{aligned} \tag{4.24}$$

We can check the PG2 algorithm with Exercise 4.3.6, the code of which is in the internet hosting for the software of this book[7].

4.3.7 Pardos-Gotor Subproblem Three (PG3) – Translation to a Distance

4.3.7.1 TRANSLATION to a Given DISTANCE Applied to a POINT

We define this movement by the translation of a point "p" along an axis "$\hat{v}$" so that the point passes to occupy the position given by points "c" or "d," complying with the condition that the distance from either of those two points to a certain point "k" in the space 3D, is given by the magnitude "δ" (Figure 4.11). The subproblem definition takes the distance "δ" (norm or module) of the exponential defined by the twists "ξ" and the magnitude "θ," applied to a point "p" and then subtracted from point "k." The expression is Equation 4.25.

Given the points "p" and "k," the translation axis "$\hat{v}$" and the distance "δ," the IK problem must determine the necessary translation magnitudes "θ" to accomplish that movement. There can be none, one, or two solutions, as we can infer from the representation (Figure 4.11).

The canonical PG3 subproblem algorithm solves the intersection between a line and a sphere in 3D. Afterward, the subproblem can be easily solved geometrically by Equation 4.26. The point "k" does not need to be on the axis "$\hat{v}$" of the twist "ξ."

$$\left\|e^{\hat{\xi}\theta}p - k\right\| = \delta \tag{4.25}$$

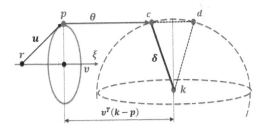

FIGURE 4.11 Pardos-Gotor IK subproblem THREE (PG3).

Inverse Kinematics

$$\theta = v^T(k-p) \pm \sqrt{\left(v^T(k-p)\right)^2 - \|k-p\|^2 + \delta^2} \quad (4.26)$$

We can check the PG3 algorithm with Exercise 4.3.7, the code of which is in the internet hosting for the software of this book[8].

4.3.8 Pardos-Gotor Subproblem Four (PG4) – Two Parallel Rotations

4.3.8.1 ROTATION around TWO Subsequent PARALLEL AXES Applied to a POINT

We define this geometry by a first rotation of a point "p" around an axis "ω_2," and then followed by a second rotation around an axis "ω_1," so that the point passes to occupy the position given by another point "k." Axes "ω_2" and "ω_1" are parallel. The expression for this subproblem is the product of two exponentials defined by twists "ξ_1" and "ξ_2," and magnitudes "θ_1" and "θ_2," which applied to a point "p" give as result point "k." The definition is Equation 4.27.

Given points "p" and "k" and the rotation axes "ω_2," and "ω_1," the IK problem must determine the necessary angle rotation "θ_2," and "θ_1," to accomplish that motion.

The subproblem can have none, one, or two solutions as it is possible to infer from the graphic representation (Figure 4.12). In existence, each solution has a couple of angles, "$\theta 2$-θ_1" (the complementary solutions to 2π not considered). To get the possible solutions, we must determine the intermediate points "c" and "d," where point "p" is after the first rotation. Afterward, we solve the subproblem by a division into two PK1 subproblems.

$$e^{\hat{\xi}_1 \theta_1} e^{\hat{\xi}_2 \theta_2} p = k \quad (4.27)$$

The IK solution needs to find points "c" and "d" in any 3D case, and we get them by geometry. The canonical PG4 subproblem implements an algorithm with the core idea of the geometrical calculation of two segments, "a" and "h" (see Figure 4.12). Knowing "o_1" by Equation 4.28, the magnitudes for "a" and "h" according to Equation 4.29, and their axes "ω_a" and "ω_h" by Equation 4.30, it is relatively straightforward to get the points "c" and "d" following Equation 4.31.

$$u = p - r_2 \; ; \; v = k - r_1 \; ; \; u_{p2} = u - \omega_2 \omega_2^T u \; ; \; v_{p1} = v - \omega_1 \omega_1^T v$$

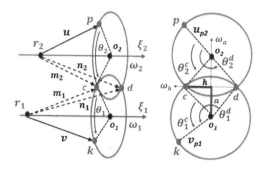

FIGURE 4.12 Pardos-Gotor IK subproblem FOUR (PG4).

$$o_2 = r_2 + \omega_2 \omega_2^T u$$

$$o_1 = r_1 + \omega_1 \omega_1^T v \qquad (4.28)$$

$$a = \frac{\|o_2 - o_1\|^2 - \|u_{p2}\|^2 + \|v_{p1}\|^2}{2\|o_2 - o_1\|} \;;\; h = \sqrt{\|v_{p1}\|^2 - a^2} \qquad (4.29)$$

$$\omega_a = \frac{(o_2 - o_1)}{\|o_2 - o_1\|} \;;\; \omega_h = \omega_1 \times \omega_a \qquad (4.30)$$

$$c = o_1 + a\omega_a + h\omega_h \;;\; d = o_1 + a\omega_a - h\omega_h \qquad (4.31)$$

Now, given points "p" and "k" and the rotation axes "ω_2" and "ω_1," we get the two possible geometric solutions which are couples "θ_2-θ_1," with two PK1 subproblems (complementary solutions to 2π not considered). The first solution develops the path "p-c-k" with the couple of magnitudes "θ^c_2-θ^c_1," as Equation 4.32. The second solution develops the path "p-d-k" with the couple of magnitudes "θ^d_2-θ^d_1," as Equation 4.33 (see Figure 4.12).

$$m_2 = c - r_2 \;;\; n_2 = d - r_2 \;;\; m_1 = c - r_1 \;;\; n_1 = d - r_1$$

$$m_{p2} = m_2 - \omega_2 \omega_2^T m_2 \;;\; n_{p2} = n_2 - \omega_2 \omega_2^T n_2$$

$$m_{p1} = m_1 - \omega_1 \omega_1^T m_1 \;;\; n_{p1} = n_1 - \omega_1 \omega_1^T n_1$$

$$\begin{aligned}\theta^c_2 &= atan2\left(\omega_2^T (u_{p2} \times m_{p2}), u_{p2}^T \cdot m_{p2}\right) \\ \theta^c_1 &= atan2\left(\omega_1^T (m_{p1} \times v_{p1}), m_{p1}^T \cdot v_{p1}\right)\end{aligned} \qquad (4.32)$$

$$\begin{aligned}\theta^d_2 &= atan2\left(\omega_2^T (u_{p2} \times n_{p2}), u_{p2}^T \cdot n_{p2}\right) \\ \theta^d_1 &= atan2\left(\omega_1^T (n_{p1} \times v_{p1}), n_{p1}^T \cdot v_{p1}\right)\end{aligned} \qquad (4.33)$$

4.3.8.2 PG4 Extension - ROTATION around TWO PARALLEL AXES Applied to a LINE

We define the motion by the first rotation around an axis "ω_2" applied to a line "ω_p," parallel to the axes of both twists "ω_2" and "ω_1." The first motion is followed by a second rotation around an axis "ω_1" so that the line passes to occupy the position "ω_k" (Figure 4.13). The expression for this subproblem uses the product of two exponentials defined by twists "ξ_1," and "ξ_2" and magnitudes "θ_1" and "θ_2," which applied to the line "ω_p" give another line "ω_k" (see Equation 4.34).

$$e^{\hat{\xi}_1 \theta_1} e^{\hat{\xi}_2 \theta_2} \omega_p = \omega_k \qquad (4.34)$$

Inverse Kinematics

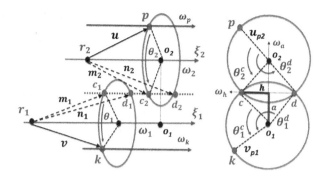

FIGURE 4.13 Pardos-Gotor IK subproblem FOUR (PG4) extension.

The only difference with the leading formulation of the PG4 subproblem is the way to calculate the values for "m_2," "n_2," "m_1," and "n_1" (see Equation 4.35). The rest of the formulation for the algorithm is the same.

$$m_2 = c_2 - r_2 \; ; \; n_2 = d_2 - r_2 \; ; \; m_1 = c_1 - r_1 \; ; \; n_1 = d_1 - r_1 \qquad (4.35)$$

The solution is also the same as for PG4 (see the right-hand side of Figure 4.13). The first couple of magnitudes "$\theta^c_2 - \theta^c_1$" by Equation 4.32, and the second "$\theta^d_2 - \theta^d_1$" by Equation 4.33.

We can check the PG4 algorithm with Exercise 4.3.8, the code of which is in the internet hosting for the software of this book[9].

4.3.9 Pardos-Gotor Subproblem Five (PG5) – Rotation of a Line or Plane

4.3.9.1 ROTATION around ONE Single AXIS Applied to a Perpendicular LINE or PLANE

This problem is an extension of the PK1. The motion to solve is the orientation of a line (Figure 4.14a) or a plane (Figure 4.14b) around some axis. The reason to propose this subproblem is to get a convenient solution for some robot geometries, where the critical information to the IK is to find out the plane on which a joint orientation must

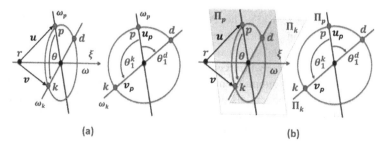

FIGURE 4.14 (a) Pardos-Gotor IK subproblem FIVE (PG5) applied to a line. (b) PG5 applied to a plane.

lie. Sometimes, it is beneficial to know the plane where a manipulator's arm is moving. In those cases, only the control of one joint is orienting the plane. Therefore, two complementary values for the rotation of that joint comply with the same pose of the plane. This PG5 permits to get those two viable magnitudes of the joint at once. These are the solutions to this canonical IK subproblem.

We can define a line and a plane by any point "p" being part of them and the perpendicular to the screw axis through that point. This IK problem is defined by the rotation of point "p" around "ω," to make the point occupy the position "k" or "d." This subproblem has the POE expression of Equation 4.36 for the line "ω_p," and the plane "Π_p," respectively.

$$e^{\hat{\xi}\theta}\omega_p = \omega_k \; ; \; e^{\hat{\xi}\theta}\Pi_p = \Pi_k \qquad (4.36)$$

Given the axis "ω" of the screw, the point "p" and "d" or "k," the IK problem must determine the rotation angle "θ_1," to accomplish the movement. There are two geometric solutions "θ_1^k" and "θ_1^d" for any 3D problem, which we straightforwardly define as Equation 4.37.

$$u = p - r \; ; \; u_p = u - \omega\omega^T u \; ; \; v = k - r \; ; \; v_p = v - \omega\omega^T v$$
$$\theta_1^k = \operatorname{atan2}\left(\omega^T\left(u_p \times v_p\right), u_p^T \cdot v_p\right) \qquad (4.37)$$
$$\theta_1^d = \theta_1^k - \pi.$$

We can check the PG5 algorithm with Exercise 4.3.9, the code of which is in the internet hosting for the software of this book[10].

4.3.10 Pardos-Gotor Subproblem Six (PG6) – Two Skewed Rotations

4.3.10.1 ROTATION around TWO Subsequent SKEW AXES Applied to a POINT

We have seen before two IK canonical subproblems related to two subsequent rotations, PK2 and PG4, which applied to a point move it to a different place in SE(3). The former is classical that solves the case for crossing axes. The latter is new for parallel axes. We present a generalization to find the IK of two consecutive rotations with disjoint axes (Figure 4.15). This PG6 is more generic than PK2, as it solves the two consecutive rotations problem irrespective of the axes' characteristics. It works equally well for crossing and skew axes. However, this PG6 does not work for parallel axes, which we can solve with PG4.

The expression for this subproblem in terms of the screw theory is Equation 4.38, as the product of two exponentials defined by twists "ξ_1" and "ξ_2" and magnitudes "θ_1" and "θ_2," which applied to a point "p" give as result point "k." The characteristic magnitudes of both screws can be easily obtained by Equation 4.39.

$$e^{\hat{\xi}_1\theta_1}e^{\hat{\xi}_2\theta_2}p = k \qquad (4.38)$$

Inverse Kinematics

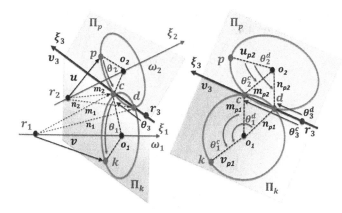

FIGURE 4.15 Pardos-Gotor IK subproblem SIX (PG6).

$$
\begin{aligned}
u &= p - r_2 \;\; ; \;\; v = k - r_1 \\
u_{p2} &= u - \omega_2 \omega_2^T u \;\; ; \;\; v_{p1} = v - \omega_1 \omega_1^T v \\
o_2 &= r_2 + \omega_2 \omega_2^T u \;\; ; \;\; o_1 = r_1 + \omega_1 \omega_1^T v
\end{aligned}
\tag{4.39}
$$

Given points "p" and "k" and the rotation axes "ω_2" and "ω_1," the IK problem must determine the necessary angle rotation "θ_2" and "θ_1" to accomplish that motion.

The subproblem can have none, one, or two solutions, as we can infer from the graphic representation (Figure 4.15). In existence, each solution definition includes a couple of angles, "θ_2-θ_1" (the complementary solutions to 2π not considered).

To get the possible solutions, we must be able to determine the intermediate points "c" and "d," where point "p" is after the first rotation. Afterward, we solve the subproblem by a division into two PK1 subproblems, one for each screw.

This IK canonical subproblem has some solutions in the literature. For instance, there is an algorithm (Yue-Sheng and Ai-ping, 2008) based on introducing the vector "r_2-r_1" (Figure 4.15). Our proposal takes a different path. The idea behind this new canonical PG6 subproblem is to solve the intersecting line between the two planes "Π_p" and "Π_k," where the two rotations develop (i.e., the path of rotation for "p" and "k"). Afterward, it is relatively easy to get the points "c" and "d." Pay attention because these points might be different on each plane. Then, we solve the canonical subproblem by a division into two PK1 subproblems.

The cornerstone of this new PG6 algorithm is the introduction of a pure translational screw. The twist of this new screw "ξ_3" has an axis defined by Equation 4.40, which is the intersecting line between the two planes where the points "p" and "k" rotate. A point "r_3" on this twist axis is available, knowing the information from the twists "ξ_1" and "ξ_2" with Equation 4.41. Now, it is possible to propose the same approach given for the Pardos-Gotor Three subproblem (i.e., translation to a given distance PG3), to find out the intersection between the axis of the twist "ξ_3" and the circles described by the rotations of the point "p" and "k," if they exist, using the Equation 4.42 and Equation 4.43.

$$\xi_3 = \begin{bmatrix} \upsilon_3 \\ 0 \end{bmatrix} = \begin{bmatrix} \omega_1 \times \omega_2 \\ 0 \end{bmatrix} \quad (4.40)$$

$$r_3 = \frac{\omega_1\left(\omega_1^T o_1 - \omega_2^T o_2 \cdot \omega_1^T \omega_2\right) + \omega_2\left(\omega_2^T o_2 - \omega_1^T o_1 \cdot \omega_1^T \omega_2\right)}{1 - \omega_1^T \omega_2} \quad (4.41)$$

$$\left\| e^{\hat{\xi}_3 \theta_3} r_3 - o_2 \right\| = \left\| u_{p2} \right\| \quad (4.42)$$

$$\left\| e^{\hat{\xi}_3 \theta_3} r_3 - o_1 \right\| = \left\| v_{p1} \right\| \quad (4.43)$$

We must pay attention to the picture of the point "c" and "d" (Figure 4.15). This figure is a particular case of crossing axes for the twists "ξ_1" and "ξ_2" and the existence of two double solutions to this IK problem. In a general case, the points exist for the plane of rotation of each twist. The twist "ξ_1" will have the points (if they exist) of the intersection of the axis of "ξ_3" with the circle of rotation of "k" with "c_1" and "d_1." The second twist, "ξ_2," has the points (if they exist) of the intersection of the axis of "ξ_3" with the circle of rotation of "p" will be "c_2" and "d_2." The magnitudes for the pure translation "$\theta{c}13\text{-}\theta^{d1}{}_3$" and "$\theta{c}23\text{-}\theta^{d2}{}_3$" make the point "$r_3$" intersect the circles of "k" and "p" respectively.

We can obtain these quantities by two formulations. The first uses the direct formula for the intersection between a line and a sphere in 3D, defined by Equation 4.44.

$$\theta_3^{c2} = \upsilon_3^T(o_2 - r_3) + \sqrt{\left(\upsilon_3^T(o_2 - r_3)\right)^2 - \|o_2 - r_3\|^2 + \|u_{p2}\|^2} \quad (4.44)$$

$$\theta_3^{d2} = \upsilon_3^T(o_2 - r_3) - \sqrt{\left(\upsilon_3^T(o_2 - r_3)\right)^2 - \|o_2 - r_3\|^2 + \|u_{p2}\|^2}$$

$$\theta_3^{c1} = \upsilon_3^T(o_1 - r_3) + \sqrt{\left(\upsilon_3^T(o_1 - r_3)\right)^2 - \|o_1 - r_3\|^2 + \|v_{p1}\|^2}$$

$$\theta_3^{d1} = \upsilon_3^T(o_1 - r_3) - \sqrt{\left(\upsilon_3^T(o_1 - r_3)\right)^2 - \|o_1 - r_3\|^2 + \|v_{p1}\|^2}$$

As we pointed to before, another way to obtain these magnitudes for the pure translation screw of "ξ_3" might be to recur directly to the canonical PG3 subproblem for a translation to a given distance, using Equation 4.45 and Equation 4.46.

$$\left\| e^{\hat{\xi}_3 \theta_3} r_3 - o_2 \right\| = |u_{p2}| \rightarrow \begin{bmatrix} \theta_3^{c2} & \theta_3^{d2} \end{bmatrix} \quad (4.45)$$

Inverse Kinematics

$$\left\|e^{\hat{\xi}_3\theta_3}r_3 - o_1\right\| = \left\|v_{p1}\right\| \rightarrow \begin{bmatrix} \theta_3^{c1} & \theta_3^{d1} \end{bmatrix} \tag{4.46}$$

Then we obtain the points "c_2" and "d_2" (if they exist) in the circle of rotation of "p" by Equation 4.47, and the points "c_1" and "d_1" (if they exist) in the circle of rotation of "k" by Equation 4.48.

$$\begin{aligned} c_2 &= r_3 + \theta_3^{c2} v_3 \\ d_2 &= r_3 + \theta_3^{d2} v_3 \end{aligned} \tag{4.47}$$

$$\begin{aligned} c_1 &= r_3 + \theta_3^{c1} v_3 \\ d_1 &= r_3 + \theta_3^{d1} v_3 \end{aligned} \tag{4.48}$$

It is necessary to analyze the possible outcomes of the PG6 algorithm. They are heavily dependent on the twists "ξ_1" and "ξ_2" geometry and the relative position of their axes. We identify the alternatives by the values of the just calculated points "c_2," "d_2," "c_1," and "d_1." The possible causes of the solutions are:

- **TWO double solutions:** which can happen when the axes "ω_2" and "ω_1" cross in a point and "$c_1 = c_2 = c$" and "$d_1 = d_2 = d$."
- **ONE double solution:** which might happen when the axes "ω_2" and "ω_1" are skewed and "$c_1 = c_2 = c$" or "$d_1 = d_2 = d$." It also comes off when the axes "ω_2" and "ω_1" cross in a point and "$c_1 = c_2 = c = d_1 = d_2 = d$."
- **NO solution:** which might happen for skewed and crossing axes, but with "$c_1 <> c_2$" and "$d_1 <> d_2$."

Having arrived here, with the input points "p" and "k," the rotation axes "ω_2-ω_1," the points "r_2-r_1" of the twists "ξ_1" and "ξ_2," and the points "c_1, c_2, d_1, d_2," we are in conditions to get the two possible geometric double solutions which are couples "θ_2-θ_1" with the application twice of PK1 subproblem (complementary solutions to 2π not considered). First, we calculate the points "m_1, m_2, n_1, n_2" according to Equation 4.49 (see Figure 4.15)

$$m_2 = c_2 - r_2 \; ; \; n_2 = d_2 - r_2 \; ; \; m_1 = c_1 - r_1 \; ; \; n_1 = d_1 - r_1 \tag{4.49}$$

$$m_{p2} = m_2 - \omega_2 \omega_2^T m_2 \; ; \; n_{p2} = n_2 - \omega_2 \omega_2^T n_2$$

$$m_{p1} = m_1 - \omega_1 \omega_1^T m_1 \; ; \; n_{p1} = n_1 - \omega_1 \omega_1^T n_1$$

Finally, we obtain the two possible geometric double solutions "θ^c_2-θ^c_1" by Equation 4.50 and "θ^d_2-θ^d_1" by Equation 4.51, following the PK1 concept.

$$\begin{aligned} \theta_2^c &= atan2\left(\omega_2^T\left(u_{p2} \times m_{p2}\right), u_{p2}^T \cdot m_{p2}\right) \\ \theta_1^c &= atan2\left(\omega_1^T\left(m_{p1} \times v_{p1}\right), m_{p1}^T \cdot v_{p1}\right) \end{aligned} \tag{4.50}$$

$$\theta_2^d = atan2\left(\omega_2^T\left(u_{p2} \times n_{p2}\right), u_{p2}^T \cdot n_{p2}\right)$$
$$\theta_1^d = atan2\left(\omega_1^T\left(n_{p1} \times v_{p1}\right), n_{p1}^T \cdot v_{p1}\right) \tag{4.51}$$

It is essential to recall the "ST24R" programming philosophy when implementing the PG6. The function always gives this couple of doubles solutions, even though they might not exist sometimes. In those situations, the PG6 provides only an approximation that we find helpful for applications and simulations.

We can check the PG6 algorithm with Exercise 4.3.10, the code of which is in the internet hosting for the software of this book[11].

4.3.11 Pardos-Gotor Subproblem Seven (PG7) – Three Rotations to a Point

4.3.11.1 ROTATION around THREE Subsequent AXES (ONE SKEW + TWO PARALLEL) Applied to a POINT

This subproblem comprises a rotation of a point "p" about three successive axes, so this point moves to coincide with another point "k." The axes of twists "ξ_2" and "ξ_3" are parallel to each other, and the first axis of a twist "ξ_1" is not parallel to the last two axes (Figure 4.16). We must determine the angles "θ_1," "θ_2," and "θ_3" to solve the new subproblem.

The expression for this subproblem in terms of the screw theory is Equation 4.52, which is the product of three exponentials defined by twists "ξ_1," "ξ_2," and "ξ_3" and magnitudes "θ_1," "θ_2," and "θ_3," applied to a point "p" give as result point "k." The characteristic magnitudes of the first and third screws can be easily obtained by Equation 4.53.

$$e^{\hat{\xi}_1\theta_1}e^{\hat{\xi}_2\theta_2}e^{\hat{\xi}_3\theta_3}p = k \tag{4.52}$$

$$u = p - r_3 \; ; \; v = k - r_1 \; ; \; u_{p3} = u - \omega_3\omega_3^T u \; ; \; v_{p1} = v - \omega_1\omega_1^T v$$
$$o_3 = r_3 + \omega_3\omega_3^T u \; ; \; o_1 = r_1 + \omega_1\omega_1^T v \tag{4.53}$$

Given points "p" and "k" and the axes "ω_3," "ω_2," and "ω_1," the IK problem must determine the necessary angle rotation "θ_3," "θ_2," and "θ_1" to accomplish that motion.

The subproblem can have none, one, two, or four solutions, as we can infer from the graphic representation (Figure 4.16). Each solution (if it exists) contains a triplet of angles "θ_3-θ_2-θ_1" (we do not consider the complementary solutions to 2π).

To get the possible solutions, we must determine the intermediate points "c" and "d," where point "p" is after the second rotation. Afterward, we solve the subproblem by one PK1 subproblem plus two canonical PG4 subproblems of parallel rotations, one for each point.

This canonical subproblem has some solutions in the literature (Chen et al., 2015; Dimovski et al., 2018) that proposed algorithms to solve the same geometry. Our proposal takes a different path. The idea for this new approach is to use even more the screw theory to solve the intersecting line between the two planes "Π_p" and "Π_k"

Inverse Kinematics

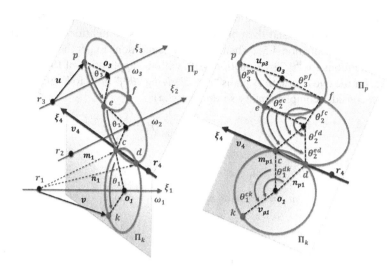

FIGURE 4.16 Pardos-Gotor IK subproblem SEVEN (PG7).

where are the two rotations (i.e., the paths for "p" and "k"). Afterward, it is easier to get the points "c" and "d." Pay attention because these points might be different on each plane. Then, the canonical subproblem can be solved by first a PK1 subproblem to get the possible solutions for "θ_1," plus a division into two PG4 subproblems to get the possible solutions for "θ_3" and "θ_2."

This new PG7 algorithm's cornerstone introduces a pure translational screw (Figure 4.16). The axis of the new twist "ξ_4," given by Equation 4.54, is the intersecting line between the two planes "Π_p" and "Π_k." We obtain the point "r_4" on this twist axis from the twists "ξ_1" and "ξ_3" following Equation 4.55.

$$\xi_4 = \begin{bmatrix} v_4 \\ 0 \end{bmatrix} = \begin{bmatrix} \omega_1 \times \omega_3 \\ 0 \end{bmatrix}. \tag{4.54}$$

$$r_4 = \frac{\omega_1 \left(\omega_1^T o_1 - \omega_3^T o_3 \cdot \omega_1^T \omega_3 \right) + \omega_3 \left(\omega_3^T o_3 - \omega_1^T o_1 \cdot \omega_1^T \omega_3 \right)}{1 - \omega_1^T \omega_3} \tag{4.55}$$

We use the canonical PG3 for translation to a given distance to determine the intersection between the axis of "ξ_4" and the circle described by the rotation of "k." Solving this motion with PG3 by Equation 4.56, we get the magnitudes for the pure translation screw "θ_4." Then, we get the points "c" and "d" (if they exist) in the circle of "k" by Equation 4.57.

$$\left\| e^{\hat{\xi}_4 \theta_4} r_4 - o_1 \right\| = \left\| v_{p1} \right\| \rightarrow \begin{bmatrix} \theta_4^c & \theta_4^d \end{bmatrix} \tag{4.56}$$

$$\begin{aligned} c &= r_4 + \theta_4^c v_4 \\ d &= r_4 + \theta_4^d v_4 \end{aligned} \tag{4.57}$$

Now, with the inputs of the rotation axis "ω_1," the point "r_1" of the twists "ξ_1," the point "k" to obtain "v_{p1}," and the points "c" and "d," is straightforward to get the initial values with Equation 4.58. Then, we are in conditions to get the two possible geometric solutions for "θ_1" by Equation 4.59, with the application of PK1 subproblem twice, for the two alternatives paths "c to k" giving "θ_1^{ck}," or "d to k" giving "θ_1^{dk}," (we do not consider complementary solutions to 2π).

$$v = k - r_1 \;\; ; \;\; v_{p1} = v - \omega_1 \omega_1^T v \tag{4.58}$$

$$m_1 = c - r_1 \;\; ; \;\; n_1 = d - r_1$$
$$m_{p1} = m_1 - \omega_1 \omega_1^T m_1 \;\; ; \;\; n_{p1} = n_1 - \omega_1 \omega_1^T n_1$$

$$\theta_1^{ck} = atan2\left(\omega_1^T \left(m_{p1} \times v_{p1}\right), m_{p1}^T \cdot v_{p1}\right)$$
$$\theta_1^{dk} = atan2\left(\omega_1^T \left(n_{p1} \times v_{p1}\right), n_{p1}^T \cdot v_{p1}\right) \tag{4.59}$$

After solving the first unknown "θ_1," applying the PG4 subproblem for two consecutive parallel rotations is straightforward to solve the unknowns "θ_2" and "θ_3." It is clear (Figure 4.16) that we must apply PG4 for two cases, the motion from the point "p" to "c" by Equation 4.60 and the movement from "p" to "d" by Equation 4.61. As we know, the PG4 provides none, one, or two double solutions for the IK. These are, respectively, their cases for no intersection of the motions, "e" and "f" coincident points and "e" and "f" different points.

$$e^{\hat{\xi}_2 \theta_2} e^{\hat{\xi}_3 \theta_3} p = c \rightarrow \begin{bmatrix} \theta_3^{pe} & \theta_2^{ec} ; \theta_3^{pf} & \theta_2^{fc} \end{bmatrix} \tag{4.60}$$

$$e^{\hat{\xi}_2 \theta_2} e^{\hat{\xi}_3 \theta_3} p = d \rightarrow \begin{bmatrix} \theta_3^{pe} & \theta_2^{ed} ; \theta_3^{pf} & \theta_2^{fd} \end{bmatrix} \tag{4.61}$$

Putting together all possible solutions for the set "θ_1-θ_2-θ_3," the final solution for the IK subproblem PG7 has up to four triple solutions, as shown in Equation 4.62.

$$\begin{bmatrix} \theta_1^{ck} & \theta_2^{ec} & \theta_3^{pe} ; \theta_1^{ck} & \theta_2^{fc} & \theta_3^{pf} ; \theta_1^{dk} & \theta_2^{ed} & \theta_3^{pe} ; \theta_1^{dk} & \theta_2^{fd} & \theta_3^{pf} \end{bmatrix} \tag{4.62}$$

Nevertheless, it is necessary to analyze with more detail the possible outcomes of the PG7 algorithm. The alternatives depend on the values of the points: "c," "d," "e," and "f." The possible causes of the solutions are:

- FOUR triple solutions, which happen when the intersecting points "c," "d," "e," and "f" exist, and all of them are different.
- TWO triple solutions happen when the intersecting points "$c = d$" or "$e = f$," but the two conditions do not exist simultaneously.
- ONE triple solution happens when the intersecting points "$c = d$" and "$e = f$."

Inverse Kinematics

- NO solution, which can happen when the intersecting points "c" and "d" or "e" and "f" do not exist, and there cannot be continuity from the point "p" to "k."

The implementation of the PG7 follows the "ST24R" philosophy and always gives the four triple solutions. If some of them do not exist, the PG7 provides a good approximation, which we find helpful for applications and simulations.

Once we have understood the geometric solution for this canonical PG7 subproblem, we would like to introduce some examples of robot mechanisms where we can apply this IK solution. The typical configuration is the one we find in manipulators of type PUMA or Bending Backwards. It is possible to appreciate the first three DoF for these robots, with two commercial manipulators ABB IRB120 and ABB IRB1600. The first joint axis is not parallel (in fact, it is perpendicular) to the second and third joint axes, parallel to each other. In the Bending Backwards robot, the first joint axis does not cross the others. In both cases, they comply with the definition of the PG7, and this IK can be solved geometrically, giving up to four different closed-form triple solutions for the magnitudes of the first three joint motions. We will see practical cases for these robots in several exercises.

Of course, there are many more robots and mechanisms for which PG7 can be helpful. For instance, the robots: "ABB IRB7600," "Qianjiang I," or "KUKA KR 360 R2830."

The following sections of this chapter will show how to combine this PG7 canonical subproblem with others to solve the complete kinematics of commercial robots.

We can check the PG7 algorithm with Exercise 4.3.11, the code of which is in the internet hosting for the software of this book[12].

4.3.12 Pardos-Gotor Subproblem Eight (PG8) – Three Rotations to a Pose

4.3.12.1 ROTATION around THREE Subsequent PARALLEL AXES Applied to a POSE (Position Plus Orientation) or COORDINATE SYSTEM

This canonical problem raises a geometry with three consecutive parallel rotations applied to a coordinate system. It is somewhat an extension of the classic planar arm problem with three rotational joints but generalized and extended to 3D (see Figure 4.17).

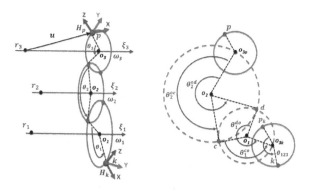

FIGURE 4.17 Pardos-Gotor IK subproblem EIGHT (PG8).

PG8 is a unique canonical problem because the simplification techniques for the kinematic equation do not apply the POE to a point but a pose or coordinate system (i.e., some configuration including position and orientation). This pose includes the translation and rotation information expressed as a homogeneous matrix. The PG8 canonical subproblem may have no solution, one or two triple solutions for the set "θ_3-θ_2-θ_1."

The PG8 subproblem expression in terms of screw theory comes as the POE of the three consecutive parallel rotations, applied to a homogeneous matrix "H_p," and the result is another homogeneous matrix "H_k." This formulation gives Equation 4.63. Both homogeneous matrices represent the coordinate system of a pose (position and orientation). We represent those configurations "H_p" and "H_k" with more detail as Equation 4.64 because we use some of the internal components for the next steps.

$$e^{\hat{\xi}_1\theta_1} e^{\hat{\xi}_2\theta_2} e^{\hat{\xi}_3\theta_3} H_p = H_k \tag{4.63}$$

$$e^{\hat{\xi}_1\theta_1} e^{\hat{\xi}_2\theta_2} e^{\hat{\xi}_3\theta_3} \begin{bmatrix} Xp_x & Yp_x & Zp_x & Pp_x \\ Xp_y & Yp_y & Zp_y & Pp_y \\ Xp_z & Yp_z & Zp_z & Pp_z \\ 0 & 0 & 0 & 1 \end{bmatrix} = \begin{bmatrix} Xk_x & Yk_x & Zk_x & Pk_x \\ Xk_y & Yk_y & Zk_y & Pk_y \\ Xk_z & Yk_z & Zk_z & Pk_z \\ 0 & 0 & 0 & 1 \end{bmatrix} \tag{4.64}$$

The core of the proposed algorithm to solve this PG8 subproblem is a divide and conquer approach. In the first place, we reduce the problem to one with only two parallel rotations. For doing so, we pass "H_p" to the right-hand side of the kinematics problem equation and apply both sides to the center of the third screw, and this is the point "o_{3p}," defined as Equation 4.65. Then, the right-hand side of Equation 4.66 is another point, "o_{3k}," which is the center of the third screw affected by the first and second rotations (see Figure 4.17). On the left-hand side of Equation 4.66, we can cancel the third exponential by PK1 simplification (i.e., the third rotation does not affect a point "o_{3p}" on its axis). Consequently, the outcome is the expression of Equation 4.67.

$$p = \begin{bmatrix} Pp_x \\ Pp_y \\ Pp_z \end{bmatrix} \ ; \ u = p - r_3 \ ; \ o_{3p} = r_3 + \omega_3 \omega_3^T u \tag{4.65}$$

$$e^{\hat{\xi}_1\theta_1} e^{\hat{\xi}_2\theta_2} e^{\hat{\xi}_3\theta_3} o_{3p} = H_k H_p^{-1} o_{3p} = o_{3k} \tag{4.66}$$

$$e^{\hat{\xi}_1\theta_1} e^{\hat{\xi}_2\theta_2} o_{3p} = o_{3k} \rightarrow \begin{bmatrix} \theta_1^{co} & \theta_2^{oc} ; \theta_1^{do} & \theta_2^{od} \end{bmatrix} \tag{4.67}$$

Now, we only need to realize that Equation 4.67 is exactly the expression for the PG4 subproblem of two consecutive parallel rotations applied to a point. Therefore, the

Inverse Kinematics

solution of PG4 can give none, one, or two double solutions for "θ_2-θ_1" as a function of the existence of the point "c" and "d" (for more details, see Figure 4.17).

Once we knew the final position for the center of the third screw, "o_{3k}" we can reevaluate the third twist on "o_{3k}" and translate the point "p" on this new twist "ξ_{3k}" to get "p_k" by Equation 4.68. Jointly with the value for the position "k" of "H_k," we are in the condition to calculate the full rotation "θ_{123}" employing the subproblem PK1 with Equation 4.69. Afterward, we get the values for "θ_3" only subtracting from the total rotation the calculated values for "θ_2-θ_1."

$$\xi_{3k} = \begin{bmatrix} -\omega_3 \times o_{3k} \\ \omega_3 \end{bmatrix} \tag{4.68}$$

$$p_k = o_{3k} + (p - o_{3p})$$

$$k = \begin{bmatrix} Pk_x \\ Pk_y \\ Pk_z \end{bmatrix}$$

$$e^{\hat{\xi}_{3k}\theta_{123}} p_k = k \rightarrow \theta_{123} \tag{4.69}$$

We get the final solution with the two sets of triple solutions by Equation 4.70.

$$\theta_{123} = \theta_1 + \theta_2 + \theta_3 \tag{4.70}$$

$$\left[\theta_3^{01} = \theta_{123} - \theta_1^{co} - \theta_2^{oc} \; ; \; \theta_3^{02} = \theta_{123} - \theta_1^{do} - \theta_2^{od}\right]$$

$$\left[\theta_1^{co} \quad \theta_2^{oc} \quad \theta_3^{01} \; ; \; \theta_1^{do} \quad \theta_2^{od} \quad \theta_3^{02}\right]$$

We can check the PG8 algorithm with Exercise 4.3.12, the code of which is in the internet hosting for the software of this book[13].

4.4 PRODUCT OF EXPONENTIALS APPROACH

The mathematical formulation of the kinematics using the screw theory notation provides numerous advantages, particularly for IK. The POE (Brockett, 1983) offers the possibility to obtain closed-form geometric solutions suitable for real-time applications. The payoff is multiple advantages for solving both kinematics and dynamics problems. Besides, it allows an optimal kinematic design of mechanisms (Park, 1991).

This treatment is somewhat a different path from many textbooks, which prefer a DH formulation. Nevertheless, among other advantages, the twists' geometric significance makes the POE a superior alternative to using the DH parameters, as will be demonstrated along with more complex problems of inverse kinematics (Davidson and Hunt, 2004).

This chapter provides examples to apply this kinematics formulation for manipulators and robots with different architectures and many DoF (Brockett et al., 1993). The inverse kinematics problem remains as simple as possible for practical application with these geometrical considerations (Li, 1990). Afterward, we can extend this approach to other robot configurations and mechanisms.

For a robot manipulator, the IK problem aims to calculate the joint magnitudes in the given tool pose. In other words, given the desired position and orientation for the tool (i.e., end-effector pose), we must get the robot joint (i.e., θ_1-θ_n) magnitudes, which comply with the desired motion. The IK problems might have no solution, one or multiple solutions. Thus, solving IK is not a trivial problem at all.

The complexity of the robot IK problem increases dramatically with the number of joints. For example, in a Puma robot architecture with six DoF, we must solve a system of 12 nonlinear, coupled equations with six unknowns and zero to eight solutions. All in all, this is genuinely a challenging mathematical problem.

There is a new IK kinematics treatment with screw theory and Lie algebras. They formalize the POE mathematically to represent a manipulator's kinematics with an exquisite treatment. The POE combines a great deal of analytical sophistication to becoming a successful approach.

To illustrate the comparison between the use of the POE instead of other methods, we can make a simile using the formula of Euler as Equation 4.71. It is possible to work with the arithmetic of sines and cosines to deal with complex numbers. Using the exponential expression allows a much more powerful calculation tool when working with complexes. Thus, similarly, the use of screw theory and its canonical POE expression will enable us to tackle better rigid body chains' complex problems. We will confirm the highest mathematical performance provided by the exponential of matrices along the following sections.

$$e^{i\omega} = \cos\omega + i\sin\omega \qquad (4.71)$$

The POE provides a truly geometric and global representation of kinematics, which dramatically simplifies robotics analysis using twists. Because the essential mathematical expression is the exponential of a matrix, we can quickly treat the resulting equations of motion for the robot. The manipulation of the exponentials expressions is very convenient to solve IK problems (Lynch and Park, 2017).

4.4.1 General Solution to Inverse Kinematics

Under the screw theory formalism and using the POE, the IK problem must obtain the joints' motion (i.e., **magnitudes $\theta1$-θ_n**), knowing the configuration for **the tool pose (i.e., orientation and position)**.

The general solution to any manipulator IK problem follows this algorithm:

- Select the **Spatial coordinate system "S"** (usually stationary and the base) and the **Mobile coordinate system "T"** (typically the tool). There is no rule

Inverse Kinematics

and complete flexibility to make this definition, and we can choose the most convenient frames according to the application.

- Define the **Axis of each joint**, which are the axes "ω_i" for revolute joints and "v_i" for the prismatic joints, and a point "q_i" on those axes.
- Obtain the **Twists** "ξ_i" for the joints, knowing for each revolute joint its axis and a point on that axis (see Equation 4.72), and for each prismatic joint its axis (see Equation 4.73).

$$\xi_i = \begin{bmatrix} v_i \\ \omega_i \end{bmatrix} = \begin{bmatrix} -\omega_i \times q_i \\ \omega_i \end{bmatrix} \qquad (4.72)$$

$$\xi_i = \begin{bmatrix} v_i \\ 0 \end{bmatrix} \qquad (4.73)$$

- Get **the pose of the tool at the reference "$H_{ST}(0)$"** (home) robot position. This configuration of the end-effect happens when all the joint magnitudes are zero. This tool pose comes as a homogeneous matrix.
- Express the **kinematics mapping "$H_{ST}(\theta)$," with the product of all joint screw exponentials (POE) and "$H_{ST}(0)$."** In this equation, the input to the IK problem is the homogeneous matrix (i.e., **noap**) which represents the desired tool pose or configuration (i.e., orientation and position in 3D). Simultaneously, the searched problem output must be the joints' magnitudes "θ_i" to comply with the IK map.

$$H_{ST}(\theta) = \prod_{i=1}^{n} e^{\hat{\xi}_i \theta_i} H_{ST}(0) = \begin{bmatrix} n\ o\ a\ p \\ 0\ 0\ 0\ 1 \end{bmatrix}$$

- **Analyze the robot mechanics to recognize patterns of Canonical Subproblems inside the POE expression** for different sets of joints (i.e., one, two, or three) as introduced in Section 4.3 of this chapter.

$$H_{ST}(\theta) = e^{\hat{\xi}_1 \theta_1} \cdots e^{\hat{\xi}_n \theta_n} H_{ST}(0) = \begin{bmatrix} n\ o\ a\ p \\ 0\ 0\ 0\ 1 \end{bmatrix}$$

- **Manipulate the POE to cancel some of the unknown variables** (i.e., joint magnitudes "θ_i"), introducing some elements (e.g., points, lines, planes, coordinate systems) on both sides of the kinematics equation. For doing that, to move the screw exponential conveniently between the expression left and right side. This operation aims to convert the complete IK map into a simpler POE, corresponding with one of the identified canonical subproblems, whose geometric solutions we know have an exact meaning and are numerically stable. Afterward, we introduce the solved joint magnitudes in the IK map and repeat the exponential manipulation for some other variables. This process ends when we get the set of all feasible joint magnitudes "$\theta_i \ldots \theta_n$" which can be none, one, or multiple solutions.

This IK general approach is not unique nor completely systematic for all types of mechanisms, as the procedure depends very much on identifying the applicable canonical subproblems. We need the ability and intelligence of a human being to analyze the different mechanisms. However, this method produces geometric algorithms very efficient and effective in comparison con other numeric and algebraic alternatives.

To learn how to solve IK problems is necessary to practice the proposed procedure with several examples. There are exercises with typical robot architectures in the following sections. They will permit us to consolidate the skills to manipulate the POE expressions. These examples have different joints, configurations, and a diverse selection of base and tool frames. The solved robot architectures are Puma (e.g., ABB IRB120), Bending Backwards (ABB IRB1600), Gantry (e.g., ABB 6620LX), Scara (e.g., ABB IRB910SC), Collaborative (e.g., UNIVERSAL UR16e), and a Redundant manipulator (e.g., KUKA IIWA). We briefly introduce an example with a humanoid of 21 DoF (i.e., UC3M RH0). After practicing these manipulators, we realize how a powerful tool is the POE to solve much more complex IK for challenging robotics structures.

It is advisable to review the Screw Theory Toolbox for Robotics "ST24R" (Pardos-Gotor, 2021a) to understand better the concepts working in the examples of the following sections.

This general formulation of the solution for IK problems also represents a very nice collaboration between mathematics and engineering subjects. It all starts with the paradox of using screw theory, a 19th-century mathematical approach, to better solve the IK problems addressed with the DH convention, which is an eminently 20th-century engineering approach. Then, we can solve the most abstract formulation of the kinematic map according to the POE of screw theory with the help of the engineer's vision, who knows the robot's mechanics, solving nonlinear equations with a powerful geometric sense.

What a beautiful cooperation between the disciplines of mathematics and engineering!

4.4.2 Puma Robots (e.g., ABB IRB120)

It is clear to appreciate how beautiful and elegant it is to get the problem's definition with the POE of the joints' motion. We only need to apply several canonical subproblems in different ways. It is essential to highlight the flexibility of the IK screw theory approach because different algorithms can solve a problem. To illustrate this fact, in this section, we solve this ABB IRB120 IK in four different ways, all of them providing closed-form geometric solutions. The same can happen with other robot architectures included in the following sections. For them, we will show fewer algorithms and leave it up for further exercises other alternatives, which are easy grasping the insights of the IK algorithms of this first example.

This section results descriptive to solve the IK for the typical Puma robot mechanics. The problem definition for the manipulator configuration is helpful to analyze all the robot joints with the screw theory approach (see Figure 4.18).

Inverse Kinematics

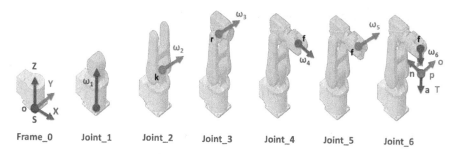

FIGURE 4.18 Puma robot with its screw inverse kinematics analysis.

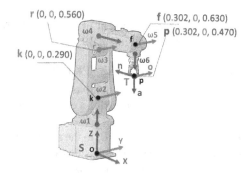

FIGURE 4.19 Puma ABB IRB120 inverse kinematics POE parameters.

4.4.2.1 Inverse Kinematics Puma Robot ABB IRB120 Problem Definition

We obtain the equation that relates the joints' motion ($\theta 1$-θ_6) to the pose for the tool (noap), with the POE of the joints' twists, as defined by the screw theory (Figure 4.19).

- Define the axis of each joint "$\omega_1...\omega_6$" Attention to the sign for ω_6 because of home position.

$$\omega_1 = \begin{bmatrix} 0 \\ 0 \\ 1 \end{bmatrix} \quad \omega_2 = \begin{bmatrix} 0 \\ 1 \\ 0 \end{bmatrix} \quad \omega_3 = \begin{bmatrix} 0 \\ 1 \\ 0 \end{bmatrix} \quad \omega_4 = \begin{bmatrix} 1 \\ 0 \\ 0 \end{bmatrix} \quad \omega_5 = \begin{bmatrix} 0 \\ 1 \\ 0 \end{bmatrix} \quad \omega_6 = \begin{bmatrix} 0 \\ 0 \\ -1 \end{bmatrix}$$

- Obtain the **Twists** ($\xi_1...\xi_6$), knowing each joint's axis and a point on that axis.

$$\xi_1 = \begin{bmatrix} -\omega_1 \times o \\ \omega_1 \end{bmatrix} \quad \xi_2 = \begin{bmatrix} -\omega_2 \times k \\ \omega_2 \end{bmatrix} \quad \xi_3 = \begin{bmatrix} -\omega_3 \times r \\ \omega_3 \end{bmatrix}$$

$$\xi_4 = \begin{bmatrix} -\omega_4 \times f \\ \omega_4 \end{bmatrix} \quad \xi_5 = \begin{bmatrix} -\omega_5 \times f \\ \omega_5 \end{bmatrix} \quad \xi_6 = \begin{bmatrix} -\omega_6 \times f \\ \omega_6 \end{bmatrix}$$

- Get $H_{st}(0)$, the pose of the tool at the reference (home) robot position.

$$H_{ST}(0) = T_{xyz}\begin{bmatrix} p_x \\ p_y \\ p_z \end{bmatrix} R_Y(\pi) = \begin{bmatrix} -1 & 0 & 0 & p_x \\ 0 & 1 & 0 & 0 \\ 0 & 0 & -1 & p_z \\ 0 & 0 & 0 & 1 \end{bmatrix}$$

- Define the problem in terms of the kinematics **POE map**.

$$H_{ST}(\theta) = e^{\hat{\xi}_1\theta_1}e^{\hat{\xi}_2\theta_2}e^{\hat{\xi}_3\theta_3}e^{\hat{\xi}_4\theta_4}e^{\hat{\xi}_5\theta_5}e^{\hat{\xi}_6\theta_6}H_{ST}(0) = \begin{bmatrix} n & o & a & p \\ 0 & 0 & 0 & 1 \end{bmatrix}$$

For the rest of this section, the development of the four algorithms refers to the robot parameters, which are explicit in Figure 4.19.

4.4.2.2 First Algorithm for ABB IRB120 IK "PK3+PK2+PK2+PK1"

The first approach presents the classical solution based on the PK subproblems. The canonical subproblems used are the PK3, the PK2 a couple of times, and finally PK1.

- **Solution for θ_3 with PK3:** We pass $H_{ST}(0)$ to the right-hand side of the problem definition equation. Then we apply both sides of the equation to point "f" (crossing of the last three axes: "ω_4," "ω_5," "ω_6"), subtract the result from point "k" (crossing of the first two axes: "ω_1," and "ω_2"), and calculate the norm of both sides (see Equation 4.74). We can apply the PK1 and PK3 simplification cases to the left-hand side of the equation to cancel the unknowns: "θ_6," "θ_5," and "θ_4" by PK1, and "θ_2" and "θ_1" by PK3. The right-hand side of the equation is now a known value "δ." The resulting equation is precisely the definition for the canonical PK3. We know the geometric IK solutions for PK3, none, one, or two. Finally, the motion for the third joint of the robot is obtained as the rotation magnitudes (if they exist) for the "θ_3" double solution by Equation 4.75.

$$\left\| e^{\hat{\xi}_1\theta_1}e^{\hat{\xi}_2\theta_2}e^{\hat{\xi}_3\theta_3}e^{\hat{\xi}_4\theta_4}e^{\hat{\xi}_5\theta_5}e^{\hat{\xi}_6\theta_6}f - k \right\| = \left\| H_{ST}(\theta)H_{ST}(0)^{-1}f - k \right\| \quad (4.74)$$

$$\left\| e^{\hat{\xi}_3\theta_3}f - k \right\| = \delta \rightarrow \begin{bmatrix} \theta_3^{01} & \theta_3^{02} \end{bmatrix} \quad (4.75)$$

- **Solutions for θ_2 & θ_1 with PK2:** We pass $H_{ST}(0)$ to the right-hand side of the problem definition equation. Then, we apply both sides of the equation to the point "f" (crossing of the last three axes: "ω_4," "ω_5," "ω_6") (see Equation 4.76). We can cancel the unknowns "θ_6," "θ_5," and "θ_4" by PK1 simplification. Besides, we already know "θ_3" from the previous step of the algorithm, and by applying the exponential of this third screw to point "f," we get another point "f_1." The right-hand side of the equation is now a known value "k_1." The

Inverse Kinematics

resulting equation is the definition for the canonical PK2. Moreover, we know the geometric IK solutions for PK2, none, one, or two double. Finally, for each "θ_3" value, we solve the motion for the first two joints, obtaining the magnitudes (if they exist) for "θ_2-θ_1" by Equation 4.77.

$$e^{\hat{\xi}_1\theta_1}e^{\hat{\xi}_2\theta_2}e^{\hat{\xi}_3\theta_3}e^{\hat{\xi}_4\theta_4}e^{\hat{\xi}_5\theta_5}e^{\hat{\xi}_6\theta_6}f = H_{ST}(\theta)H_{ST}(0)^{-1}f \qquad (4.76)$$

$$e^{\hat{\xi}_1\theta_1}e^{\hat{\xi}_2\theta_2}f_1 = k_1 \begin{cases} \theta_3^{01} \Rightarrow \begin{bmatrix} \theta_1^{01} & \theta_2^{01} & ; & \theta_1^{02} & \theta_2^{02} \end{bmatrix} \\ \theta_3^{02} \Rightarrow \begin{bmatrix} \theta_1^{03} & \theta_2^{03} & ; & \theta_1^{04} & \theta_2^{04} \end{bmatrix} \end{cases} \qquad (4.77)$$

- **Solutions for θ_5 & θ_4 with PK2**: We pass $H_{ST}(0)$ and the exponentials of the first three screws, which we already know from the previous steps of the algorithm (i.e., we got "θ_1," "θ_2," and "θ_3") to the right-hand side of the problem definition equation. Then, we apply both sides to the point "p" (see Equation 4.78). The right-hand side of the equation is now a known value "k_2." We can cancel the unknown "θ_6" by PK1 simplification. The resulting equation is the canonical PK2. We know the geometric IK solutions for PK2, none, one, or two double. Finally, for each set of θ_1-θ_2-θ_3" values, we obtained the rotation magnitudes (if they exist) for "θ_5-θ_4" double solutions by Equation 4.79.

$$e^{\hat{\xi}_4\theta_4}e^{\hat{\xi}_5\theta_5}e^{\hat{\xi}_6\theta_6}p = e^{-\hat{\xi}_3\theta_3}e^{-\hat{\xi}_2\theta_2}e^{-\hat{\xi}_1\theta_1}H_{ST}(\theta)H_{ST}(0)^{-1}p \qquad (4.78)$$

$$e^{\hat{\xi}_4\theta_4}e^{\hat{\xi}_5\theta_5}p = k_2 \begin{cases} \begin{bmatrix} \theta_1^{01} & \theta_2^{01} & \theta_3^{01} \end{bmatrix} \Rightarrow \begin{bmatrix} \theta_4^{01} & \theta_5^{01} & ; & \theta_4^{02} & \theta_5^{02} \end{bmatrix} \\ \begin{bmatrix} \theta_1^{02} & \theta_2^{02} & \theta_3^{01} \end{bmatrix} \Rightarrow \begin{bmatrix} \theta_4^{03} & \theta_5^{03} & ; & \theta_4^{04} & \theta_5^{04} \end{bmatrix} \\ \begin{bmatrix} \theta_1^{03} & \theta_2^{03} & \theta_3^{02} \end{bmatrix} \Rightarrow \begin{bmatrix} \theta_4^{05} & \theta_5^{05} & ; & \theta_4^{06} & \theta_5^{06} \end{bmatrix} \\ \begin{bmatrix} \theta_1^{04} & \theta_2^{04} & \theta_3^{02} \end{bmatrix} \Rightarrow \begin{bmatrix} \theta_4^{07} & \theta_5^{07} & ; & \theta_4^{08} & \theta_5^{08} \end{bmatrix} \end{cases} \qquad (4.79)$$

- **Solution for θ_6 with PK1**: We pass $H_{ST}(0)$ and the exponentials of the first five screws, which we already know from the previous steps of the algorithm (i.e., we got "θ_1," "θ_2," "θ_3," "θ_4," and "θ_5") to the right-hand side of the problem definition equation. Then, we apply both sides of the equation to the point "o" (see Equation 4.80). The right-hand side of the equation is now a known value "k_3." The resulting equation is precisely the definition for the canonical PK1. Furthermore, we know the geometric IK solutions for PK1. Finally, for each set of "θ_1-θ_2-θ_3-θ_4-θ_5" values, we solve the motion for the sixth joint of the robot, obtaining the rotation for "θ_6" solution by Equation 4.81.

$$e^{\hat{\xi}_6\theta_6}o = e^{-\hat{\xi}_5\theta_5}e^{-\hat{\xi}_4\theta_4}e^{-\hat{\xi}_3\theta_3}e^{-\hat{\xi}_2\theta_2}e^{-\hat{\xi}_1\theta_1}H_{ST}(\theta)H_{ST}(0)^{-1}o \qquad (4.80)$$

$$e^{\hat{\xi}_6\theta_6}o = k_3 \begin{cases} \begin{bmatrix} \theta_1^{01} & \theta_2^{01} & \theta_3^{01} & \theta_4^{01} & \theta_5^{01} \end{bmatrix} \Rightarrow \theta_6^{01} \\ \begin{bmatrix} \theta_1^{01} & \theta_2^{01} & \theta_3^{01} & \theta_4^{02} & \theta_5^{02} \end{bmatrix} \Rightarrow \theta_6^{02} \\ \begin{bmatrix} \theta_1^{02} & \theta_2^{02} & \theta_3^{01} & \theta_4^{03} & \theta_5^{03} \end{bmatrix} \Rightarrow \theta_6^{03} \\ \begin{bmatrix} \theta_1^{02} & \theta_2^{02} & \theta_3^{01} & \theta_4^{04} & \theta_5^{04} \end{bmatrix} \Rightarrow \theta_6^{04} \\ \begin{bmatrix} \theta_1^{03} & \theta_2^{03} & \theta_3^{02} & \theta_4^{05} & \theta_5^{05} \end{bmatrix} \Rightarrow \theta_6^{05} \\ \begin{bmatrix} \theta_1^{03} & \theta_2^{03} & \theta_3^{02} & \theta_4^{06} & \theta_5^{06} \end{bmatrix} \Rightarrow \theta_6^{06} \\ \begin{bmatrix} \theta_1^{04} & \theta_2^{04} & \theta_3^{02} & \theta_4^{07} & \theta_5^{07} \end{bmatrix} \Rightarrow \theta_6^{07} \\ \begin{bmatrix} \theta_1^{04} & \theta_2^{04} & \theta_3^{02} & \theta_4^{08} & \theta_5^{08} \end{bmatrix} \Rightarrow \theta_6^{08} \end{cases} \quad (4.81)$$

Exercise 4.4.2a implements this first algorithm. The code is in the internet hosting for the software of this book[14].

4.4.2.3 Second Algorithm for ABB IRB120 IK "PG7+PK2+PK1"

This approach uses the PG7 to get the solutions for the first three robot DoF, instead of "PK3 and PK2." We solve the last three DoF like in the previous algorithm.

- **Solutions for θ_3, θ_2 & θ_1 with PG7**: We pass $H_{ST}(0)$ to the right-hand side of the problem definition equation and apply both sides of the equation to the point "f" (crossing of the last three axes: "ω_4," "ω_5," "ω_6"). Then we can apply the PK1 simplification on the left-hand side to cancel the unknowns "θ_6," "θ_5," and "θ_4" (see Equation 4.82). The right-hand side of the equation is now a known value "k_1." The resulting equation is the canonical PG7 for three consecutive rotations about one skewed and two parallel axes. We know the geometric IK solution for PG7, which can give none, two, or four triple solutions for "θ_1-θ_2-θ_3" by Equation 4.83.

$$e^{\hat{\xi}_1\theta_1}e^{\hat{\xi}_2\theta_2}e^{\hat{\xi}_3\theta_3}e^{\hat{\xi}_4\theta_4}e^{\hat{\xi}_5\theta_5}e^{\hat{\xi}_6\theta_6}f = H_{ST}(\theta)H_{ST}(0)^{-1}f \quad (4.82)$$

$$e^{\hat{\xi}_1\theta_1}e^{\hat{\xi}_2\theta_2}e^{\hat{\xi}_3\theta_3}f = k_1 \begin{cases} \begin{bmatrix} \theta_1^{01} & \theta_2^{01} & \theta_3^{01} \end{bmatrix} \\ \begin{bmatrix} \theta_1^{01} & \theta_2^{02} & \theta_3^{02} \end{bmatrix} \\ \begin{bmatrix} \theta_1^{02} & \theta_2^{03} & \theta_3^{01} \end{bmatrix} \\ \begin{bmatrix} \theta_1^{02} & \theta_2^{04} & \theta_3^{02} \end{bmatrix} \end{cases} \quad (4.83)$$

- **Solutions for θ_6, θ_5 & θ_4**: We can calculate the magnitudes for the last three joints with the same approach as the first algorithm. With PK2, we obtain "θ_4--θ_5" (see Equation 4.78 and Equation 4.79). We got "θ_6" with PK1: (see Equation 4.80 and Equation 4.81).

Inverse Kinematics

Exercise 4.4.2b implements this second algorithm. The code is in the internet hosting for the software of this book[15].

4.4.2.4 Third Algorithm for ABB IRB120 IK "PG5+PG4+PK2+PK1"

This approach uses the PG5 and PG4 subproblems to get the solutions for the first three robot DoF, instead of "PK3 and PK2" or "PG7." The last three DoF are solved precisely like for the first algorithm.

- **Solutions for θ_1 with PG5:** We pass $H_{ST}(0)$ to the right-hand side of the problem definition equation and apply both sides of the equation to the point "f" (crossing of the last three axes: "ω_4," "ω_5," "ω_6") (see Equation 4.84). Then we can apply the PK1 simplification in the left-hand side of the equation to cancel "θ_6," "θ_5," and "θ_4." The second and third rotations do not change the plane where "f" moves, so they do not affect the "θ_1" value. In other words, "θ_1" is the only DoF used to orientate the robot body plane, and for calculating it, we can cancel the unknowns "θ_2" and "θ_3." The right-hand side of the equation is now a known value "k_1". The resulting equation is the canonical PG5, for one rotation about one single axis, applied to a plane. We know the geometric IK solutions can be two for "θ_1" by Equation 4.85.

$$e^{\hat{\xi}_1\theta_1}e^{\hat{\xi}_2\theta_2}e^{\hat{\xi}_3\theta_3}e^{\hat{\xi}_4\theta_4}e^{\hat{\xi}_5\theta_5}e^{\hat{\xi}_6\theta_6}f = H_{ST}(\theta)H_{ST}(0)^{-1}f \quad (4.84)$$

$$e^{\hat{\xi}_1\theta_1}f = k_1 \cdot \begin{bmatrix} \theta_1^{01} & \theta_1^{02} \end{bmatrix} \quad (4.85)$$

- **Solutions for θ_3 & θ_2 with PG4:** We pass $H_{ST}(0)$ and the exponential of the first screw, which we already know from the previous step of the algorithm (i.e., we got "θ_1"), to the right-hand side of the problem definition equation. Then, we apply both sides of the equation to point "f" (see Equation 4.86). Therefore, we can cancel the unknowns "θ_6," "θ_5," and "θ_4" by PK1 simplification. The right-hand side of the equation is a known value "k_4." The result is the PG4 subproblem, for two consecutive rotations about parallel axes. We know the IK solutions for PG4, none, one, or two double solutions. With the "θ_1" value, we solve the motion of the second and third joints of the robot, obtaining the rotation magnitudes (if they exist) for "θ_3-θ_2" double solutions by Equation 4.87.

$$e^{\hat{\xi}_2\theta_2}e^{\hat{\xi}_3\theta_3}e^{\hat{\xi}_4\theta_4}e^{\hat{\xi}_5\theta_5}e^{\hat{\xi}_6\theta_6}f = e^{-\hat{\xi}_1\theta_1}H_{ST}(\theta)H_{ST}(0)^{-1}f \quad (4.86)$$

$$e^{\hat{\xi}_2\theta_2}e^{\hat{\xi}_3\theta_3}f = k_4 \begin{cases} \theta_1^{01} \Rightarrow \begin{bmatrix} \theta_2^{01} & \theta_3^{01} & ; & \theta_2^{02} & \theta_3^{02} \end{bmatrix} \\ \theta_1^{02} \Rightarrow \begin{bmatrix} \theta_2^{03} & \theta_3^{03} & ; & \theta_2^{04} & \theta_3^{04} \end{bmatrix} \end{cases} \quad (4.87)$$

- **Solutions for θ_6, θ_5 & θ_4:** We get the magnitudes for the last three joints like the first algorithm. With PK2, it is possible to get "θ_4-θ_5" (see Equation 4.78 and Equation 4.79). The "θ_6" got with PK1 (see Equation 4.80 and Equation 4.81).

Exercise 4.4.2c implements this third algorithm. The code is in the internet hosting for the software of this book[16].

4.4.2.5 Fourth Algorithm for ABB IRB120 IK "PG5+PG4+PG6+PK1"

This approach uses the PG6 subproblem to get the fourth and fifth robot DoF solutions instead of "PK2." The rest of the DoF is solved precisely like for the third algorithm.

- **Solutions for θ_5 & θ_4 with PG6**: We pass $H_{ST}(0)$ and the exponentials of the first three screws, which we already know from the previous steps (i.e., we got "θ_1," "θ_2," and "θ_3") to the right-hand side of the problem definition equation. Then, we apply both sides to the point "p" as Equation 4.88. We can use the PK1 simplification case to the equation's left-hand side to cancel the unknown "θ_6." The right-hand side of the equation is a known value "k_2." The resulting equation corresponds with the definition of the canonical PG6 for two consecutive rotations about skewed axes. Moreover, we know the geometric solutions for PG6, none, one, or two double. For each set of "θ_1-θ_2-θ_3," we solve the motion for the fourth and fifth joints of the robot, obtaining the rotation magnitudes (if they exist) for "θ_5-θ_4" by Equation 4.89.

$$e^{\hat{\xi}_4\theta_4} e^{\hat{\xi}_5\theta_5} e^{\hat{\xi}_6\theta_6} p = e^{-\hat{\xi}_3\theta_3} e^{-\hat{\xi}_2\theta_2} e^{-\hat{\xi}_1\theta_1} H_{ST}(\theta) H_{ST}(0)^{-1} p \qquad (4.88)$$

$$e^{\hat{\xi}_4\theta_4} e^{\hat{\xi}_5\theta_5} p = k_2 \begin{cases} \begin{bmatrix} \theta_1^{01} & \theta_2^{01} & \theta_3^{01} \end{bmatrix} \Rightarrow \begin{bmatrix} \theta_4^{01} & \theta_5^{01} & ; & \theta_4^{02} & \theta_5^{02} \end{bmatrix} \\ \begin{bmatrix} \theta_1^{01} & \theta_2^{02} & \theta_3^{02} \end{bmatrix} \Rightarrow \begin{bmatrix} \theta_4^{03} & \theta_5^{03} & ; & \theta_4^{04} & \theta_5^{04} \end{bmatrix} \\ \begin{bmatrix} \theta_1^{02} & \theta_2^{03} & \theta_3^{03} \end{bmatrix} \Rightarrow \begin{bmatrix} \theta_4^{05} & \theta_5^{05} & ; & \theta_4^{06} & \theta_5^{06} \end{bmatrix} \\ \begin{bmatrix} \theta_1^{02} & \theta_2^{04} & \theta_3^{04} \end{bmatrix} \Rightarrow \begin{bmatrix} \theta_4^{07} & \theta_5^{07} & ; & \theta_4^{08} & \theta_5^{08} \end{bmatrix} \end{cases} \qquad (4.89)$$

- **Solutions for θ_1, θ_2, θ_3 & θ_6**: The details for calculating these magnitudes are in the third algorithm. The "θ_1" comes from the PG5 application (see Equation 4.84 and Equation 4.85). We solve the "θ_2-θ_3" with PG4 (see Equation 4.86 and Equation 4.87). The "θ_6" comes from the PK1 application (see Equation 4.80 and Equation 4.81).

Exercise 4.4.2d implements this fourth algorithm. The code is in the internet hosting for the software of this book[17].

Finally, we have got a set of eight exact geometric solutions for each of the four algorithms applied to this Puma robot.

4.4.2.6 Comparison between the Four Algorithms for ABB IRB120 IK

A good exercise is to check the algorithms' performance developed in this section, working together to solve the same IK targets of the ABB IRB120 manipulator.

Inverse Kinematics

We can check that the effectiveness of all these algorithms is exact. This fact is evident given the geometric nature of all these methods. In terms of efficiency, the differences come from two sources, the canonical subproblems used and the targets' characteristics. Anyhow is instructive to play with some examples to learn more about these implementations.

We hope it is clear the high flexibility provided by the screw theory approach to work with mechanics after these four examples to solve this Puma architecture with the POE. Besides, the POE offers several algorithms to solve any IK problem.

Exercise 4.4.2e implements the comparison between the different algorithms. The code is in the internet hosting for the software of this book[18].

4.4.2.7 Comment on the Implementation of the Algorithms for ABB IRB120 IK

The geometric approach for solving this problem by POE implies that we first solve the "θ_1," "θ_2," and "θ_3" magnitudes for positioning the "f" wrist point in space, and afterward we solve "θ_4," "θ_5," and "θ_6." This algorithm uses PK and PG subproblems solutions for positioning "f," if the robot has a spherical wrist (last three DoF with crossing point axes). It does not matter how the first joints move the robot if they reach the pose for the correct "f" position because from thereon, the last three joints can get any necessary orientation for the TCP. Be careful not to confuse this algorithm with the approach of "kinematics uncoupling." It is evident (see Figure 4.19) that "θ_4" and "θ_5" also imply TCP translation in space, and not only rotation, as it happens with the kinematics uncoupling. Here, there is no kinematics simplification, and that is why the algorithm results so elegant.

The "ST24R" (Screw Theory Toolbox for Robotics) code gives some solution for the IK canonical problems, for whatever input. The algorithm's results could be inexact when the TCP target is out of the robot's workspace, but the output is always a fair approximation even in that event. In any case, this proves to be very useful for most applications. Suppose we need another behavior for the robot. It is straightforward to reprogram the IK canonical problems with another approach. For instance, a convenient development method can sometimes freeze the robot whenever there is no exact solution for the IK.

It is possible to observe the robot configurations for eight different IK solutions for a tool target. Check the values for the magnitudes of the joints. It might seem that some of them are identical, but they are not. What happens is that some configurations show a complementary magnitude as a solution for the same joint. In those cases, two configurations look the same.

4.4.2.8 Performance Contrast for Both Numeric and Geometric ABB IRB120 IK Algorithms

The goal is to compare the computational performance between a numeric algorithm based on optimization methods, a standard approach, and a geometric algorithm based on screw theory for robotics. We will use a Puma-type robot (e.g., ABB IRB120) to test both IK algorithms' performance.

We use the very well-known MATLAB® "RST" (Robotics System Toolbox) to build the IK function on a "BFGS" iterative gradient-based optimization method. The quality of "RST" is unquestionable, but it cannot compete with a screw theory geometric approach's performance. To implement the screw theory POE algorithm, we use the "ST24R" (Screw Theory Toolbox for Robotics). We code these POE examples with interpreted MATLAB® language (MathWorks, 2021a). While the numeric method only gives one approximate solution, the screw theory leads to 8 EXACT SOLUTIONS for the Puma robot's IK. The geometric POE algorithm is several orders of magnitude faster. Furthermore, this great advantage is found even without an optimized implementation of the software.

4.4.2.9 RST - Robotics System Toolbox™

The "RST" provides software and hardware connectivity for developing robotics applications. The toolbox includes algorithms for IK, constraints, and robot dynamics. The library uses a rigid body tree model representation as a portrayal of a robot structure. It can render robot models such as manipulators using the "RigidBodyTree" class.

MATLAB® supports the Broyden-Fletcher-Goldfarb-Shanno (BFGS) projection algorithm to solve IK mapping. It is an iterative, gradient-based optimization method that starts from an initial guess at the solution and seeks to minimize a specific cost function. For some combinations of initial guesses and desired end-effector poses, the algorithm may finish without any correct configuration solution for the manipulator.

This exercise can test different pose targets for the Puma robot tool (even random) and check the results. The computational cost of "RST" is in the order of milliseconds if the algorithm converges to a solution. Even worse is its effectiveness, which allows for only an approximate solution. On top of this, we cannot obtain the set of all possible IK solutions and, if it exists, we get only one solution.

4.4.2.10 ST24R - Screw Theory Toolbox for Robotics

This software library supports the mechanics presented in this book. We have seen exercises and examples to understand better and comprehend the concepts introduced regarding screw theory for robotics. The idea behind this "ST24R" development is to support clarity and simplicity to the teaching task and a more comfortable grasp of the concepts. The programming of this toolbox has a didactical purpose and has not a commercial scope.

The "ST24R" includes some screw theory functions, which you can also find in other very well-known libraries, like the MATLAB® Robotics Toolbox (RTB) (Corke, 2017). However, for manipulator robots, the "ST24R" includes some new algorithms, like those presented in this chapter for the new canonical IK subproblems or some innovative dynamics functions. The algorithms provide closed-form solutions and avoid the need for iteration, initial guesses, or error tolerance, as is evident in the implementation. One can realize how elegant the implementation is, as the only argument for the function is the target pose.

We can test different pose targets for the Puma robot tool (even random) and check the results. The computational cost of "ST24R" is in the order of microseconds, and the algorithm always converges to a closed-form solution. Its effectiveness allows us to get exact answers for all configurations inside the workspace. Besides, there is a set of all possible solutions (eight in this case), which permits choosing the most suitable outcome for the application.

This contrast between numeric and geometric algorithms is seen in Exercise 4.4.2f. The code is in the internet hosting for the software of this book[19].

Inverse Kinematics

4.4.3 PUMA ROBOTS (E.G., ABB IRB120) "TOOL-UP."

The motion representation of the POE is relative to an initial pose (reference or home) of the robot (Figure 4.20). This reference pose can be chosen arbitrarily in the most convenient way for the robotics application, depending on the manipulator's mechanical configuration. This feature is another advantage of the screw theory.

For example, we develop another exercise for the Puma here, but in this case, the tool oriented upward in the reference home position. The definition of the twists and the initial posture of the TCP change (see formulations below). The resolution of the IK problem is as easy as presented in the previous example. We use the actual geometry of the ABB IRB120 for exercising some practice application with this robot.

We only need to apply an algorithm using the PG and PK subproblems in this way: "**PG7+PG6+PK1.**" However, as we showed in the previous section, there are many more algorithms to solve this problem, left it up as a possible extra practice.

4.4.3.1 Inverse Kinematics PUMA ABB IRB120 "Tool-Up" Problem Definition

We obtain the kinematics equation that relates the motion of the joints ($\theta_1...\theta_6$) to the pose for the tool (noap), with the POE of the joints' screws, defined by the twists and magnitudes. The problem definition is easy, seeing Figure 4.20.

- Define the axis of each joint "$\omega_1...\omega_6$." We must pay attention to this definition based on the spatial frame, which is oriented differently from the typical commercial with the "Z" axis upward.

$$\omega_1 = \begin{bmatrix} 0 \\ 1 \\ 0 \end{bmatrix} \quad \omega_2 = \begin{bmatrix} 0 \\ 0 \\ 1 \end{bmatrix} \quad \omega_3 = \begin{bmatrix} 0 \\ 0 \\ 1 \end{bmatrix} \quad \omega_4 = \begin{bmatrix} 1 \\ 0 \\ 0 \end{bmatrix} \quad \omega_5 = \begin{bmatrix} 0 \\ 0 \\ 1 \end{bmatrix} \quad \omega_6 = \begin{bmatrix} 0 \\ 1 \\ 0 \end{bmatrix}$$

- Obtain the Twists ($\xi_1...\xi_6$), knowing each joint's axis and a point on those axes.

$$\xi_1 = \begin{bmatrix} -\omega_1 \times o \\ \omega_1 \end{bmatrix} \quad \xi_2 = \begin{bmatrix} -\omega_2 \times k \\ \omega_2 \end{bmatrix} \quad \xi_3 = \begin{bmatrix} -\omega_3 \times r \\ \omega_3 \end{bmatrix}$$

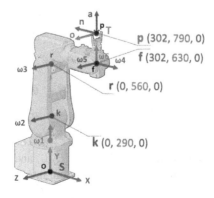

p (302, 790, 0)
f (302, 630, 0)
r (0, 560, 0)
k (0, 290, 0)

FIGURE 4.20 Puma ABB IRB120 "Tool-Up" inverse kinematics POE parameters.

$$\xi_4 = \begin{bmatrix} -\omega_4 \times f \\ \omega_4 \end{bmatrix} \quad \xi_5 = \begin{bmatrix} -\omega_5 \times f \\ \omega_5 \end{bmatrix} \quad \xi_6 = \begin{bmatrix} -\omega_6 \times f \\ \omega_6 \end{bmatrix}$$

- Get $H_{ST}(0)$, the pose of the tool at the reference (home) robot position.

$$H_{ST}(0) = T_{xyz}\begin{bmatrix} p_x \\ p_y \\ p_z \end{bmatrix} R_X\left(\frac{-\pi}{2}\right) R_Z(\pi) = \begin{bmatrix} -1 & 0 & 0 & p_x \\ 0 & 0 & 1 & p_y \\ 0 & 1 & 0 & 0 \\ 0 & 0 & 0 & 1 \end{bmatrix}$$

- Define the kinematics problem in terms of the POE as Equation 4.90.

$$H_{ST}(\theta) = e^{\hat{\xi}_1\theta_1} e^{\hat{\xi}_2\theta_2} e^{\hat{\xi}_3\theta_3} e^{\hat{\xi}_4\theta_4} e^{\hat{\xi}_5\theta_5} e^{\hat{\xi}_6\theta_6} H_{ST}(0) = \begin{bmatrix} n\,o\,a\,p \\ 0\,0\,0\,1 \end{bmatrix} \quad (4.90)$$

We know that different algorithms provide closed-form geometric solutions based on the application of different canonical subproblems. Next, we present only one of them, but it is also possible to develop others, following the previous section's example.

4.4.3.2 First Algorithm for ABB IRB120 "Tool-Up" IK "PG7+PG6+PK1"

- **Solutions for θ_3, θ_2 & θ_1 with PG7**: We pass $H_{ST}(0)$ to the right-hand side of the problem definition equation and apply both sides of Equation 4.90 to the point "f," resulting in Equation 4.91. Then, we can use the PK1 simplification case on the left-hand side of the equation to cancel the unknowns "θ_6," "θ_5," and "θ_4." The right-hand side of the equation is now a known value "k_1." The resulting Equation 4.92 defines the canonical PG7 for three consecutive rotations about one skewed and two parallel axes. We know the geometric IK for PG7, which can be none, two, or four triple solutions for "θ_3-θ_2-θ_1."

$$e^{\hat{\xi}_1\theta_1} e^{\hat{\xi}_2\theta_2} e^{\hat{\xi}_3\theta_3} e^{\hat{\xi}_4\theta_4} e^{\hat{\xi}_5\theta_5} e^{\hat{\xi}_6\theta_6} f = H_{ST}(\theta) H_{ST}(0)^{-1} f \quad (4.91)$$

$$e^{\hat{\xi}_1\theta_1} e^{\hat{\xi}_2\theta_2} e^{\hat{\xi}_3\theta_3} f = k_1 \begin{cases} \begin{bmatrix} \theta_1^{01} & \theta_2^{01} & \theta_3^{01} \end{bmatrix} \\ \begin{bmatrix} \theta_1^{01} & \theta_2^{02} & \theta_3^{02} \end{bmatrix} \\ \begin{bmatrix} \theta_1^{02} & \theta_2^{03} & \theta_3^{01} \end{bmatrix} \\ \begin{bmatrix} \theta_1^{02} & \theta_2^{04} & \theta_3^{02} \end{bmatrix} \end{cases} \quad (4.92)$$

- **Solutions for θ_4 & θ_5 with PG6**: We pass $H_{ST}(0)$ and the exponentials of the first three screws, which we already know from the previous steps of the algorithm (i.e., we got "θ_1," "θ_2," and "θ_3") to the right-hand side of the problem

Inverse Kinematics

definition. Then, we apply both sides of the equation to the point "p," getting Equation 4.93. We can use the PK1 simplification case to the equation's left-hand side to cancel the unknown "θ_6." The right-hand side of the equation is a known value "k_2." The resulting equation corresponds with the canonical PG6 subproblem for two consecutive rotations about skewed axes. Furthermore, we know the geometric IK solutions for PG6. For each set of "θ_1-θ_2-θ_3" values, we solve the motion for the fourth and fifth joints of the robot, obtaining the rotation magnitudes (if they exist) for "θ_5-θ_4" by Equation 4.94.

$$e^{\hat{\xi}_4\theta_4} e^{\hat{\xi}_5\theta_5} e^{\hat{\xi}_6\theta_6} p = e^{-\hat{\xi}_3\theta_3} e^{-\hat{\xi}_2\theta_2} e^{-\hat{\xi}_1\theta_1} H_{ST}(\theta) H_{ST}(0)^{-1} p \tag{4.93}$$

$$e^{\hat{\xi}_4\theta_4} e^{\hat{\xi}_5\theta_5} p = k_2 \begin{cases} \begin{bmatrix} \theta_1^{01} & \theta_2^{01} & \theta_3^{01} \end{bmatrix} \Rightarrow \begin{bmatrix} \theta_4^{01} & \theta_5^{01} & ; & \theta_4^{02} & \theta_5^{02} \end{bmatrix} \\ \begin{bmatrix} \theta_1^{01} & \theta_2^{02} & \theta_3^{02} \end{bmatrix} \Rightarrow \begin{bmatrix} \theta_4^{03} & \theta_5^{03} & ; & \theta_4^{04} & \theta_5^{04} \end{bmatrix} \\ \begin{bmatrix} \theta_1^{02} & \theta_2^{03} & \theta_3^{01} \end{bmatrix} \Rightarrow \begin{bmatrix} \theta_4^{05} & \theta_5^{05} & ; & \theta_4^{06} & \theta_5^{06} \end{bmatrix} \\ \begin{bmatrix} \theta_1^{02} & \theta_2^{04} & \theta_3^{02} \end{bmatrix} \Rightarrow \begin{bmatrix} \theta_4^{07} & \theta_5^{07} & ; & \theta_4^{08} & \theta_5^{08} \end{bmatrix} \end{cases} \tag{4.94}$$

- **Solution for θ_6 with PK1**: We pass $H_{ST}(0)$, and the exponentials of the first five screws, which we already know from the previous steps of the algorithm (i.e., we got "θ_1," "θ_2," "θ_3," "θ_4," and "θ_5") to the right-hand side of the problem definition of Equation 4.90. Then, we apply both sides of the equation to the point "o" (origin of the spatial reference frame), getting Equation 4.95. The right-hand side of the equation is now a known value "k_3." The resulting equation is precisely the canonical PK1, and we know its geometric IK solutions. Finally, for each set of "θ_1-θ_2-θ_3-θ_4-θ_5" values, we solve the motion for the sixth joint of the Puma robot, obtaining the magnitude for "θ_6" by Equation 4.96.

$$e^{\hat{\xi}_6\theta_6} o = e^{-\hat{\xi}_5\theta_5} e^{-\hat{\xi}_4\theta_4} e^{-\hat{\xi}_3\theta_3} e^{-\hat{\xi}_2\theta_2} e^{-\hat{\xi}_1\theta_1} H_{ST}(\theta) H_{ST}(0)^{-1} o \tag{4.95}$$

$$e^{\hat{\xi}_6\theta_6} o = k_3 \begin{cases} \begin{bmatrix} \theta_1^{01} & \theta_2^{01} & \theta_3^{01} & \theta_4^{01} & \theta_5^{01} \end{bmatrix} \Rightarrow \theta_6^{01} \\ \begin{bmatrix} \theta_1^{01} & \theta_2^{01} & \theta_3^{01} & \theta_4^{02} & \theta_5^{02} \end{bmatrix} \Rightarrow \theta_6^{02} \\ \begin{bmatrix} \theta_1^{01} & \theta_2^{02} & \theta_3^{02} & \theta_4^{03} & \theta_5^{03} \end{bmatrix} \Rightarrow \theta_6^{03} \\ \begin{bmatrix} \theta_1^{01} & \theta_2^{02} & \theta_3^{02} & \theta_4^{04} & \theta_5^{04} \end{bmatrix} \Rightarrow \theta_6^{04} \\ \begin{bmatrix} \theta_1^{02} & \theta_2^{03} & \theta_3^{01} & \theta_4^{05} & \theta_5^{05} \end{bmatrix} \Rightarrow \theta_6^{05} \\ \begin{bmatrix} \theta_1^{02} & \theta_2^{03} & \theta_3^{01} & \theta_4^{06} & \theta_5^{06} \end{bmatrix} \Rightarrow \theta_6^{06} \\ \begin{bmatrix} \theta_1^{02} & \theta_2^{04} & \theta_3^{02} & \theta_4^{07} & \theta_5^{07} \end{bmatrix} \Rightarrow \theta_6^{07} \\ \begin{bmatrix} \theta_1^{02} & \theta_2^{04} & \theta_3^{02} & \theta_4^{08} & \theta_5^{08} \end{bmatrix} \Rightarrow \theta_6^{08} \end{cases} \tag{4.96}$$

Finally, we have got a set of eight exact geometric solutions for this Puma robot.

The code for this Exercise 4.4.3 is in the internet hosting for the software of this book[20].

4.4.4 Bending Backwards Robots (e.g., ABB IRB1600)

This exercise addresses the IK for a Bending Backwards robot (Figure 4.21). We can solve the problem again with multiple IK algorithms. For instance, the same algorithm used previously for robots with Puma configuration works fine, even though, for this case, the two first joint axes do not cross. We present the new algorithm "PG7+PG6+PK1" and do not consider it necessary to specify other solutions because they were given before for the ABB IRB120. Nonetheless, we can check different algorithms like "PG5+PG4+PK2+PK1" or "PG5+PG4+PG6+PK1."

4.4.4.1 Inverse Kinematics ABB IRB1600 Problem Definition

We obtain the equation that relates the motion of the joints $(\theta_1...\theta_6)$ to the pose for the tool (noap), with the POE as defined by the screw theory. The problem definition seems clear from Figure 4.21.

- Define the axis of each joint "$\omega_1...\omega_6$."

$$\omega_1 = \begin{bmatrix} 0 \\ 0 \\ 1 \end{bmatrix} \quad \omega_2 = \begin{bmatrix} 0 \\ 1 \\ 0 \end{bmatrix} \quad \omega_3 = \begin{bmatrix} 0 \\ 1 \\ 0 \end{bmatrix} \quad \omega_4 = \begin{bmatrix} 1 \\ 0 \\ 0 \end{bmatrix} \quad \omega_5 = \begin{bmatrix} 0 \\ 1 \\ 0 \end{bmatrix} \quad \omega_6 = \begin{bmatrix} 1 \\ 0 \\ 0 \end{bmatrix}$$

- Obtain the Twists $(\xi_1... \xi_6)$, knowing each joint's axis and a point on those axes.

$$\xi_1 = \begin{bmatrix} -\omega_1 \times o \\ \omega_1 \end{bmatrix} \quad \xi_2 = \begin{bmatrix} -\omega_2 \times k \\ \omega_2 \end{bmatrix} \quad \xi_3 = \begin{bmatrix} -\omega_3 \times r \\ \omega_3 \end{bmatrix}$$

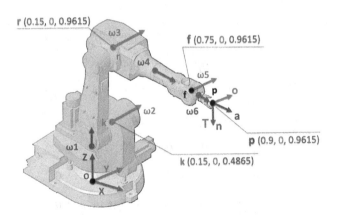

FIGURE 4.21 Bending Backwards ABB IRB1600 inverse kinematics POE parameters.

Inverse Kinematics

$$\xi_4 = \begin{bmatrix} -\omega_4 \times f \\ \omega_4 \end{bmatrix} \quad \xi_5 = \begin{bmatrix} -\omega_5 \times f \\ \omega_5 \end{bmatrix} \quad \xi_6 = \begin{bmatrix} -\omega_6 \times f \\ \omega_6 \end{bmatrix}$$

- Get $H_{ST}(0)$, the pose of the tool at the reference (home) robot position.

$$H_{ST}(0) = T_{xyz}\begin{bmatrix} p_x \\ p_y \\ p_z \end{bmatrix} R_Y\left(\frac{\pi}{2}\right) = \begin{bmatrix} 0 & 0 & 1 & p_x \\ 0 & 1 & 0 & 0 \\ -1 & 0 & 0 & p_z \\ 0 & 0 & 0 & 1 \end{bmatrix}$$

- Define the problem in terms of the POE as Equation 4.97.

$$H_{ST}(\theta) = e^{\hat{\xi}_1\theta_1}e^{\hat{\xi}_2\theta_2}e^{\hat{\xi}_3\theta_3}e^{\hat{\xi}_4\theta_4}e^{\hat{\xi}_5\theta_5}e^{\hat{\xi}_6\theta_6}H_{ST}(0) = \begin{bmatrix} n\,o\,a\,p \\ 0\,0\,0\,1 \end{bmatrix} \quad (4.97)$$

4.4.4.2 First Algorithm for ABB IRB1600 IK "PG7+PG6+PK1"

- **Solutions for θ_3, θ_2 & θ_1 with PG7**: Manipulating Equation 4.97 and applying both sides of the equation to "f," we get the Equation 4.98, which is precisely the canonical subproblem PG7 of Equation 4.99, for three consecutive rotations about one skewed and two parallel axes applied to a point. We know the possible solutions for the PG7 algorithm, which gives the values for "θ_3-θ_3-θ_1."

$$e^{\hat{\xi}_1\theta_1}e^{\hat{\xi}_2\theta_2}e^{\hat{\xi}_3\theta_3}e^{\hat{\xi}_4\theta_4}e^{\hat{\xi}_5\theta_5}e^{\hat{\xi}_6\theta_6}f = H_{ST}(\theta)H_{ST}(0)^{-1}f \quad (4.98)$$

$$e^{\hat{\xi}_1\theta_1}e^{\hat{\xi}_2\theta_2}e^{\hat{\xi}_3\theta_3}f = k_1 \begin{cases} \begin{bmatrix} \theta_1^{01} & \theta_2^{01} & \theta_3^{01} \end{bmatrix} \\ \begin{bmatrix} \theta_1^{01} & \theta_2^{02} & \theta_3^{02} \end{bmatrix} \\ \begin{bmatrix} \theta_1^{02} & \theta_2^{03} & \theta_3^{01} \end{bmatrix} \\ \begin{bmatrix} \theta_1^{02} & \theta_2^{04} & \theta_3^{02} \end{bmatrix} \end{cases} \quad (4.99)$$

- **Solutions for θ_4 & θ_5 with PG6**: We demonstrated how to get Equation 4.100, from the manipulation of Equation 4.97. The result is the definition of the canonical subproblem PG6 as Equation 4.101 for two consecutive rotations about skewed axes. The rotation magnitudes for "θ_5-θ_4" are the result of getting the PG6 solutions for each set of the already known values "θ_1-θ_2-θ_3."

$$e^{\hat{\xi}_4\theta_4}e^{\hat{\xi}_5\theta_5}e^{\hat{\xi}_6\theta_6}p = e^{-\hat{\xi}_3\theta_3}e^{-\hat{\xi}_2\theta_2}e^{-\hat{\xi}_1\theta_1}H_{ST}(\theta)H_{ST}(0)^{-1}p \quad (4.100)$$

$$e^{\hat{\xi}_4\theta_4}e^{\hat{\xi}_5\theta_5}p = k_2 \begin{cases} \begin{bmatrix} \theta_1^{01} & \theta_2^{01} & \theta_3^{01} \end{bmatrix} \Rightarrow \begin{bmatrix} \theta_4^{01} & \theta_5^{01} & ; & \theta_4^{02} & \theta_5^{02} \end{bmatrix} \\ \begin{bmatrix} \theta_1^{01} & \theta_2^{02} & \theta_3^{02} \end{bmatrix} \Rightarrow \begin{bmatrix} \theta_4^{03} & \theta_5^{03} & ; & \theta_4^{04} & \theta_5^{04} \end{bmatrix} \\ \begin{bmatrix} \theta_1^{02} & \theta_2^{03} & \theta_3^{01} \end{bmatrix} \Rightarrow \begin{bmatrix} \theta_4^{05} & \theta_5^{05} & ; & \theta_4^{06} & \theta_5^{06} \end{bmatrix} \\ \begin{bmatrix} \theta_1^{02} & \theta_2^{04} & \theta_3^{02} \end{bmatrix} \Rightarrow \begin{bmatrix} \theta_4^{07} & \theta_5^{07} & ; & \theta_4^{08} & \theta_5^{08} \end{bmatrix} \end{cases} \quad (4.101)$$

- **Solution for θ_6 with PK1**: Finally, applying PK1 to Equation 4.102, we obtain the result of "θ_6" by Equation 4.103.

$$e^{\hat{\xi}_6\theta_6}o = e^{-\hat{\xi}_5\theta_5}e^{-\hat{\xi}_4\theta_4}e^{-\hat{\xi}_3\theta_3}e^{-\hat{\xi}_2\theta_2}e^{-\hat{\xi}_1\theta_1}H_{ST}(\theta)H_{ST}(0)^{-1}o \quad (4.102)$$

$$e^{\hat{\xi}_6\theta_6}o = k_3 \begin{cases} \begin{bmatrix} \theta_1^{01} & \theta_2^{01} & \theta_3^{01} & \theta_4^{01} & \theta_5^{01} \end{bmatrix} \Rightarrow \theta_6^{01} \\ \begin{bmatrix} \theta_1^{01} & \theta_2^{01} & \theta_3^{01} & \theta_4^{02} & \theta_5^{02} \end{bmatrix} \Rightarrow \theta_6^{02} \\ \begin{bmatrix} \theta_1^{01} & \theta_2^{02} & \theta_3^{02} & \theta_4^{03} & \theta_5^{03} \end{bmatrix} \Rightarrow \theta_6^{03} \\ \begin{bmatrix} \theta_1^{01} & \theta_2^{02} & \theta_3^{02} & \theta_4^{04} & \theta_5^{04} \end{bmatrix} \Rightarrow \theta_6^{04} \\ \begin{bmatrix} \theta_1^{02} & \theta_2^{03} & \theta_3^{01} & \theta_4^{05} & \theta_5^{05} \end{bmatrix} \Rightarrow \theta_6^{05} \\ \begin{bmatrix} \theta_1^{02} & \theta_2^{03} & \theta_3^{01} & \theta_4^{06} & \theta_5^{06} \end{bmatrix} \Rightarrow \theta_6^{06} \\ \begin{bmatrix} \theta_1^{02} & \theta_2^{04} & \theta_3^{02} & \theta_4^{07} & \theta_5^{07} \end{bmatrix} \Rightarrow \theta_6^{07} \\ \begin{bmatrix} \theta_1^{02} & \theta_2^{04} & \theta_3^{02} & \theta_4^{08} & \theta_5^{08} \end{bmatrix} \Rightarrow \theta_6^{08} \end{cases} \quad (4.103)$$

Finally, we have got a set of eight exact geometric solutions for this Bending Backwards robot.

The code for this Exercise 4.4.4 is in the internet hosting for the software of this book[21].

4.4.5 GANTRY ROBOTS (E.G., ABB IRB6620LX)

This exercise is for the ABB IRB6620LX, a very illustrative typical Gantry robot geometry (Figure 4.22). The selection of the coordinate frames for the spatial and tool systems is entirely arbitrary, as the screw theory does allow this flexibility.

The POE approach provides elegant results applying PG and PK subproblems. Pay attention to the several possible algorithms to solve this problem: "PG1+PG4+PG6+PK1," "PG1+PK3+PK1+PG6+PK1," "PG1+PG4+PK2+PK1," "PG1+PK3+PK1+PK2+PK1." All of them are alternative methods of equal effectiveness. Hereafter, we present the first of those algorithms in detail, but the others can be easily implemented, following the previous exercises of this book.

4.4.5.1 Inverse Kinematics ABB IRB6620LX Problem Definition

We obtain the equation that relates the motion of the joints ($\theta_1...\theta_6$) to the pose for the tool (noap), with the screw theory POE. The problem definition emerges from Figure 4.22.

Inverse Kinematics

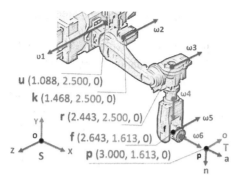

FIGURE 4.22 Gantry ABB IRB6620LX inverse kinematics POE parameters.

- Define the axis of each joint "$v_1, \omega_2...\omega_6$."

$$v_1 = \begin{bmatrix} 0 \\ 0 \\ 1 \end{bmatrix} \quad \omega_2 = \begin{bmatrix} 0 \\ 0 \\ -1 \end{bmatrix} \quad \omega_3 = \begin{bmatrix} 0 \\ 0 \\ -1 \end{bmatrix} \quad \omega_4 = \begin{bmatrix} 0 \\ -1 \\ 0 \end{bmatrix} \quad \omega_5 = \begin{bmatrix} 0 \\ 0 \\ -1 \end{bmatrix} \quad \omega_6 = \begin{bmatrix} 1 \\ 0 \\ 0 \end{bmatrix}$$

- Obtain the Twists ($\xi_1 ... \xi_6$), knowing each joint's axis and a point on the axes of the revolute joints.

$$\xi_1 = \begin{bmatrix} v_1 \\ 0 \end{bmatrix} \quad \xi_2 = \begin{bmatrix} -\omega_2 \times k \\ \omega_2 \end{bmatrix} \quad \xi_3 = \begin{bmatrix} -\omega_3 \times r \\ \omega_3 \end{bmatrix}$$

$$\xi_4 = \begin{bmatrix} -\omega_4 \times f \\ \omega_4 \end{bmatrix} \quad \xi_5 = \begin{bmatrix} -\omega_5 \times f \\ \omega_5 \end{bmatrix} \quad \xi_6 = \begin{bmatrix} -\omega_6 \times f \\ \omega_6 \end{bmatrix}$$

- Get $H_{ST}(0)$, the pose of the tool at the reference (home) robot position.

$$H_{ST}(0) = T_{xyz}\begin{bmatrix} p_x \\ p_y \\ p_z \end{bmatrix} R_Y\left(\frac{\pi}{2}\right) R_Z\left(\frac{-\pi}{2}\right) = \begin{bmatrix} 0 & 0 & 1 & p_X \\ -1 & 0 & 0 & p_Y \\ 0 & -1 & 0 & 0 \\ 0 & 0 & 0 & 1 \end{bmatrix}$$

- Define the problem in terms of the POE by Equation 4.104.

$$H_{ST}(\theta) = e^{\hat{\xi}_1\theta_1} e^{\hat{\xi}_2\theta_2} e^{\hat{\xi}_3\theta_3} e^{\hat{\xi}_4\theta_4} e^{\hat{\xi}_5\theta_5} e^{\hat{\xi}_6\theta_6} H_{ST}(0) = \begin{bmatrix} n & o & a & p \\ 0 & 0 & 0 & 1 \end{bmatrix} \quad (4.104)$$

4.4.5.2 First Algorithm for ABB IRB6620LX IK "PG1+PG4+PG6+PK1"

- **Solution for θ_1 with PG1:** We pass $H_{ST}(0)$ to the right-hand side of the problem definition Equation 4.104. Then, we apply both sides of the equation to point "f" to get Equation 4.105. We can use the PK1 simplification case to the left-hand side of the equation and cancel the unknowns "θ_6," "θ_5," and "θ_4." We must realize that the screw rotations θ_2 and θ_3 (for whatever magnitude) do not change the plane where "f" moves, which is perpendicular to the axis of $\hat{v}_1$ and therefore do not affect the calculation for θ_1. The right-hand side of the equation is now a known value "k_1." The resulting equation is the canonical Pardos-Gotor subproblem One (PG1) for one translation along a single axis. We know the geometric IK solution for PG1, and solving the motion for the first joint, we get the translation magnitude for "θ_1" by Equation 4.106.

$$e^{\hat{\xi}_1\theta_1}e^{\hat{\xi}_2\theta_2}e^{\hat{\xi}_3\theta_3}e^{\hat{\xi}_4\theta_4}e^{\hat{\xi}_5\theta_5}e^{\hat{\xi}_6\theta_6}f = H_{ST}(\theta)H_{ST}(0)^{-1}f \qquad (4.105)$$

$$e^{\hat{\xi}_1\theta_1}f = k_1\left\{\left[\theta_1^{01}\right]\right\} \qquad (4.106)$$

- **Solutions for θ_3 & θ_2 with PG4**: We pass $H_{ST}(0)$ and the exponential of the first screw, which we already know from the previous step of the algorithm (i.e., we got "θ_1"), to the right-hand side of the problem definition equation. Then, we apply both sides of the equation to point "f" to get Equation 4.107. Therefore, we can cancel the unknowns "θ_6," "θ_5," and "θ_4" by PK1 simplification case. The right-hand side of the equation is a value "k_2." The resulting equation is the canonical PG4 subproblem for two consecutive parallel rotations. Moreover, we know the geometric solutions for PG4. Then, knowing "θ_1," we solve the motion for the second and third joints of the robot, obtaining the rotation magnitudes (if they exist) for "θ_3-θ_2" double solutions by Equation 4.108.

$$e^{\hat{\xi}_2\theta_2}e^{\hat{\xi}_3\theta_3}e^{\hat{\xi}_4\theta_4}e^{\hat{\xi}_5\theta_5}e^{\hat{\xi}_6\theta_6}f = e^{-\hat{\xi}_1\theta_1}H_{ST}(\theta)H_{ST}(0)^{-1}f \qquad (4.107)$$

$$e^{\hat{\xi}_2\theta_2}e^{\hat{\xi}_3\theta_3}f = k_2\left\{\theta_1^{01} \Rightarrow \left[\theta_2^{01} \quad \theta_3^{01} \quad ; \quad \theta_2^{02} \quad \theta_3^{02}\right]\right\} \qquad (4.108)$$

- **Solutions for θ_4 & θ_5 with PG6**: We pass $H_{ST}(0)$ and the exponentials of the first three screws, which we already know from the previous steps of the algorithm (i.e., we got "θ_1," "θ_2," and "θ_3") to the right-hand side of the problem definition Equation 4.104. Then, we apply both sides of the equation to the point "p," getting Equation 4.109. We can use the PK1 simplification case to the equation's left-hand side to cancel the unknown "θ_6." The right-hand side of the equation is a value "k_3." The resulting equation corresponds with the canonical PG6 subproblem for two consecutive rotations about skewed axes. The geometric solutions for PG6 can be none, one, or two double. Finally, for each set of "θ_1-θ_2-θ_3" values, we solve the motion for the fourth and fifth joints, obtaining the rotation magnitudes (if they exist) for "θ_5-θ_4" by Equation 4.110.

$$e^{\hat{\xi}_4\theta_4}e^{\hat{\xi}_5\theta_5}e^{\hat{\xi}_6\theta_6}p = e^{-\hat{\xi}_3\theta_3}e^{-\hat{\xi}_2\theta_2}e^{-\hat{\xi}_1\theta_1}H_{ST}(\theta)H_{ST}(0)^{-1}p \qquad (4.109)$$

Inverse Kinematics

$$e^{\hat{\xi}_4\theta_4}e^{\hat{\xi}_5\theta_5}p = k_3 \begin{cases} \begin{bmatrix} \theta_1^{01} & \theta_2^{01} & \theta_3^{01} \end{bmatrix} \Rightarrow \begin{bmatrix} \theta_4^{01} & \theta_5^{01} & ; & \theta_4^{02} & \theta_5^{02} \end{bmatrix} \\ \begin{bmatrix} \theta_1^{01} & \theta_2^{02} & \theta_3^{02} \end{bmatrix} \Rightarrow \begin{bmatrix} \theta_4^{03} & \theta_5^{03} & ; & \theta_4^{04} & \theta_5^{04} \end{bmatrix} \end{cases} \quad (4.110)$$

- **Solution for θ_6 with PK1**: We pass $H_{ST}(0)$, and the exponentials of the first five screws, which we already know from the previous steps of the algorithm (i.e., we already got "θ_1," "θ_2," "θ_3," "θ_4," and "θ_5") to the right-hand side of the problem definition equation. Then, we apply both sides of the equation to the point "o," the origin of the spatial reference frame, to get Equation 4.111. The right-hand side of the equation is now a known value "k_4." The resulting equation is precisely the definition for the canonical subproblem PK1. And we know the geometric IK solutions for PK1. Finally, for each set of "θ_1-θ_2-θ_3-θ_4-θ_5" values, we solve the motion for the sixth joint of the robot, obtaining the rotation magnitude for "θ_6" solution by Equation 4.112.

$$e^{\hat{\xi}_6\theta_6}o = e^{-\hat{\xi}_5\theta_5}e^{-\hat{\xi}_4\theta_4}e^{-\hat{\xi}_3\theta_3}e^{-\hat{\xi}_2\theta_2}e^{-\hat{\xi}_1\theta_1}H_{ST}(\theta)H_{ST}(0)^{-1}o \quad (4.111)$$

$$e^{\hat{\xi}_6\theta_6}o = k_4 \begin{cases} \begin{bmatrix} \theta_1^{01} & \theta_2^{01} & \theta_3^{01} & \theta_4^{01} & \theta_5^{01} \end{bmatrix} \Rightarrow \theta_6^{01} \\ \begin{bmatrix} \theta_1^{01} & \theta_2^{01} & \theta_3^{01} & \theta_4^{02} & \theta_5^{02} \end{bmatrix} \Rightarrow \theta_6^{02} \\ \begin{bmatrix} \theta_1^{01} & \theta_2^{02} & \theta_3^{02} & \theta_4^{03} & \theta_5^{03} \end{bmatrix} \Rightarrow \theta_6^{03} \\ \begin{bmatrix} \theta_1^{01} & \theta_2^{02} & \theta_3^{02} & \theta_4^{04} & \theta_5^{04} \end{bmatrix} \Rightarrow \theta_6^{04} \end{cases} \quad (4.112)$$

Finally, we have got a set of four exact geometric solutions for this Gantry-type robot.

The code for this Exercise 4.4.5 is in the internet hosting for the software of this book[22].

4.4.6 Scara Robots (e.g., ABB IRB910SC)

This exercise is for any Scara robot, exemplified with the ABB IRB910SC (Figure 4.23). We can also apply the POE approach to this mechanism with only four DoF, using PG and PK subproblems to get the solution. There are several possible ways to solve this problem, for example, "PG1+PG4+PK1" or "PG1+PK3+PK1+PK1." Both are equivalent alternatives. Hereafter, we present in detail those algorithms.

4.4.6.1 Inverse Kinematics ABB IRB910SC Problem Definition

We obtain the equation that relates the motion of the joints (θ_1...θ_4) to the pose for the tool (noap), with the screw theory POE. The problem definition is evident from Figure 4.23.

- Define the axis of each joint "ω_1...ω_4."

$$\omega_1 = \begin{bmatrix} 0 \\ 1 \\ 0 \end{bmatrix} \quad \omega_2 = \begin{bmatrix} 0 \\ 1 \\ 0 \end{bmatrix} \quad v_3 = \begin{bmatrix} 0 \\ 1 \\ 0 \end{bmatrix} \quad \omega_4 = \begin{bmatrix} 0 \\ -1 \\ 0 \end{bmatrix}$$

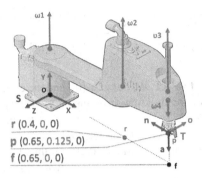

FIGURE 4.23 Scara ABB IRB910SC inverse kinematics POE parameters.

- Obtain the Twists (ξ_1... ξ_4), knowing each joint's axis and a point on those axes.

$$\xi_1 = \begin{bmatrix} -\omega_1 \times o \\ \omega_1 \end{bmatrix} \quad \xi_2 = \begin{bmatrix} -\omega_2 \times r \\ \omega_2 \end{bmatrix} \quad \xi_3 = \begin{bmatrix} v_3 \\ 0 \end{bmatrix} \quad \xi_4 = \begin{bmatrix} -\omega_4 \times f \\ \omega_4 \end{bmatrix}$$

- Get $H_{ST}(0)$, the pose of the tool at the reference (home) robot position.

$$H_{ST}(0) = T_{xyz}\begin{bmatrix} p_x \\ p_y \\ p_z \end{bmatrix} R_X\left(\frac{\pi}{2}\right) R_Z(\pi) = \begin{bmatrix} -1 & 0 & 0 & p_X \\ 0 & 0 & -1 & p_Y \\ 0 & -1 & 0 & 0 \\ 0 & 0 & 0 & 1 \end{bmatrix}$$

- Define the problem in terms of the POE as Equation 4.113.

$$H_{ST}(\theta) = e^{\hat{\xi}_1 \theta_1} e^{\hat{\xi}_2 \theta_2} e^{\hat{\xi}_3 \theta_3} e^{\hat{\xi}_4 \theta_4} H_{ST}(0) = \begin{bmatrix} n\,o\,a\,p \\ 0\,0\,0\,1 \end{bmatrix} \quad (4.113)$$

4.4.6.2 First Algorithm for ABB IRB910SC IK "PG1+PG4+PK1"

- **Solution for θ_3 with PG1:** We pass $H_{ST}(0)$ to the right-hand side of the problem definition of Equation 4.113. Then, we apply both sides of the equation to point "f" to get Equation 4.114 and apply the PK1 simplification case to the equation's left-hand side to cancel the unknown "θ_4." We must realize that the screw rotations θ_1 and θ_2 do not change the plane where "f" moves, which is perpendicular to the axis of $\check{v}_3$ and therefore do not affect θ_3. The right-hand side of the equation is now a known value "k_1." The resulting equation is the canonical Pardos-Gotor subproblem One (PG1) for one translation along a single axis as Equation 4.115. And we know the geometric IK solutions for PG1, obtaining the translation magnitude for "θ_3."

$$e^{\hat{\xi}_1 \theta_1} e^{\hat{\xi}_2 \theta_2} e^{\hat{\xi}_3 \theta_3} e^{\hat{\xi}_4 \theta_4} f = H_{ST}(\theta) H_{ST}(0)^{-1} f \quad (4.114)$$

Inverse Kinematics

$$e^{\hat{\xi}_3\theta_3}f = k_1\left\{\begin{bmatrix}\theta_3^{01}\end{bmatrix}\right\} \tag{4.115}$$

- **Solutions for θ_2 & θ_1 with PG4**: We pass $H_{ST}(0)$ to the right-hand side of the problem definition. Then, we apply both sides of the equation to point "f," getting Equation 4.116. The right-hand side of the equation is now a known value "k_1." On the left-hand side, we can cancel the unknown "θ_4" by PK1 simplification case. We know "θ_3," and applying the exponential of this third screw to point "f," we get another point, "f_1." The resulting equation is the definition for the canonical subproblem PG4, for two consecutive rotations about the parallel axis as Equation 4.117. Moreover, we know the solutions for PG4, which can be none, one, or two double. Understanding the "θ_3" value, we then solve the motion, obtaining the rotation magnitudes (if they exist) for "θ_2-θ_1" double solutions.

$$e^{\hat{\xi}_1\theta_1}e^{\hat{\xi}_2\theta_2}e^{\hat{\xi}_3\theta_3}e^{\hat{\xi}_4\theta_4}f = H_{ST}(\theta)H_{ST}(0)^{-1}f \tag{4.116}$$

$$e^{\hat{\xi}_1\theta_1}e^{\hat{\xi}_2\theta_2}f_1 = k_1\left\{\theta_3^{01} \Rightarrow \begin{bmatrix}\theta_1^{01} & \theta_2^{01} & ; & \theta_1^{02} & \theta_2^{02}\end{bmatrix}\right\} \tag{4.117}$$

- **Solutions for θ_4 with PK1**: We pass $H_{ST}(0)$ and the exponentials of the first three screws, which we already know from the algorithm's previous steps to the right-hand side of the problem definition equation. Then, we apply both sides of the equation to point "o" to get Equation 4.118. The right-hand side of the equation is now a known value "k_2." The resulting Equation 4.119 is precisely the definition for the canonical PK1. Finally, for each set of "θ_1-θ_2-θ_3," we get the magnitude for "θ_4."

$$e^{\hat{\xi}_4\theta_4}o = e^{-\hat{\xi}_3\theta_3}e^{-\hat{\xi}_2\theta_2}e^{-\hat{\xi}_1\theta_1}H_{ST}(\theta)H_{ST}(0)^{-1}o \tag{4.118}$$

$$e^{\hat{\xi}_4\theta_4}o = k_2\begin{cases}\begin{bmatrix}\theta_1^{01} & \theta_2^{01} & \theta_3^{01}\end{bmatrix} \Rightarrow \theta_4^{01} \\ \begin{bmatrix}\theta_1^{02} & \theta_2^{02} & \theta_3^{01}\end{bmatrix} \Rightarrow \theta_4^{02}\end{cases} \tag{4.119}$$

Finally, we have got a set of two exact geometric solutions for this Scara type robot.
The code for this first algorithm is in Exercise 4.4.6a is in the internet hosting for the software of this book[23].

4.4.6.3 Second Algorithm for ABB IRB910SC IK "PG1+PK3+PK1+PK1"

- **Solution for θ_3 with PG1**: We use the previous resolution of Equation 4.114 and Equation 4.115.
- **Solutions for θ_2 with PK3**: We pass $H_{ST}(0)$ to the right-hand side of the problem definition equation. Then, we apply both sides of the equation to point "f" and subtract the result from the point "o" and calculate both sides' norm, as shown in Equation 4.120. We can cancel the unknown "θ_4" by PK1 and "θ_1" by

PK3 simplification cases. We know "θ_3," and applying the exponential of this third screw to point "f," we get another point, "f_1." The right-hand side of the equation is now a value "δ." The resulting Equation 4.121 is precisely the definition for the canonical PK3 to get "θ_2."

$$\left\| e^{\hat{\xi}_1\theta_1} e^{\hat{\xi}_2\theta_2} e^{\hat{\xi}_3\theta_3} e^{\hat{\xi}_4\theta_4} f - o \right\| = \left\| H_{ST}(\theta) H_{ST}(0)^{-1} f - o \right\| \quad (4.120)$$

$$\left\| e^{\hat{\xi}_2\theta_2} f_1 - o \right\| = \delta \left\{ \theta_3^{01} \rightarrow \begin{bmatrix} \theta_2^{01} & \theta_2^{02} \end{bmatrix} \right. \quad (4.121)$$

- **Solutions for θ_1 with PK1**: We pass $H_{ST}(0)$ to the right-hand side of the problem definition equation. Then, we apply both sides of the equation to point "f" to get Equation 4.122. We can use the PK1 simplification case to the equation's left-hand side to cancel "θ_4." Besides, we already know "θ_2" and "θ_3" from the algorithm's previous steps. Applying the exponential of these screws to point "f," we get another point, "f_2." The right-hand side of the equation is a known value "k_1." The resulting equation is precisely the definition for the canonical PK1 with Equation 4.123, which gives "θ_1."

$$e^{\hat{\xi}_1\theta_1} e^{\hat{\xi}_2\theta_2} e^{\hat{\xi}_3\theta_3} e^{\hat{\xi}_4\theta_4} f = H_{ST}(\theta) H_{ST}(0)^{-1} f \quad (4.122)$$

$$e^{\hat{\xi}_1\theta_1} f_2 = k_1 \begin{cases} \begin{bmatrix} \theta_2^{01} & \theta_3^{01} \end{bmatrix} \Rightarrow \theta_1^{01} \\ \begin{bmatrix} \theta_2^{02} & \theta_3^{01} \end{bmatrix} \Rightarrow \theta_1^{02} \end{cases} \quad (4.123)$$

- **Solutions for θ_4 with PK1**: We use the previous solution of Equation 4.118 and Equation 4.119.

Finally, we have got a set of two exact geometric solutions for this Scara type robot.

4.4.6.4 Comments on the SCARA Robot (ABB IRB910SC) IK Implementation

We must pay attention to detail that demonstrates the elegance and freedom that the screw theory provides. The manipulator's kinematics scheme (Figure 4.23) presents a careful selection of the points on the axes of the twists that are different. The points "r" and "f" are out of the mechanism, but of course, they work perfectly to the definition of the twists. This more abstract aspect of the screw theory opens more possibilities for the IK analysis.

Exercise 4.4.6c is in the internet hosting for the software of this book[24] and includes a comparison between the two algorithms presented for this Scara robot.

4.4.7 COLLABORATIVE ROBOTS (E.G., UNIVERSAL UR16E)

This exercise is for the UNIVERSAL UR16e, a very illustrative cobot system (Figure 4.24). Other robots of the same family can benefit from the same algorithm.

This exercise shows the flexibility of the POE canonical problems seamlessly with other IK techniques. We see with this robot how algebraic and geometric methods work

Inverse Kinematics

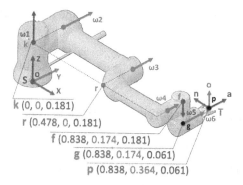

FIGURE 4.24 Collaborative robot UNIVERSAL UR16e inverse kinematics POE parameters.

together with screw theory to solve the IK of this robot. Nonetheless, the trade-off is that the complete solution ends up being less elegant than those presented in the previous sections using PK and PG canonical subproblems exclusively. We can pay attention to the several possible ways to solve this problem with alternative algorithms, such as "PG5+PG3+PK1+PG8," "PK1+PG3+PK1+PG8," or "PK1+PG3+PK1+PG4+PK1." Hereafter, we present in detail the first of those algorithms.

4.4.7.1 Inverse Kinematics UNIVERSAL UR16e Problem Definition

We obtain the equation that relates the motion of the joints ($\theta_1 \ldots \theta_6$) to the pose for the tool (noap), with the screw theory POE. We make the problem definition only by watching Figure 4.24.

- Define the axis of each joint "$\omega_1 \ldots \omega_6$."

$$\omega_1 = \begin{bmatrix} 0 \\ 0 \\ 1 \end{bmatrix} \quad \omega_2 = \begin{bmatrix} 0 \\ 1 \\ 0 \end{bmatrix} \quad \omega_3 = \begin{bmatrix} 0 \\ 1 \\ 0 \end{bmatrix} \quad \omega_4 = \begin{bmatrix} 0 \\ 1 \\ 0 \end{bmatrix} \quad \omega_5 = \begin{bmatrix} 0 \\ 0 \\ -1 \end{bmatrix} \quad \omega_6 = \begin{bmatrix} 0 \\ 1 \\ 0 \end{bmatrix}$$

- Obtain the Twists ($\xi_1 \ldots \xi_6$), knowing each joint's axis and a point on those axes.

$$\xi_1 = \begin{bmatrix} -\omega_1 \times o \\ \omega_1 \end{bmatrix} \quad \xi_2 = \begin{bmatrix} -\omega_2 \times k \\ \omega_2 \end{bmatrix} \quad \xi_3 = \begin{bmatrix} -\omega_3 \times r \\ \omega_3 \end{bmatrix}$$

$$\xi_4 = \begin{bmatrix} -\omega_4 \times f \\ \omega_4 \end{bmatrix} \quad \xi_5 = \begin{bmatrix} -\omega_5 \times g \\ \omega_5 \end{bmatrix} \quad \xi_6 = \begin{bmatrix} -\omega_6 \times p \\ \omega_6 \end{bmatrix}$$

- Get $H_{ST}(0)$, the pose of the tool at the reference (home) robot position.

$$H_{ST}(0) = T_{xyz}\begin{bmatrix} p_x \\ p_y \\ p_z \end{bmatrix} R_X\left(\frac{-\pi}{2}\right) R_Z(\pi) = \begin{bmatrix} -1 & 0 & 0 & p_X \\ 0 & 0 & 1 & p_Y \\ 0 & 1 & 0 & p_Z \\ 0 & 0 & 0 & 1 \end{bmatrix}$$

- Formulate the problem definition in terms of the POE as Equation 4.124.

$$H_{ST}(\theta) = e^{\hat{\xi}_1\theta_1} e^{\hat{\xi}_2\theta_2} e^{\hat{\xi}_3\theta_3} e^{\hat{\xi}_4\theta_4} e^{\hat{\xi}_5\theta_5} e^{\hat{\xi}_6\theta_6} H_{ST}(0) = \begin{bmatrix} n & o & a & p \\ 0 & 0 & 0 & 1 \end{bmatrix} \quad (4.124)$$

4.4.7.2 First Algorithm for UNIVERSAL UR16e IK "PG5+PG3+PK1+PG8"

- **Solutions for θ_1 with PG5**: We pass $H_{ST}(0)$ to the right-hand side of the problem definition of Equation 4.124. Then, we apply both sides of the equation to "g" (crossing of the last two axes: "ω_5," and "ω_6") to get Equation 4.125. We can apply the PK1 simplification case to cancel the unknowns "θ_6" and "θ_5." Pay attention to the fact that the screw rotations "θ_2, θ_3 and θ_4" (for whatever magnitude) do not change the plane where "g" moves "Π_g," which is oriented around the "Z" axis only by the effect of the first joint and therefore, "θ_2, θ_3, and θ_4," do not affect the calculation for "θ_1." The right-hand side of the equation is a known value "k_1" on a plane "Π_{k1}." The result is equivalent to the canonical PG5 subproblem, which evaluates a plane's rotation. The rotation "θ_1" applies to the plane "Π_g," containing the points "o" and "g" and the axis "ω_1." The resulting moving plane is "Π_{k1}," which includes the points "o" and "k_1" and the axis "ω_1." We get two values for this "θ_1" solving this PG5 with Equation 4.126. There are generally two solutions, which correspond to configurations where the robot shoulder is left or right. We must appreciate that because the point "g" does not lie in the manipulator plane of motion (i.e., the plane defined by points "o," and "r"), it is necessary to introduce some geometric corrections by Equation 4.127. In that way, we get the first joint feasible magnitudes "θ_1^{01}" and "θ_1^{02}" by equation. It is not difficult to understand the trigonometrical solution for looking at the robot geometry in the "X-Y" plane and considering the possible shoulder change of configuration.

$$e^{\hat{\xi}_1\theta_1} e^{\hat{\xi}_2\theta_2} e^{\hat{\xi}_3\theta_3} e^{\hat{\xi}_4\theta_4} e^{\hat{\xi}_5\theta_5} e^{\hat{\xi}_6\theta_6} g = H_{ST}(\theta) H_{ST}(0)^{-1} g \quad (4.125)$$

$$e^{\hat{\xi}_1\theta_1} \Pi_g = \Pi_{k1} \begin{cases} \theta_1^1 \\ \theta_1^2 \end{cases} \quad (4.126)$$

$$u = g - o \; ; \; \|u_p\| = \|u - \omega_1\omega_1^T u\| \; ; \; v = k_1 - o \; ; \; \|v_p\| = \|v - \omega_1\omega_1^T v\| \quad (4.127)$$

$$\theta_1^{01} = \theta_1^1 - \sin^{-1}\left(\frac{g_y}{\|v_p\|}\right) + \sin^{-1}\left(\frac{g_y}{\|u_p\|}\right)$$

$$\theta_1^{02} = \theta_1^2 + \sin^{-1}\left(\frac{g_y}{\|v_p\|}\right) + \sin^{-1}\left(\frac{g_y}{\|u_p\|}\right) \quad (1.128)$$

Inverse Kinematics

- **Solutions for θ_5 with PG3 and PK1**: The solution to this magnitude is a little bit difficult to comprehend. Therefore, we are going to explain it in detail. We pass $H_{ST}(0)$ and the exponential of the first four screws to the right-hand side of the problem definition of Equation 4.124. Then, we apply both sides of the equation to point "p" to get Equation 4.129. On the left-hand side of the equation, we cancel "θ_6" by PK1 simplification. In the right-hand side of the equation, the product of the inverted first exponential by $H_{ST}(\theta)$ and the inverted $H_{ST}(0)$ is a specific value "k_{2p}" (Equation 4.130). On this right-hand side, the product of the inverted three parallel exponentials (i.e., second, third, and four) translate the point "k_{2p}" along the "X" axis a certain amount. To evaluate this magnitude, we employ the PG3 subproblem, which calculates the translation to a given distance. PG3 moves the point "k_{2p}," along the axis "X" to a certain point, which is at magnitude given by the "θ_5" radius of rotation, to the "θ_5" rotation center. We identify this translation as "θ_7" and evaluate this magnitude by PG3 with Equation 4.130. The equation's right-hand side is then completely defined and equivalent to a certain point, "k_2," as shown in Equation 4.131. The resulting final Equation 4.132 is precisely the definition for the canonical PK1. We know the geometric IK solution for PK1, none, or one solution. With the "θ_1" value, we solve the motion for the fifth joint of the robot, obtaining two possible solutions with that magnitude and the negative of that quantity (Equation 4.132).

$$e^{\hat{\xi}_5\theta_5}e^{\hat{\xi}_6\theta_6}p = e^{-\hat{\xi}_4\theta_4}e^{-\hat{\xi}_3\theta_3}e^{-\hat{\xi}_2\theta_2}e^{-\hat{\xi}_1\theta_1}H_{ST}(\theta)H_{ST}(0)^{-1}p \qquad (4.129)$$

$$e^{-\hat{\xi}_1\theta_1}H_{ST}(\theta)H_{ST}(0)^{-1}p = k_{2p} \; ; \; \left\|e^{\hat{\xi}_7\theta_7}k_{2p} - g\right\| = \|p-g\| \qquad (4.130)$$

$$k_2 = k_{2p} + \omega_7^T\theta_7 \qquad (4.131)$$

$$e^{\hat{\xi}_5\theta_5}p = k_2 \begin{cases} \theta_1^{01} \Rightarrow \left[\theta_5^{01} \; ; \; \theta_5^{02} = -\theta_5^{01}\right] \\ \theta_1^{02} \Rightarrow \left[\theta_5^{03} \; ; \; \theta_5^{04} = -\theta_5^{03}\right] \end{cases} \qquad (4.132)$$

- **Solutions for θ_6 with an algebraic approach**: From the expression of the robot kinematics Equation 4.133, it is possible to quickly solve the sixth joint's value with a purely geometric approach. The sixth joint's magnitude depends only on the robot's first and fifth joints magnitudes, which form a spherical coordinate expression. There can be none or one solution for "θ_6" for each set of "θ_1-θ_5" by Equation 4.134.

$$H_{ST}(\theta) = e^{\hat{\xi}_1\theta_1}e^{\hat{\xi}_2\theta_2}e^{\hat{\xi}_3\theta_3}e^{\hat{\xi}_4\theta_4}e^{\hat{\xi}_5\theta_5}e^{\hat{\xi}_6\theta_6}H_{ST}(0) = \begin{bmatrix} n_x & o_x & a_x & p_x \\ n_y & o_y & a_y & p_y \\ n_z & o_z & a_z & p_z \\ 0 & 0 & 0 & 1 \end{bmatrix} \qquad (4.133)$$

$$\theta_6 = atan2\left(\frac{o_x \sin\theta_1 - o_y \cos\theta_1}{\sin\theta_5}, \frac{n_y \cos\theta_1 - n_x \sin\theta_1}{\sin\theta_5}\right) \begin{cases} \left[\theta_1^{01} \quad \theta_5^{01}\right] \Rightarrow \left[\theta_6^{01}\right] \\ \left[\theta_1^{01} \quad \theta_5^{02}\right] \Rightarrow \left[\theta_6^{02}\right] \\ \left[\theta_1^{02} \quad \theta_5^{03}\right] \Rightarrow \left[\theta_6^{03}\right] \\ \left[\theta_1^{02} \quad \theta_5^{04}\right] \Rightarrow \left[\theta_6^{04}\right] \end{cases} \quad (4.134)$$

- **Solutions for θ_2, θ_3 & θ_4 with PG8**: From the POE kinematics expression of Equation 4.124, we pass the first exponential to the right-hand side to get Equation 4.135. In doing so, we get a known pose (i.e., homogeneous transformation matrix) in the right-hand side "H_k." On the other hand, we already know the magnitudes for "θ_5" and "θ_6" and then the product of the exponential fifth and sixth by $H_{ST}(0)$, which is a particular pose "H_p." Therefore, the resulting Equation 4.136 is precisely the definition of the canonical Pardos-Gotor subproblem Eight (PG8) for three consecutive parallel rotations applied to a coordinate frame. We know the geometric IK solutions for PG8, which can be none, one, or two triple solutions for the set "θ_2-θ_3-θ_4." We must solve PG8 for each set of solutions "θ_1-θ_5-θ_6."

$$e^{\hat{\xi}_2\theta_2} e^{\hat{\xi}_3\theta_3} e^{\hat{\xi}_4\theta_4} e^{\hat{\xi}_5\theta_5} e^{\hat{\xi}_6\theta_6} H_{ST}(0) = e^{-\hat{\xi}_1\theta_1} H_{ST}(\theta) \quad (4.135)$$

$$e^{\hat{\xi}_2\theta_2} e^{\hat{\xi}_3\theta_3} e^{\hat{\xi}_4\theta_4} H_p = H_k \begin{cases} \left[\theta_1^{01} \quad \theta_5^{01} \quad \theta_6^{01}\right] \Rightarrow \left[\theta_2^{01} \quad \theta_3^{01} \quad \theta_4^{01}\right]\left[\theta_2^{02} \quad \theta_3^{02} \quad \theta_4^{02}\right] \\ \left[\theta_1^{01} \quad \theta_5^{02} \quad \theta_6^{02}\right] \Rightarrow \left[\theta_2^{03} \quad \theta_3^{03} \quad \theta_4^{03}\right]\left[\theta_2^{04} \quad \theta_3^{04} \quad \theta_4^{04}\right] \\ \left[\theta_1^{02} \quad \theta_5^{03} \quad \theta_6^{03}\right] \Rightarrow \left[\theta_2^{05} \quad \theta_3^{05} \quad \theta_4^{05}\right]\left[\theta_2^{06} \quad \theta_3^{06} \quad \theta_4^{06}\right] \\ \left[\theta_1^{02} \quad \theta_5^{04} \quad \theta_6^{04}\right] \Rightarrow \left[\theta_2^{07} \quad \theta_3^{07} \quad \theta_4^{07}\right]\left[\theta_2^{08} \quad \theta_3^{08} \quad \theta_4^{08}\right] \end{cases}$$

(4.136)

Finally, the outcome is the set of eight possible geometric solutions for the IK of the UR16e.

4.4.7.3 Comments on the UNIVERSAL UR16e IK Complete Solution Implementation

The IK of this UR16e, without kinematics decoupling (i.e., system of 12 nonlinear coupled equations with six unknowns: θ_1, θ_2, θ_3, θ_4, θ_5, θ_6) is solved with eight geometric solutions.

Perhaps this algorithm is not so precise and elegant as those of the previous exercises with other robot configurations. However, it demonstrates how well the screw theory tools can work intertwined with other methodologies for solving IK from a different point of view. In this case, we have mixed the POE with some geometric solutions. In fact, in the literature, there are several solutions for the IK of this robot. However, we have wanted to introduce this new algorithm based on the application of screw theory canonical subproblems.

The dexterous workspace and the singularities of the robot can limit the set of solutions. The outcome can be none, two, four, or eight exact IK solutions. Therefore,

Inverse Kinematics

it is good practice to test each solution before using them. It is easy to do that by checking the solutions with the FK map. Afterward, it is possible to choose the most suitable for each application. Because of the nature of the "ST24R" library, if the solutions are not exact, they are a good approximation, which can be very useful on most occasions. However, if the robot behavior is not adequate to the required application, it is easy to reprogram the proposed algorithm

The code for Exercise 4.4.7 is in the internet hosting for the software of this book[25].

4.4.8 Redundant Robots (e.g., KUKA IIWA)

Other kinematic mechanisms frequently occurred in robotic manipulation, such as the redundant robots. We extend the screw theory methodologies for this kind of system. A robot must have enough DoF to accomplish a given task. In the mathematics presented so far, we have concentrated most of the time on the cases in which the robot has the required DoF (i.e., six precisely for a general motion in 3D). A kinematically redundant robot has more than the minimal number of DoF required to complete a set of tasks.

A redundant robot can have an infinite number of joint configurations that give the same tool pose. The extra DoF present in redundant manipulators can be used to develop some motion strategy. These extra DoFs can help avoid obstacles and kinematic singularities or optimize the manipulator's motion relative to a cost function. If some joint limits are present, we can use the redundant robot to increase the dexterous workspace.

The FK of a redundant manipulator is an elementary problem, as we can solve it simply using the POE formula.

The IK problem for redundant robots is ill-posed, as there may exist infinite configurations of the joints, giving the desired tool target. Even if we keep the pose for the tool fixed, the robot is still free to move along some trajectories.

Since there may be an infinite number of joint magnitudes that give the requisite end-effector pose, therefore, we must introduce additional criteria to choose among them. To select a solution, we need to introduce some constraints. One standard answer is to choose the minimum joint velocity to give the desired workspace velocity. Another one is to select the minimum norm, which returns a solution where the joint vector has a minor magnitude. It is also possible to introduce a particular constraint that is better suited for other applications.

The exercise presented here is for the manipulator KUKA IIWA, with the robot's real geometry. We can realize how elegant, and practical it is to bring solutions employing the POE and the canonical subproblems for solving both constraints and IK of this redundant robot. In the literature, we have seen some solutions for this robot that neglect one DoF to eliminate the redundancy. However, these solutions are wasting the great possibilities of the robot workspace.

We propose a new approach to get closed-form solutions to this KUKA IIWA robot (Figure 4.25). We introduce a criterion for the constraints, consisting of solving

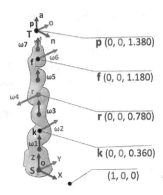

FIGURE 4.25 Redundant robot KUKA IIWA inverse kinematics POE parameters.

first "θ_1" or "θ_3" to orientate these two joints toward the tool target. Then, for each set value of these joints, we solve the IK for the rest six joints and then getting a set of sixteen exact geometric solutions. This way, the algorithm provides a natural motion of the robot and better use of dexterous workspace.

4.4.8.1 Inverse Kinematics KUKA IIWA Problem Definition

We obtain the equation that relates the motion of the joints ($\theta_1...\theta_7$) with the pose for the tool (noap), with the POE of the joints' twists as Equation 4.137. The problem definition is straightforward from Figure 4.25.

- Define the axis of each joint "$\omega_1...\omega_7$."

$$\omega_1 = \begin{bmatrix} 0 \\ 0 \\ 1 \end{bmatrix} \quad \omega_2 = \begin{bmatrix} 0 \\ 1 \\ 0 \end{bmatrix} \quad \omega_3 = \begin{bmatrix} 0 \\ 0 \\ 1 \end{bmatrix} \quad \omega_4 = \begin{bmatrix} 0 \\ -1 \\ 0 \end{bmatrix} \quad \omega_5 = \begin{bmatrix} 0 \\ 0 \\ 1 \end{bmatrix} \quad \omega_6 = \begin{bmatrix} 0 \\ 1 \\ 0 \end{bmatrix} \quad \omega_7 = \begin{bmatrix} 0 \\ 0 \\ 1 \end{bmatrix}$$

- Obtain the Twists ($\xi_1... \xi_7$), knowing each joint's axis and a point on those axes.

$$\xi_1 = \begin{bmatrix} -\omega_1 \times o \\ \omega_1 \end{bmatrix} \quad \xi_2 = \begin{bmatrix} -\omega_2 \times k \\ \omega_2 \end{bmatrix} \quad \xi_3 = \begin{bmatrix} -\omega_3 \times o \\ \omega_3 \end{bmatrix} \quad \xi_4 = \begin{bmatrix} -\omega_4 \times r \\ \omega_4 \end{bmatrix}$$

$$\xi_5 = \begin{bmatrix} -\omega_5 \times o \\ \omega_5 \end{bmatrix} \quad \xi_6 = \begin{bmatrix} -\omega_6 \times f \\ \omega_6 \end{bmatrix} \quad \xi_7 = \begin{bmatrix} -\omega_7 \times o \\ \omega_7 \end{bmatrix}$$

- Get $H_{ST}(0)$, the pose of the tool at the reference (home) robot position.

$$H_{ST}(0) = T_{xyz} \begin{bmatrix} p_x \\ p_y \\ p_z \end{bmatrix} = \begin{bmatrix} 1 & 0 & 0 & 0 \\ 0 & 1 & 0 & 0 \\ 0 & 0 & 1 & p_z \\ 0 & 0 & 0 & 1 \end{bmatrix}$$

Inverse Kinematics

- Then we get the problem definition in terms of the POE.

$$H_{ST}(\theta) = e^{\hat{\xi}_1\theta_1} e^{\hat{\xi}_2\theta_2} e^{\hat{\xi}_3\theta_3} e^{\hat{\xi}_4\theta_4} e^{\hat{\xi}_5\theta_5} e^{\hat{\xi}_6\theta_6} e^{\hat{\xi}_7\theta_7} H_{ST}(0) = \begin{bmatrix} n_x & o_x & a_x & p_x \\ n_y & o_y & a_y & p_y \\ n_z & o_z & a_z & p_z \\ 0 & 0 & 0 & 1 \end{bmatrix} \quad (4.137)$$

4.4.8.2 First Algorithm for KUKA IIWA IK "PK1+PK3+PK2+PK2+PK2+PK1"

- **Solutions for θ_1^{in} & θ_3^{in} with PK1:** It is a redundant robot with 7 DoF, and so it can have infinite solutions for its IK problem. To limit the number of solutions, we decide to choose some θ_1 & θ_3 (input) values as a function of some criteria and then solve the IK for each input magnitudes to get 8 + 8 solutions. These values are named "θ_1^{in}" & "θ_3^{in}" after "input" constraint solutions.

  We can choose several criteria: the minimum joint velocity, the minimum norm for the joint magnitude, the secure path to avoid obstacles or kinematic singularities, or the optimum motion of the robot relative to some cost function. In general, any wise criteria must increase the dexterous workspace for the redundant robot compared to a robot with only 6 DoF. **To get "θ_1^{in} & θ_3^{in}," we propose an algorithm to orientate the robot TCP in the direction of the target position given by the desired TCP (i.e., p_x, p_y, p_z).** To get these solutions, simply apply the PK1 subproblem with Equation 4.138. The whole movement of the robot is quite natural when following a planned trajectory, and at the same time, it exploits the workspace capacities.

$$e^{\hat{\xi}_1\theta_1}\begin{bmatrix}1\\0\\0\end{bmatrix} = \begin{bmatrix}p_x\\p_y\\p_z\end{bmatrix}\{[\theta_1^{in}]\} \quad ; \quad e^{\hat{\xi}_3\theta_3}\begin{bmatrix}1\\0\\0\end{bmatrix} = \begin{bmatrix}p_x\\p_y\\p_z\end{bmatrix}\{[\theta_3^{in}]\} \quad (4.138)$$

- **Solutions for θ_4 with PK3:** We pass $H_{ST}(0)$ to the right-hand side of the problem definition of Equation 4.137. Then, we apply both sides of the equation to point "f" (crossing of the last three axes: "ω_5," "ω_6," "ω_7"), to later subtract the result from point "k" (crossing of the first three axes: "ω_1," "ω_2," "ω_3"), and calculate the norm of both sides, to get Equation 4.139. We can apply to the left-hand side of the equation the PK1 and PK3 simplification cases, and thus we can cancel the unknowns "θ_7-θ_6-θ_5" by PK1 and "θ_3-θ_2-θ_1" by PK3. The right-hand side of the equation is now a known value "δ." The resulting expression is precisely the definition for the canonical PK3. We know the geometric IK solutions for PK3, none, one, or two. We solve the motion for the fourth joint, obtaining the rotation magnitudes (if they exist) for the "θ_4" double solution by Equation 4.140.

$$\left\|e^{\hat{\xi}_1\theta_1} e^{\hat{\xi}_2\theta_2} e^{\hat{\xi}_3\theta_3} e^{\hat{\xi}_4\theta_4} e^{\hat{\xi}_5\theta_5} e^{\hat{\xi}_6\theta_6} e^{\hat{\xi}_7\theta_7} f - k\right\| = \left\|H_{ST}(\theta) H_{ST}(0)^{-1} f - k\right\| \quad (4.139)$$

$$\left\| e^{\hat{\xi}_4\theta_4} f - k \right\| = \delta \Rightarrow \begin{bmatrix} \theta_4^{01} & \theta_4^{02} \end{bmatrix} \tag{4.140}$$

- **Solutions for θ_2 & θ_1 with PK2:** We pass $H_{ST}(0)$ to the right-hand side of the problem definition of Equation 4.137. Then, we apply both sides of the equation to point "f," getting Equation 4.141. We can then use the simplification case PK1 to the left-hand side of the equation for canceling "θ_7-θ_6-θ_5." We already know "θ_3^{in}" and "θ_4," from the previous step of the algorithm, and by applying the exponential of these screws to point "f," we get another point, "f_1." The right-hand side of the equation is now a known value "k_1." The resulting Equation 4.142 is precisely the definition for the canonical PK2. Then, we know the geometric IK solutions for PK2 (none, one, or two double). Finally, for each "θ_3^{in}" and "θ_4" couple of values, we obtain the rotation magnitudes (if they exist) for "θ_2-θ_1" double solutions.

$$e^{\hat{\xi}_1\theta_1} e^{\hat{\xi}_2\theta_2} e^{\hat{\xi}_3\theta_3} e^{\hat{\xi}_4\theta_4} e^{\hat{\xi}_5\theta_5} e^{\hat{\xi}_6\theta_6} e^{\hat{\xi}_7\theta_7} f = H_{ST}(\theta) H_{ST}(0)^{-1} f \tag{4.141}$$

$$e^{\hat{\xi}_1\theta_1} e^{\hat{\xi}_2\theta_2} f_1 = k_1 \begin{cases} \begin{bmatrix} \theta_3^{in} & \theta_4^{01} \end{bmatrix} \Rightarrow \begin{bmatrix} \theta_1^{01} & \theta_2^{01} & ; & \theta_1^{02} & \theta_2^{02} \end{bmatrix} \\ \begin{bmatrix} \theta_3^{in} & \theta_4^{02} \end{bmatrix} \Rightarrow \begin{bmatrix} \theta_1^{03} & \theta_2^{03} & ; & \theta_1^{04} & \theta_2^{04} \end{bmatrix} \end{cases} \tag{4.142}$$

- **Solutions for θ_3 & θ_2 with PK2:** We pass $H_{ST}(0)$ and the exponential of the first screw to the right-hand side of the problem definition of Equation 4.137. Then, we apply both sides of the equation to point "f," getting Equation 4.143. We can use the simplification case PK1 to the left-hand side for canceling "θ_7-θ_6-θ_5." Also, we already know "θ_4" from a previous step of the algorithm, and by applying the exponential of this screw to point "f," we get another point, "f_2." The right-hand side of the equation is now a known value "k_2." The resulting equation is precisely the definition for the canonical PK2 of Equation 4.144. We know the geometric IK solutions for PK2, none, one, or two double values. Finally, for each set "θ_1^{in}-θ_4" couple of values, we solve the motion for the second and third joints of the robot to obtain the rotation magnitudes (if they exist) for "θ_3-θ_2" double solutions.

$$e^{\hat{\xi}_2\theta_2} e^{\hat{\xi}_3\theta_3} e^{\hat{\xi}_4\theta_4} e^{\hat{\xi}_5\theta_5} e^{\hat{\xi}_6\theta_6} e^{\hat{\xi}_7\theta_7} f = e^{-\hat{\xi}_1\theta_1} H_{ST}(\theta) H_{ST}(0)^{-1} f \tag{4.143}$$

$$e^{\hat{\xi}_2\theta_2} e^{\hat{\xi}_3\theta_3} f_2 = k_2 \begin{cases} \begin{bmatrix} \theta_1^{in} & \theta_4^{01} \end{bmatrix} \Rightarrow \begin{bmatrix} \theta_2^{05} & \theta_3^{01} & ; & \theta_2^{06} & \theta_3^{02} \end{bmatrix} \\ \begin{bmatrix} \theta_1^{in} & \theta_4^{02} \end{bmatrix} \Rightarrow \begin{bmatrix} \theta_2^{07} & \theta_3^{03} & ; & \theta_2^{08} & \theta_3^{04} \end{bmatrix} \end{cases} \tag{4.144}$$

- **Solutions for θ_6 & θ_5 with PK2:** We pass $H_{ST}(0)$ and the exponentials of the first four screws, which we already know from the previous steps of the

Inverse Kinematics

algorithm (i.e., we got "θ_1," "θ_2," "θ_3," and "θ_4") to the right-hand side of the problem definition of Equation 4.137. Then, we apply both sides of the equation to point "p" to get Equation 4.145. The right-hand side of the equation is now a known value "k_3." On the left-hand side of the equation, we can cancel the unknown "θ_7" by the PK1 simplification case. The resulting equation is precisely the definition for the canonical PK2 with Equation 4.146. We know how to solve the geometric IK problem for PK2, none, one, or two double solutions. Finally, for each set of "θ_1-θ_2-θ_3-θ_4" values, we solve the motion for the fifth and sixth joints of the robot to obtain the rotation magnitudes (if they exist) for "θ_6-θ_5" double solutions.

$$e^{\hat{\xi}_5\theta_5}e^{\hat{\xi}_6\theta_6}e^{\hat{\xi}_7\theta_7}p = e^{-\hat{\xi}_4\theta_4}e^{-\hat{\xi}_3\theta_3}e^{-\hat{\xi}_2\theta_2}e^{-\hat{\xi}_1\theta_1}H_{ST}(\theta)H_{ST}(0)^{-1}p \qquad (4.145)$$

$$e^{\hat{\xi}_5\theta_5}e^{\hat{\xi}_6\theta_6}p = k_3 \begin{cases} \begin{bmatrix} \theta_1^{01} & \theta_2^{01} & \theta_3^{in} & \theta_4^{01} \end{bmatrix} \Rightarrow \begin{bmatrix} \theta_5^{01} & \theta_6^{01} & ; & \theta_5^{02} & \theta_6^{02} \end{bmatrix} \\ \begin{bmatrix} \theta_1^{02} & \theta_2^{02} & \theta_3^{in} & \theta_4^{01} \end{bmatrix} \Rightarrow \begin{bmatrix} \theta_5^{03} & \theta_6^{03} & ; & \theta_5^{04} & \theta_6^{04} \end{bmatrix} \\ \begin{bmatrix} \theta_1^{03} & \theta_2^{03} & \theta_3^{in} & \theta_4^{02} \end{bmatrix} \Rightarrow \begin{bmatrix} \theta_5^{05} & \theta_6^{05} & ; & \theta_5^{06} & \theta_6^{06} \end{bmatrix} \\ \begin{bmatrix} \theta_1^{04} & \theta_2^{04} & \theta_3^{in} & \theta_4^{02} \end{bmatrix} \Rightarrow \begin{bmatrix} \theta_5^{07} & \theta_6^{07} & ; & \theta_5^{08} & \theta_6^{08} \end{bmatrix} \\ \begin{bmatrix} \theta_1^{in} & \theta_2^{05} & \theta_3^{01} & \theta_4^{01} \end{bmatrix} \Rightarrow \begin{bmatrix} \theta_5^{09} & \theta_6^{09} & ; & \theta_5^{10} & \theta_6^{10} \end{bmatrix} \\ \begin{bmatrix} \theta_1^{in} & \theta_2^{06} & \theta_3^{02} & \theta_4^{01} \end{bmatrix} \Rightarrow \begin{bmatrix} \theta_5^{11} & \theta_6^{11} & ; & \theta_5^{12} & \theta_6^{12} \end{bmatrix} \\ \begin{bmatrix} \theta_1^{in} & \theta_2^{07} & \theta_3^{03} & \theta_4^{02} \end{bmatrix} \Rightarrow \begin{bmatrix} \theta_5^{13} & \theta_6^{13} & ; & \theta_5^{14} & \theta_6^{14} \end{bmatrix} \\ \begin{bmatrix} \theta_1^{in} & \theta_2^{08} & \theta_3^{04} & \theta_4^{02} \end{bmatrix} \Rightarrow \begin{bmatrix} \theta_5^{15} & \theta_6^{15} & ; & \theta_5^{16} & \theta_6^{16} \end{bmatrix} \end{cases} \qquad (4.146)$$

- **Solutions for θ_7 with PK1**: We pass $H_{ST}(0)$ and the exponentials of the first six screws, which we already know from the previous steps of the algorithm (i.e., we got "θ_1," "θ_2," "θ_3," "θ_4," "θ_5," and "θ_6"), to the right-hand side of the problem definition of Equation 4.137. Then, we apply both sides of the equation to a reference point (e.g., unitary vector on "x") to obtain Equation 4.147. The right-hand side of the equation is now a known value "k_4." The resulting Equation 4.148 is precisely the canonical PK1, and we know the geometric IK solutions for this subproblem. Finally, for each set of "θ_1-θ_2-θ_3-θ_4-θ_5-θ_6" values, we solve the motion for the seventh joint of the robot, obtaining the rotation magnitude for "θ_7" single solution.

$$e^{\hat{\xi}_7\theta_7}\begin{bmatrix}1\\0\\0\end{bmatrix} = e^{-\hat{\xi}_6\theta_6}e^{-\hat{\xi}_5\theta_5}e^{-\hat{\xi}_4\theta_4}e^{-\hat{\xi}_3\theta_3}e^{-\hat{\xi}_2\theta_2}e^{-\hat{\xi}_1\theta_1}H_{ST}(\theta)H_{ST}(0)^{-1}\begin{bmatrix}1\\0\\0\end{bmatrix} \qquad (4.147)$$

$$e^{\hat{\xi}_7 \theta_7} \begin{bmatrix} 1 \\ 0 \\ 0 \end{bmatrix} = k_4 \begin{cases} [\theta_1^{01} \quad \theta_2^{01} \quad \theta_3^{in} \quad \theta_4^{01} \quad \theta_5^{01} \quad \theta_6^{01}] \Rightarrow \theta_7^{01} \\ [\theta_1^{01} \quad \theta_2^{01} \quad \theta_3^{in} \quad \theta_4^{01} \quad \theta_5^{02} \quad \theta_6^{02}] \Rightarrow \theta_7^{02} \\ [\theta_1^{02} \quad \theta_2^{02} \quad \theta_3^{in} \quad \theta_4^{01} \quad \theta_5^{03} \quad \theta_6^{03}] \Rightarrow \theta_7^{03} \\ [\theta_1^{02} \quad \theta_2^{02} \quad \theta_3^{in} \quad \theta_4^{01} \quad \theta_5^{04} \quad \theta_6^{04}] \Rightarrow \theta_7^{04} \\ [\theta_1^{03} \quad \theta_2^{03} \quad \theta_3^{in} \quad \theta_4^{02} \quad \theta_5^{05} \quad \theta_6^{05}] \Rightarrow \theta_7^{05} \\ [\theta_1^{03} \quad \theta_2^{03} \quad \theta_3^{in} \quad \theta_4^{02} \quad \theta_5^{06} \quad \theta_6^{06}] \Rightarrow \theta_7^{06} \\ [\theta_1^{04} \quad \theta_2^{04} \quad \theta_3^{in} \quad \theta_4^{02} \quad \theta_5^{07} \quad \theta_6^{07}] \Rightarrow \theta_7^{07} \\ [\theta_1^{04} \quad \theta_2^{04} \quad \theta_3^{in} \quad \theta_4^{02} \quad \theta_5^{08} \quad \theta_6^{08}] \Rightarrow \theta_7^{08} \\ [\theta_1^{in} \quad \theta_2^{05} \quad \theta_3^{01} \quad \theta_4^{01} \quad \theta_5^{09} \quad \theta_6^{09}] \Rightarrow \theta_7^{09} \\ [\theta_1^{in} \quad \theta_2^{05} \quad \theta_3^{01} \quad \theta_4^{01} \quad \theta_5^{10} \quad \theta_6^{10}] \Rightarrow \theta_7^{10} \\ [\theta_1^{in} \quad \theta_2^{06} \quad \theta_3^{02} \quad \theta_4^{01} \quad \theta_5^{11} \quad \theta_6^{11}] \Rightarrow \theta_7^{11} \\ [\theta_1^{in} \quad \theta_2^{06} \quad \theta_3^{02} \quad \theta_4^{01} \quad \theta_5^{12} \quad \theta_6^{12}] \Rightarrow \theta_7^{12} \\ [\theta_1^{in} \quad \theta_2^{07} \quad \theta_3^{03} \quad \theta_4^{02} \quad \theta_5^{13} \quad \theta_6^{13}] \Rightarrow \theta_7^{13} \\ [\theta_1^{in} \quad \theta_2^{07} \quad \theta_3^{03} \quad \theta_4^{02} \quad \theta_5^{14} \quad \theta_6^{14}] \Rightarrow \theta_7^{14} \\ [\theta_1^{in} \quad \theta_2^{08} \quad \theta_3^{04} \quad \theta_4^{02} \quad \theta_5^{15} \quad \theta_6^{15}] \Rightarrow \theta_7^{15} \\ [\theta_1^{in} \quad \theta_2^{08} \quad \theta_3^{04} \quad \theta_4^{02} \quad \theta_5^{16} \quad \theta_6^{16}] \Rightarrow \theta_7^{16} \end{cases} \quad (4.148)$$

The code for Exercise 4.4.8 is in the internet hosting for the software of this book[26].

4.4.8.3 Comments on the KUKA IIWA IK Complete Solution Implementation

The geometric approach for solving this problem with POE implies that we first solve the "θ_1," "θ_2," "θ_3," and "θ_4" magnitudes to position the "f" point in space, and afterward we solve "θ_5," "θ_6," and "θ_7." This algorithm uses canonical subproblems solutions for positioning "f" in the 3D space. However, be careful not to confuse this algorithm with the "kinematics uncoupling." It is evident (Figure 4.25) that "θ_5" and "θ_6" also imply translation for TCP in the 3D space, and not only rotation, as it happens with the kinematics uncoupling assumption. Here, the resulting algorithm is quite elegant and helpful, providing 16 exact geometric solutions.

With the broad set of closed-form results for this robot's IK, it is easier to implement various applications. It is possible to exploit the workspace given by this mechanism much better. The "ST24R" library offers, in any case, the most approximate solutions for the IK canonical problems. The algorithm results could be inexact (but approximate) even when the TCP target is out of the dexterous workspace. This implementation proves to be very useful for most applications. If we need another

Inverse Kinematics

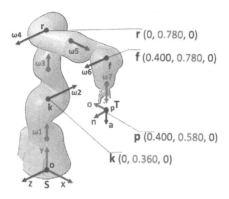

FIGURE 4.26 Redundant robot KUKA IIWA "Elbow" inverse kinematics POE parameters.

robot behavior, we only need to reprogram the IK canonical problems with a different approach according to the new application.

Other algorithms based on the POE and canonical subproblems can solve this architecture (e.g., PK1+PK3+PG6+PG6+PG6+PK1). Also, the constraints to limit the number of solutions can have another formulation matching better other specific applications.

The screw theory's flexibility allows for different robot home positions if they are more convenient for the implementation. As an example, the same algorithm works for the IIWA elbow home configuration (Figure 4.26).

4.4.9 MANY DoF ROBOTS (E.G., RH0 UC3M HUMANOID)

We can extend the advantages of the POE IK formulation to robotics mechanisms with many DoF. For instance, we present a reference to a Robot Humanoid RH0 of the UC3M (University Carlos III of Madrid).

The humanoid kinematics presents a formidable computational challenge for real-time IK applications due to the high number of DoF (e.g., 21 for the RH0) and the bipedal complexities locomotion. Nonetheless, we can express the humanoid kinematics problem with the POE for any point of the robot body, particularly for the trunk.

To apply the IK canonical subproblems to the RH0, we divide the complete analysis of the 21 DoF humanoid into different rigid-body chains (i.e., legs and arms). For instance, in the bipedal locomotion, the legs' scheme has a POE with six virtual DoF to define the feet' position and orientation about the spatial reference system, plus seven actual DoF from the mechanical joints of the leg. See these POE for the left and right legs, with Equation 4.149 and Equation 4.150. We can glimpse that the screw theory techniques can play a crucial role in solving these POE for the legs in the IK locomotion solution (Figure 4.27).

$$G_{SH}(\theta) = e^{\hat{\xi}_{VZ1}\theta_{VZ1}} \ldots e^{\hat{\xi}_{VZ6}\theta_{VZ6}} \cdot e^{\hat{\xi}_{12}\theta_{12}} \ldots e^{\hat{\xi}_{7}\theta_{7}} \cdot e^{\hat{\xi}_{13}\theta_{13}} G_{SH}(0) \qquad (4.149)$$

$$G_{SH}(\theta) = e^{\hat{\xi}_{VD1}\theta_{VD1}} \ldots e^{\hat{\xi}_{VD6}\theta_{VD6}} \cdot e^{\hat{\xi}_{1}\theta_{1}} \ldots e^{\hat{\xi}_{6}\theta_{6}} \cdot e^{\hat{\xi}_{13}\theta_{13}} G_{SH}(0) \qquad (4.150)$$

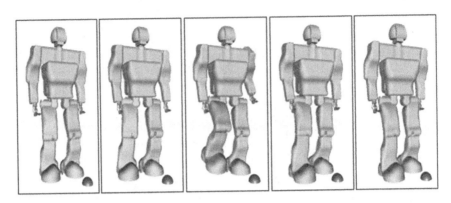

FIGURE 4.27 Humanoid robot UC3M RH0 inverse kinematics and bipedal locomotion.

Additionally, we can use the POE to solve the IK of other robots with many DoF or manipulations with kinematic loops.

From now on, we are ready to formulate the POE IK to more complex mechanisms!

4.5 SUMMARY

The IK problem tries to determine the joint magnitudes that achieve a desired robot tool configuration. The analytical difficulty of IK is quite significant. For a tool pose, multiple solutions may exist for the joints' values or no solution at all.

We reviewed the standard approach to solve the IK of robots with several DoF, which is any numerical algorithm. The result is neither very efficient nor very effective, but it has become kind of the standard.

We present **the idea of solving the IK problem using the techniques and tools from the screw theory and mainly the POE. This concept has paramount importance. It provides tools and methodologies to build geometric closed-form efficient and effective algorithms** to solve many mechanical problems, particularly robotics IK.

To show the POE benefits, we saw an example for the classic Puma robot, with numeric (i.e., the RST toolbox of MATLAB®) and geometric algorithms (i.e., the "ST24R" library). The performance is overwhelmingly better in favor of the screw theory approach.

We showed how to use POE for solving IK. It is feasible to build geometric algorithms for the IK problem of complex mechanisms with many DoF. There are some subproblems, which frequently occur in the IK solutions for many standard robot designs. These small geometric exercises constitute the cornerstone of this approach. We name these subproblems as canonical IK. They are geometrically meaningful and numerically stable. One seeks to reduce the complete IK problem into appropriate canonical subproblems whose solutions are known.

One of the more original and valuable sections of this chapter is the collection of **Canonical IK Subproblems**. The method was initially presented by Paden and built

Inverse Kinematics

on the unpublished work of Kahan. The first is the three already classical **PK canonical subproblems** (i.e., PK1, PK2, and PK3).

- **PK1**: ONE ROTATION about one single axis, applied to a POINT.
- **PK2**: TWO consecutive ROTATIONS about CROSSING axes, applied to a POINT.
- **PK3**: ONE ROTATION to a given DISTANCE, applied to a POINT.

We demonstrated that **the set of canonical subproblems is by no means exhaustive. It is possible to develop other problems to solve the IK of different robots.** This chapter showed the possibility of solving robots with prismatic joints and mechanisms with parallel or skewed rotations. We have introduced some new Pardos-Gotor (PG) canonical subproblems (i.e., PG1, PG2, PG3, PG4, PG5, PG6, PG7, and PG8). They are handy to solve the IK of many more frequent mechanical architectures. Besides, they can become the encouragement to create other canonical problems of different interests and utility.

This new set of Pardos-Gotor canonical subproblems (PG) extend the original idea of applying the POE to specific points, to other figures like lines, planes, or coordinate systems.

- **PG1**: ONE TRANSLATION along one single axis, applied to a POINT.
- **PG2**: TWO consecutive TRANSLATIONS along CROSSING axes, applied to a POINT.
- **PG3**: ONE TRANSLATION to a given DISTANCE, applied to a POINT.
- **PG4**: TWO consecutive ROTATIONS about PARALLEL axes, applied to a POINT.
- **PG5**: ONE ROTATION about one single axis, applied to a LINE or a PLANE.
- **PG6**: TWO consecutive ROTATIONS about SKEWED axes, applied to a POINT or a LINE.
- **PG7**: THREE consecutive ROTATIONS about one SKEWED and two PARALLEL axes, applied to a POINT.
- **PG8**: THREE consecutive ROTATIONS about PARALLEL axes, applied to a POSE (i.e., coordinate system, which includes rotation and translation).

There is another critical section of this chapter, where we show some systematic, elegant, and geometrically meaningful solutions for some well-known architectures. The examples use the POE for solving the IK of some robots: A Puma with six DoF (e.g., ABB IRB120), a Bending Backwards with six DoF (e.g., ABB IRB1600), a Gantry with six DoF (e.g., ABB 6620LX), a Scara with four DoF (e.g., ABB IRB910SC), a Collaborative robot (Cobot) with six DoF (e.g., UNIVERSAL UR16e) and a Redundant manipulator with seven DoF (i.e., KUKA IIWA). We present with complete detail all these algorithms, at least for one kind of implementation. For some robots, we have explained even several equivalent geometric algorithms. We hope that all these examples proved the elegance and flexibility of this new screw theory methodology.

The ideas stemmed from the exercises are helpful to develop solutions for more complicated IK, such as the example for the humanoid robot RH0 of the UC3M with 21 DoF. The goal of these exercises is to illustrate the techniques for manipulating the POE on solving IK. We are sure that they will become a significant precedent to solve other robot challenges.

The POE allows for EFFICIENT & EFFECTIVE algorithms. There are no iterations, and the calculation has convergence guaranteed. Moreover, **the geometric closed-form formulation provides the complete SET OF SOLUTIONS, if they exist, for the IK problem**. All these are significant advantages. We can choose the better algorithm for the robot's next move to follow a trajectory from the set of solutions. This POE power is remarkable compared to the typical numerical solutions, which usually provide only one approximate solution. Besides, these techniques apply to more complex mechanisms, such as robots with many DoF or robot manipulations with kinematic loops.

NOTES

1 Pardos-Gotor, J.M. (2021). *Screw Theory in Robotics*. Github. https://github.com/DrPardosGotor/Screw-Theory-in-Robotics/tree/master/Code/ST24R_MATLAB_3.70
2 Pardos-Gotor, J.M. (2021). *Screw Theory in Robotics*. Github. https://github.com/DrPardosGotor/Screw-Theory-in-Robotics/blob/master/Exercises/Exercise_4_2_5.m
3 Pardos-Gotor, J.M. (2021). *Screw Theory in Robotics*. Github. https://github.com/DrPardosGotor/Screw-Theory-in-Robotics/blob/master/Exercises/Exercise_4_3_2.m
4 Pardos-Gotor, J.M. (2021). *Screw Theory in Robotics*. Github. https://github.com/DrPardosGotor/Screw-Theory-in-Robotics/blob/master/Exercises/Exercise_4_3_3.m
5 Pardos-Gotor, J.M. (2021). *Screw Theory in Robotics*. Github. https://github.com/DrPardosGotor/Screw-Theory-in-Robotics/blob/master/Exercises/Exercise_4_3_4.m
6 Pardos-Gotor, J.M. (2021). *Screw Theory in Robotics*. Github. https://github.com/DrPardosGotor/Screw-Theory-in-Robotics/blob/master/Exercises/Exercise_4_3_5.m
7 Pardos-Gotor, J.M. (2021). *Screw Theory in Robotics*. Github. https://github.com/DrPardosGotor/Screw-Theory-in-Robotics/blob/master/Exercises/Exercise_4_3_6.m
8 Pardos-Gotor, J.M. (2021). *Screw Theory in Robotics*. Github. https://github.com/DrPardosGotor/Screw-Theory-in-Robotics/blob/master/Exercises/Exercise_4_3_7.m
9 Pardos-Gotor, J.M. (2021). *Screw Theory in Robotics*. Github. https://github.com/DrPardosGotor/Screw-Theory-in-Robotics/blob/master/Exercises/Exercise_4_3_8.m
10 Pardos-Gotor, J.M. (2021). *Screw Theory in Robotics*. Github. https://github.com/DrPardosGotor/Screw-Theory-in-Robotics/blob/master/Exercises/Exercise_4_3_9.m
11 Pardos-Gotor, J.M. (2021). *Screw Theory in Robotics*. Github. https://github.com/DrPardosGotor/Screw-Theory-in-Robotics/blob/master/Exercises/Exercise_4_3_10.m
12 Pardos-Gotor, J.M. (2021). *Screw Theory in Robotics*. Github. https://github.com/DrPardosGotor/Screw-Theory-in-Robotics/blob/master/Exercises/Exercise_4_3_11.m
13 Pardos-Gotor, J.M. (2021). *Screw Theory in Robotics*. Github. https://github.com/DrPardosGotor/Screw-Theory-in-Robotics/blob/master/Exercises/Exercise_4_3_12.m
14 Pardos-Gotor, J.M. (2021). *Screw Theory in Robotics*. Github. https://github.com/DrPardosGotor/Screw-Theory-in-Robotics/blob/master/Exercises/Exercise_4_4_2a.m
15 Pardos-Gotor, J.M. (2021). *Screw Theory in Robotics*. Github. https://github.com/DrPardosGotor/Screw-Theory-in-Robotics/blob/master/Exercises/Exercise_4_4_2b.m
16 Pardos-Gotor, J.M. (2021). *Screw Theory in Robotics*. Github. https://github.com/DrPardosGotor/Screw-Theory-in-Robotics/blob/master/Exercises/Exercise_4_4_2c.m

17 Pardos-Gotor, J.M. (2021). *Screw Theory in Robotics*. Github. https://github.com/DrPardosGotor/Screw-Theory-in-Robotics/blob/master/Exercises/Exercise_4_4_2d.m
18 Pardos-Gotor, J.M. (2021). *Screw Theory in Robotics*. Github. https://github.com/DrPardosGotor/Screw-Theory-in-Robotics/blob/master/Exercises/Exercise_4_4_2e.m
19 Pardos-Gotor, J.M. (2021). *Screw Theory in Robotics*. Github. https://github.com/DrPardosGotor/Screw-Theory-in-Robotics/blob/master/Exercises/Exercise_4_4_2f.m
20 Pardos-Gotor, J.M. (2021). *Screw Theory in Robotics*. Github. https://github.com/DrPardosGotor/Screw-Theory-in-Robotics/blob/master/Exercises/Exercise_4_4_3.m
21 Pardos-Gotor, J.M. (2021). *Screw Theory in Robotics*. Github. https://github.com/DrPardosGotor/Screw-Theory-in-Robotics/blob/master/Exercises/Exercise_4_4_4.m
22 Pardos-Gotor, J.M. (2021). *Screw Theory in Robotics*. Github. https://github.com/DrPardosGotor/Screw-Theory-in-Robotics/blob/master/Exercises/Exercise_4_4_5.m
23 Pardos-Gotor, J.M. (2021). *Screw Theory in Robotics*. Github. https://github.com/DrPardosGotor/Screw-Theory-in-Robotics/blob/master/Exercises/Exercise_4_4_6a.m
24 Pardos-Gotor, J.M. (2021). *Screw Theory in Robotics*. Github. https://github.com/DrPardosGotor/Screw-Theory-in-Robotics/blob/master/Exercises/Exercise_4_4_6c.m
25 Pardos-Gotor, J.M. (2021). *Screw Theory in Robotics*. Github. https://github.com/DrPardosGotor/Screw-Theory-in-Robotics/blob/master/Exercises/Exercise_4_4_7.m
26 Pardos-Gotor, J.M. (2021). *Screw Theory in Robotics*. Github. https://github.com/DrPardosGotor/Screw-Theory-in-Robotics/blob/master/Exercises/Exercise_4_4_8.m

5 Differential Kinematics

"Whenever the work is itself light, it becomes necessary, in order to economize time, to increase the velocity."

—Charles Babbage

5.1 PROBLEM STATEMENT IN ROBOTICS

A robot's tool moves in the 3D space with a translational and rotational velocity (Choset and Lynch, 2005). We are interested in describing this rate of change of the robot pose with a computationally efficient representation of the end-effector velocity. The relationship between the joint's velocity and the tool's velocity constitutes the concept of Differential Kinematics (DK) of the robot. The mathematical tool to relate these velocities between the operational and joint spaces is the Jacobian matrix. DK presents two fundamental problems:

- **Differential (Instantaneous) Forward Kinematics** (Equation 5.1): It gives the velocity of the tool frame as a function of the joint velocities.

$$\dot{T}(t) = J(\theta)\dot{\theta}(t) \tag{5.1}$$

- **Differential (Instantaneous) Inverse Kinematics** (Equation 5.2): It gives the joint velocities a function of the velocity of the tool frame.

$$\dot{\theta}(t) = \left[J(\theta)\right]^{-1}\dot{T}(t) \tag{5.2}$$

Carl Gustav Jacob Jacobi (1804–1851) was a Prussian Mathematician. He wrote the classic treatise on elliptic functions and described the derivative of "m" functions of "n" variables, which bears his name, as the Jacobian matrix. It maps the velocities between two spaces for the tool and the joint. The Jacobian matrix is a function of the joint coordinates, and therefore a matrix-valued function.

We can use the Jacobian to move the robot from one end-effector configuration to another without calculating the Inverse Kinematics (IK). This use of the Jacobian matrix to solve the IK is very helpful (Siciliano and Khatib, 2016). It can happen that a closed-form solution for the IK is not available or perhaps only for an uninteresting robot configuration. Using the Jacobian matrix is quite helpful for both cases, even though it is only an approximation. We will see more details in Chapter 7 for trajectory generation.

To move the end-effector from one configuration to another in the workspace, we need to know the spatial velocity between those two configurations. We obtain the new joint magnitudes following Equation 5.3 with the joint coordinates corresponding to the first end-effector pose and integrate the joint velocity over time.

$$\theta(t_2) = \theta(t_1) + \int_{t1}^{t2} \dot{\theta}(t) = \theta(t_1) + \int_{t1}^{t2} (J(\theta))^{-1} \dot{T}(t) \quad (5.3)$$

The Jacobian is also used to identify the singularities which exist in the boundaries or inside the robot workspace (Park and Kim, 1999).

Another important field when dealing with the Jacobian is the study of local manipulability (Mason, 2001). It relates infinitesimal joint motion to workspace motion, but this is an area of study beyond this book's scope. Nevertheless, there are many projects where we can explore the use of the Jacobian matrix for dexterous manipulation.

5.2 THE ANALYTIC JACOBIAN

5.2.1 A Traditional Description

The analytic Jacobian describes the transformation from the joint velocities (θ') to the tool velocities ($V^s{}_T$) of a robot (see Equation 5.4). We get this standard Jacobian (J_A) by the differentiation of the Forward Kinematics (FK) map (T) by Equation 5.5. This expression of the tool pose (i.e., position and orientation) is a function of the joint coordinates. This approach works fine when this FK mapping is uncomplicated. However, if we represent mechanisms with a complex FK mapping, like robots with many Degrees of Freedom (DoF), the analytic Jacobian is not easily obtained.

$$V_T^S = J_A(\theta)\dot{\theta} \quad (5.4)$$

$$J_A(\theta) = \frac{\partial T}{\partial \theta}(\theta) \begin{cases} T(x,y,z,\alpha,\beta,\gamma) \\ \theta_1 \cdots \theta_n \end{cases} \quad (5.5)$$

The problem is that the differentiation leads to a quantity that is not natural, and the description only holds locally. The Jacobian at a point gives the best linear approximation of the nonlinear mapping between two spaces. More importantly, choosing a local parameterization destroys the rigid body motion's natural geometric structure and has no direct geometric interpretation.

Some singularities might be an artifact due to the Jacobian mapping's parameterization. This fact leads to false conclusions about the robot's ability to reach certain poses or achieve specific velocities (Li, 1990). In the singularities, the analytic Jacobian drops rank (i.e., Jacobian's determinant is null). In those configurations, the infinitesimal tool coordinates change implies an infinite increment for joint coordinates, with some DoF loss. There might be a sharp change in the joints' speed near the singularities, with the possible collapse of the robot. Therefore, the real singularities require study and elimination for a practical application (Park, 1991).

Differential Kinematics

5.2.2 Analytic Jacobian to Forward Differential Kinematics

It relates the velocities of the robot joints (i.e., θ'_1-θ_n') with the tool velocity (i.e., V^S_T) in the spatial coordinate system. We define this concept by a six-element vector (i.e., [x', y', z', α', β', γ'). The first three elements correspond to the TCP translation velocity, and the last three refer to the rotation velocity of the coordinate tool system. **The analytic Jacobian, defined by Equation 5.6, needs to differentiate the FK map and revaluation because it is a matrix-valued function.** Then the **forward DK map** by Equation 5.7 obtains the tool velocities for any target.

$$J_A(\theta) = \begin{bmatrix} \frac{\partial T_X}{\partial \theta_1} & \cdots & \frac{\partial T_X}{\partial \theta_n} \\ \vdots & \ddots & \vdots \\ \frac{\partial T_\gamma}{\partial \theta_1} & \cdots & \frac{\partial T_\gamma}{\partial \theta_n} \end{bmatrix} \quad (5.6)$$

$$V^S_T = \begin{bmatrix} v^S_{TCP} \\ \omega^S_{ST} \end{bmatrix} = \begin{bmatrix} \dot{x} \\ \dot{y} \\ \dot{z} \\ \dot{\alpha} \\ \dot{\beta} \\ \dot{\gamma} \end{bmatrix} = J_A(\theta)\dot{\theta} = J_A(\theta)\begin{bmatrix} \dot{\theta}_1 \\ \vdots \\ \dot{\theta}_n \end{bmatrix} \quad (5.7)$$

5.2.3 Analytic Jacobian for Inverse Differential Kinematics

It relates the tool velocity with the velocities of the robot joints. **The analytic Jacobian, defined by Equation 5.8, needs the differentiation of the IK map.** Of course, calculating this inverse Jacobian is more complicated, as we know that a robot can have several and different solutions for IK. If we obtain the IK with a numeric approach, we do not have a precise map to be differentiated. Therefore, we can apply different methods to get the inverse analytic Jacobian, for instance, by the symbolic inversion of the direct Jacobian or by valuating the direct Jacobian's numeric inversion. **The inverse DK map** allows obtaining the joint velocities necessary to get any desired tool velocity by Equation 5.9 for a particular point of a trajectory.

$$[J_A(\theta)]^{-1} = \begin{bmatrix} \frac{\partial \theta_1}{\partial T_x} & \cdots & \frac{\partial \theta_1}{\partial T_\gamma} \\ \vdots & \ddots & \vdots \\ \frac{\partial \theta_n}{\partial T_x} & \cdots & \frac{\partial \theta_n}{\partial T_\gamma} \end{bmatrix} \quad (5.8)$$

$$\dot{\theta} = \begin{bmatrix} \dot{\theta}_1 \\ \vdots \\ \dot{\theta}_n \end{bmatrix} = \left[J_A(\theta) \right]^{-1} V_T^S = \left[J_A(\theta) \right]^{-1} \begin{bmatrix} \dot{x} \\ \dot{y} \\ \dot{z} \\ \dot{\alpha} \\ \dot{\beta} \\ \dot{\gamma} \end{bmatrix} = \left[J_A(\theta) \right]^{-1} \begin{bmatrix} v_{TCP}^S \\ \omega_{ST}^S \end{bmatrix} \quad (5.9)$$

5.2.4 Scara Robot (e.g., ABB IRB910SC)

To consolidate the concepts of DK presented so far, we offer some exercises. Additionally, we will use these examples to contrast and compare the approaches of analytic and geometric Jacobian.

We begin with this well-known Scara robot mechanics by its simplicity. It has only four DoF, which allows a straightforward presentation of analytic Jacobian. Later, it will be very illustrative to compare this approach with the geometric Jacobian for the same manipulator to appreciate the advantages of the screw theory on working with velocities. Figure 5.1 shows the robot dimensions for the exercises of this section.

5.2.4.1 Forward Differential Kinematics with Analytic Jacobian

First, we need an easy mapping, suitable to differentiation, to represent the FK in the stationary or spatial coordinate system (see Equation 5.10). This expression of the tool pose (i.e., position "x-y-z" and orientation "α-β-γ") is a function of the joint coordinates.

$$T^S(x,y,z,\alpha,\beta,\gamma) = \begin{cases} x = p_x = l_2 C_{12} + l_1 C_1 \\ y = p_y = l_3 + \theta_3 \\ z = p_z = -l_2 S_{12} - l_1 S_1 \\ \alpha = R_x = \dfrac{\pi}{2} \\ \beta = R_y = \theta_1 + \theta_2 - \theta_4 \\ \gamma = R_z = 0 \end{cases} \quad (5.10)$$

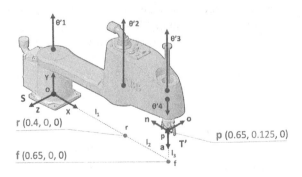

FIGURE 5.1 Scara ABB IRB910SC differential kinematics parameters.

Differential Kinematics

Then we get the direct analytic Jacobian by differentiation of the previous FK map by Equation 5.11.

$$J_A(\theta) = \begin{bmatrix} \frac{\partial T_x}{\partial \theta_1} & \frac{\partial T_x}{\partial \theta_2} & 0 & 0 \\ 0 & 0 & \frac{\partial T_y}{\partial \theta_3} & 0 \\ \frac{\partial T_z}{\partial \theta_1} & \frac{\partial T_z}{\partial \theta_2} & 0 & 0 \\ 0 & 0 & 0 & 0 \\ \frac{\partial T_\beta}{\partial \theta_1} & \frac{\partial T_\beta}{\partial \theta_2} & 0 & \frac{\partial T_\beta}{\partial \theta_4} \\ 0 & 0 & 0 & 0 \end{bmatrix} = \begin{bmatrix} -l_2 S_{12} - l_1 S_1 & -l_2 S_{12} & 0 & 0 \\ 0 & 0 & 1 & 0 \\ -l_2 C_{12} - l_1 C_1 & -l_2 C_{12} & 0 & 0 \\ 0 & 0 & 0 & 0 \\ 1 & 1 & 0 & -1 \\ 0 & 0 & 0 & 0 \end{bmatrix} \quad (5.11)$$

We can now calculate the tool's velocity in the spatial "S" frame as a function of the analytic Jacobian matrix and the joint velocities expressed as Equation 5.12.

$$\begin{bmatrix} \dot{x} \\ \dot{y} \\ \dot{z} \\ \dot{\alpha} \\ \dot{\beta} \\ \dot{\gamma} \end{bmatrix} = J_A(\theta) \begin{bmatrix} \dot{\theta}_1 \\ \dot{\theta}_2 \\ \dot{\theta}_3 \\ \dot{\theta}_4 \end{bmatrix} = \begin{bmatrix} -l_2 S_{12} - l_1 S_1 & -l_2 S_{12} & 0 & 0 \\ 0 & 0 & 1 & 0 \\ -l_2 C_{12} - l_1 C_1 & -l_2 C_{12} & 0 & 0 \\ 0 & 0 & 0 & 0 \\ 1 & 1 & 0 & -1 \\ 0 & 0 & 0 & 0 \end{bmatrix} \begin{bmatrix} \dot{\theta}_1 \\ \dot{\theta}_2 \\ \dot{\theta}_3 \\ \dot{\theta}_4 \end{bmatrix} \quad (5.12)$$

We recall that the analytic Jacobian is a matrix-valued function, and therefore it gets a different value for each tool configuration. So, to illustrate this forward DK, we propose an exercise with a pose (t_1) for the Scara robot in a specific desired trajectory for the manipulator tool (see Figure 5.2).

The code for this Exercise 5.2.4a is in the internet hosting for the software of this book[1]. It is advisable to review the details of the Screw Theory Toolbox for Robotics (ST24R) to understand better the concepts that worked in this example (Pardos-Gotor, 2021a).

5.2.4.2 Inverse Differential Kinematics with Analytic Jacobian

Here we analyze the inverse DK, but only for the TCP translation, to simplify the exercise and better focus on the Scara type singularities.

We need an IK mapping easy to differentiate in the spatial coordinate system, to get the inverse Jacobian. The expression of the joint magnitudes is a function of the tool pose (i.e., position "x-y-z"). As we have already said, this is not easy, as there might be several solutions for IK (i.e., two for this robot). One approach to solve this

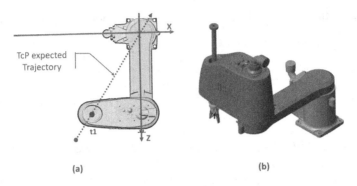

FIGURE 5.2 (a) Scara ABB IRB910SC DK trajectory top view. (b) ABB IRB910SC 3D view.

issue is to invert the symbolic expression of the TCP translation direct analytic Jacobian of Equation 5.13. It is essential to analyze the result, where it is relevant the value for the Jacobian determinant (see Equation 5.14).

$$J_a(P_{XYZ}) = \begin{bmatrix} -l_2 S_{12} - l_1 S_1 & -l_2 S_{12} & 0 \\ 0 & 0 & 1 \\ -l_2 C_{12} - l_1 C_1 & -l_2 C_{12} & 0 \end{bmatrix} \quad (5.13)$$

$$J_a(P_{XYZ})^{-1} = \frac{1}{l_1 l_2 S_2} \begin{bmatrix} l_2 C_{12} & 0 & -l_2 S_{12} \\ -l_2 C_{12} - l_1 C_1 & 0 & l_2 S_{12} + l_1 S_1 \\ 0 & l_1 l_2 S_2 & 0 \end{bmatrix} \quad (5.14)$$

Now we can aim for the joint velocities by Equation 5.15, having the desired tool velocities considered in the spatial "S" frame and the inverse analytic Jacobian matrix.

$$\begin{bmatrix} \dot{\theta}_1 \\ \dot{\theta}_2 \\ \dot{\theta}_3 \end{bmatrix} = J_a(P_{XYZ})^{-1} \begin{bmatrix} \dot{x} \\ \dot{y} \\ \dot{z} \end{bmatrix} = \frac{1}{l_1 l_2 S_2} \begin{bmatrix} l_2 C_{12} & 0 & -l_2 S_{12} \\ -l_2 C_{12} - l_1 C_1 & 0 & l_2 S_{12} + l_1 S_1 \\ 0 & l_1 l_2 S_2 & 0 \end{bmatrix} \begin{bmatrix} \dot{x} \\ \dot{y} \\ \dot{z} \end{bmatrix} \quad (5.15)$$

The Inverse analytic Jacobian is a matrix-valued function, and therefore it gets a different value for each joint configuration.

To illustrate this Inverse DK, we propose an exercise with four poses (t_1-t_2-t_3-t_4) for the Scara robot in a specific desired linear trajectory for the manipulator tool (see details in Figure 5.3). We get the robot joint velocities with Equation 5.15 for making the TCP follow a linear trajectory parallel to the floor. The tool must develop the expected trajectory with a constant velocity.

We apply the inverse DK formula only for the TCP translation to four targets in the trajectory to show the results. The joint velocities remain inside acceptable limits

Differential Kinematics

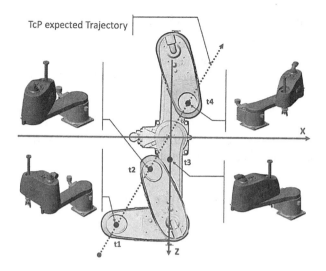

FIGURE 5.3 Scara ABB IRB910SC inverse differential kinematics.

for three of the targets (i.e., t_1-t_2-t_4). However, something different happens around the target "t_3." Or in other words, what happens between configurations "t_2" and "t_4"?

Taking a closer look at Figure 5.3, we can realize that following the desired complete trajectory is impossible. It would need the tool to go through the robot's base just by the origin of the spatial system, which is not possible. The robot could approximate the trajectory but avoiding the mechanical constraint that means a collision with the base. The robot must make a very abrupt change of configuration between points "t_2" and "t_4" of the trajectory to reasonably perform this path. It is possible to get a hint from the graphic since the first joint must complete a rotation of value "π" in very little time. Intuitively, we see that this need would require a formidable acceleration for this joint to reach the necessary velocity to comply with the motion.

Analytically, this situation translates into considering a target "t_3" located in the path between "t_2" and "t_4." For instance, we take a point "t_3," which makes "θ_2" pretty close to "π." Then, the inverse of the analytic Jacobian becomes almost infinite, and therefore it results in enormous speeds for the joints (see Figure 5.3).

These calculations around "t_3" give us speeds for the joints so big that they cannot be by the limitations of the robot mechanics. In sum, we are just in one configuration of singularity for this Scara robot. Mathematically, we can directly appreciate this fact from the inverse analytic Jacobian of Equation 5.14, as the determinant makes the expression infinite for values "$\sin\theta_2$" equal to zero.

The code for this Exercise 5.2.4b is in the internet hosting for the software of this book[2]. It is advisable to review the details of the Screw Theory Toolbox for Robotics (ST24R) to understand better the concepts that worked in this example (Pardos-Gotor, 2021a).

5.2.5 Puma Robot (e.g., ABB IRB120)

The complexity of the robot DK problem increases dramatically with the number of DoF (Figure 5.4). Here, there is only the introduction to an example with six DoF to illustrate how difficult it is from the kinematics mapping of Equation 5.16 to obtain the analytic Jacobian by differentiation as Equation 5.17. We do not complete the exercise, as it is out of this book's scope to further develop the analytic approach. We move in the next section to the DK approach based on screw theory, where we solve the robot of this example with the se(3) Lie algebra tools. Working with velocities using geometric Jacobian has many advantages. It is much more meaningful and more accessible.

$$\left.\begin{aligned}
n_x &= C_6\left(S_5\left(S_1S_4 + C_4\left(C_1C_2S_3 + C_1C_3S_2\right)\right) + C_5\left(C_1S_2S_3 - C_1C_2C_3\right)\right) \\
&\quad - S_6\left(C_4S_1 - S_4\left(C_1C_2S_3 + C_1C_3S_2\right)\right) \\
o_x &= -C_6\left(C_4S_1 - S_4\left(C_1C_2S_3 + C_1C_3S_2\right)\right) \\
&\quad - S_6\left(S_5\left(S_1S_4 + C_4\left(C_1C_2S_3 + C_1C_3S_2\right)\right) + C_5\left(C_1S_2S_3 - C_1C_2C_3\right)\right) \\
a_x &= -C_5\left(S_1S_4 + C_4\left(C_1C_2S_3 + C_1C_3S_2\right)\right) + S_5\left(C_1S_2S_3 - C_1C_2C_3\right) \\
p_x &= \frac{27}{100}C_1C_2 - \frac{151}{500}\left(C_1S_2S_3 - C_1C_2C_3\right) + \frac{7}{100}\left(C_1C_2S_3 - C_1C_3S_2\right) \\
&\quad - \frac{4}{25}\left(C_5S_1S_4 + C_1C_2C_3S_5 + C_1S_2S_3S_5 + C_1C_2C_4C_5S_3 + C_1C_3C_4C_5S_2\right) \\
n_y &= -C_6\left(S_5\left(C_1S_4 - C_4\left(C_sS_1S_3 + C_3S_1S_2\right)\right) - C_5\left(S_1S_2S_3 - C_2C_3S_1\right)\right) \\
&\quad + S_6\left(C_1C_4 + S_4\left(C_2S_1S_3 + C_3S_1S_2\right)\right) \\
o_y &= C_6\left(C_1C_4 + S_4\left(C_2S_1S_3 + C_3S_1S_2\right)\right) \\
&\quad + S_6\left(S_5\left(C_1S_4 - C_4\left(C_2S_1S_3 + C_3S_1S_2\right)\right) - C_5\left(S_1S_2S_3 - C_2C_3S_1\right)\right) \\
a_y &= C_5\left(C_1S_4 - C_4\left(C_2S_1S_3 + C_3S_1S_2\right)\right) + S_5\left(S_1S_2S_3 - C_2C_3S_1\right) \\
p_y &= \frac{27}{100}S_1S_2 - \frac{151}{500}\left(S_1S_2S_3 - C_2C_3S_1\right) + \frac{7}{100}\left(C_2S_1S_3 - C_3S_1S_2\right) \\
&\quad + \frac{4}{25}\left(C_1C_5S_4 - C_2C_3S_1S_5 + S_1S_2S_3S_5 - C_2C_4C_5S_1S_3 - C_3C_4C_5S_1S_2\right) \\
n_z &= C_6\left(S_{23}C_5 + C_{23}C_4S_5\right) + C_{23}S_4S_6 \\
o_z &= -S_6\left(S_{23}C_5 + C_{23}C_4S_5\right) + C_{23}C_6S_4 \\
a_z &= S_{23}S_5 - C_{23}C_4C_5 \\
p_z &= \frac{27}{100}C_2 - \frac{151}{500}\left(C_2S_3 + C_3S_2\right) + \frac{7}{100}\left(C_2C_3 - S_2S_3\right) \\
&\quad + \frac{4}{25}\left(C_2S_3S_5 + C_3S_2S_5 - C_2C_3C_4C_5 + C_4C_5S_2S_3\right) + \frac{29}{100}
\end{aligned}\right\} T(x,y,z,\alpha,\beta,\gamma)$$

(5.16)

Differential Kinematics

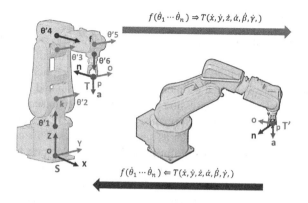

FIGURE 5.4 Puma ABB IRB120 differential kinematics concept.

$$J_A(\theta) = \begin{bmatrix} \dfrac{\partial T_x}{\partial \theta_1} & \dfrac{\partial T_x}{\partial \theta_2} & \dfrac{\partial T_x}{\partial \theta_3} & \dfrac{\partial T_x}{\partial \theta_4} & \dfrac{\partial T_x}{\partial \theta_5} & \dfrac{\partial T_x}{\partial \theta_6} \\ \dfrac{\partial T_y}{\partial \theta_1} & \dfrac{\partial T_y}{\partial \theta_2} & \dfrac{\partial T_y}{\partial \theta_3} & \dfrac{\partial T_y}{\partial \theta_4} & \dfrac{\partial T_y}{\partial \theta_5} & \dfrac{\partial T_y}{\partial \theta_6} \\ \dfrac{\partial T_z}{\partial \theta_1} & \dfrac{\partial T_z}{\partial \theta_2} & \dfrac{\partial T_z}{\partial \theta_3} & \dfrac{\partial T_z}{\partial \theta_4} & \dfrac{\partial T_z}{\partial \theta_5} & \dfrac{\partial T_z}{\partial \theta_6} \\ \dfrac{\partial T_\alpha}{\partial \theta_1} & \dfrac{\partial T_\alpha}{\partial \theta_2} & \dfrac{\partial T_\alpha}{\partial \theta_3} & \dfrac{\partial T_\alpha}{\partial \theta_4} & \dfrac{\partial T_\alpha}{\partial \theta_5} & \dfrac{\partial T_\alpha}{\partial \theta_6} \\ \dfrac{\partial T_\beta}{\partial \theta_1} & \dfrac{\partial T_\beta}{\partial \theta_2} & \dfrac{\partial T_\beta}{\partial \theta_3} & \dfrac{\partial T_\beta}{\partial \theta_4} & \dfrac{\partial T_\beta}{\partial \theta_5} & \dfrac{\partial T_\beta}{\partial \theta_6} \\ \dfrac{\partial T_\gamma}{\partial \theta_1} & \dfrac{\partial T_\gamma}{\partial \theta_2} & \dfrac{\partial T_\gamma}{\partial \theta_3} & \dfrac{\partial T_\gamma}{\partial \theta_4} & \dfrac{\partial T_\gamma}{\partial \theta_5} & \dfrac{\partial T_\gamma}{\partial \theta_6} \end{bmatrix} \quad (5.17)$$

5.3 THE GEOMETRIC JACOBIAN

The essential mathematical tool for this concept of geometric Jacobian is the Product of Exponentials (POE) (Brockett, 1983). This idea is fundamental in screw theory and not only for DK but also for robot dynamics (Davidson and Hunt, 2004). For instance, we can also use the geometric Jacobian to describe the relationship between wrenches applied at the robot links or tool and joint torques. We will see these applications in further chapters.

One can use the relationship between the robot joint and end-effector velocities for many applications. This section introduces this relationship and studies its structure and properties based on the geometric Jacobian (Millman and Parker, 1997).

Next, we present some new concepts related to the geometric Jacobian, which are fundamental to advancing the screw theory description of DK problems.

5.3.1 Robot Spatial Geometric Jacobian

The proper representation of rigid body velocity is with the twists (Murray et al., 2017), and then we give the Jacobian of the DK map in terms of twists as Equation 5.18. We shall see that the POE leads to a very natural and explicit description of the robot Jacobian, emphasizing the geometry of the mechanism and having none of the drawbacks of a local analytic expression (Selig, 2005). The consequence is the availability of better computational performance (Park, 1994). It avoids the difficulties given by local parameterization, such as some singularities. This concept is called **the robot geometric, spatial, or manipulator Jacobian**. We must bear in mind that this expression is a configuration-dependent matrix, which changes for each joint configuration (as also happened for the analytic Jacobian).

The geometric Jacobian "J^S_{ST}" has a unique structure (Lynch and Park, 2017). The contribution of each joint velocity to the tool velocity is independent of later joints' configuration in the mechanical chain, as seen in Equation 5.19. Each column of the Spatial Geometric Jacobian is the mobile twist for each joint, which depends only on the previous joints' twists. Each column of the geometric Jacobian corresponds to its joint twist, transformed into the current robot configuration. To do so, we use the adjoint transformation. All these expressions are in the Spatial frame.

For most of the developments in this book, we will use this definition of the geometric Jacobian, as the velocities refer to the spatial (i.e., inertial) coordinate system.

$$J^S_{ST}(\theta) = \begin{bmatrix} \xi'_1 & \xi'_2 \cdots \xi'_n \end{bmatrix} \tag{5.18}$$

$$\xi'_i = Ad_{\left(e^{\hat{\xi}_1 \theta_1} \cdots e^{\hat{\xi}_{i-1} \theta_{i-1}}\right)} \xi_i \tag{5.19}$$

This robust structure means that **we can calculate the geometric Jacobian even "by inspection."** It is essential to note that **we can calculate the entire robot spatial Jacobian "by definition" without explicitly differentiating the FK map.**

5.3.2 The Classical Adjoint Transformation (Ad)

It is the operation that transforms twist from one frame to another by Equation 5.20. All these expressions are in the Spatial (stationary) frame. It is also possible to convert other screw theory concepts with this expression, as we will make in the dynamics chapter.

$$Ad_H = \begin{bmatrix} R & \hat{p}R \\ 0 & R \end{bmatrix} \tag{5.20}$$

$$H = e^{\hat{\xi}_1 \theta_1} \cdots e^{\hat{\xi}_{i-1} \theta_{i-1}} = \begin{bmatrix} R & p \\ 0 & 1 \end{bmatrix}$$

$$\hat{p} = \begin{bmatrix} 0 & -p_z & p_y \\ p_z & 0 & -p_x \\ -p_y & p_x & 0 \end{bmatrix}$$

Differential Kinematics

In the same way as the analytic Jacobian, the geometric Jacobian maps joint velocities to end-effector velocities. The forward DK expression will be Equation 5.21, and for the inverse DK Equation 5.22. Nevertheless, we must be cautious because this end-effector velocity is different from the one used with the classical analytic approach. The geometric Jacobian relates the joint velocities with the end-effector velocities expressed in terms of a new concept defined as Twist Velocity "V^S_{ST}."

$$V^S_{ST} = J^S_{ST}(\theta)\dot{\theta} \tag{5.21}$$

$$\dot{\theta} = \left[J^S_{ST}(\theta)\right]^{-1} V^S_{ST} \tag{5.22}$$

5.3.3 Twist Velocity Concept

The definition of twist velocity "V^S_{ST}" (also known as Spatial Velocity) must be related to our informal understanding of rotational and translational velocity. We have already introduced the proper representation of a rigid body velocity with the twists. Consequently, this twist velocity is used in the screw theory to define the velocity of a coordinate system associated with a rigid body (e.g., a robot's tool frame).

The twist velocity of Equation 5.23 has rotational and translational components. The rotational component "ω^S_{ST}" is the instantaneous angular velocity of the body in the spatial frame. It has the classical meaning used with the analytical Jacobian. However, the interpretation of the translational component "v^S_{ST}" is somewhat unintuitive, as it is not the velocity of the origin of the tool frame "v^S_{TCP}" (i.e., TCP velocity). It is the velocity of a point (ordinarily imaginary) attached to the tool, which is traveling through the origin of the spatial frame at a given time.

$$V^S_{ST} = \begin{bmatrix} v^S_{ST} \\ \omega^S_{ST} \end{bmatrix} \tag{5.23}$$

5.3.4 Trajectory Generation

We can also use the geometric Jacobian to move the robot from one end-effector configuration to another without calculating the IK. The previous chapter solved numerous IK problems geometrically with the cornerstone of the canonical subproblems of Paden-Kahan (Kahan, 1983) and Pardos-Gotor. Nevertheless, it can be the case where there is no closed-form solution for the IK, or there is a robot configuration that is not suitable for the task trajectory because any constraint (Liu and Li, 2002). In both cases, the approach of using the geometric Jacobian matrix is quite helpful. The attention must be for not confusing the two kinds of end-effector velocity. These are the intuitive velocity of the tool system "V^S_T" and the screw theory twist velocity "V^S_{ST}."

We can move the tool from one configuration to another in the workspace if we deal well with the different definitions of the end-effector velocities. We must know the spatial velocity between those two configurations. Then, with the joint coordinates corresponding to the first end-effector configuration and the integration of the joint velocity over the time interval, we get the new joints coordinates by Equation 5.24.

$$\theta(t_2) = \theta(t_1) + \int_{t1}^{t2} \dot{\theta}(t) = \theta(t_1) + \int_{t1}^{t2} \left(J_{st}^S(\theta)\right)^{-1} V_{ST}^S(t) \quad (5.24)$$

5.3.5 Robot Tool Geometric Jacobian

Another advantage of the screw theory approach for DK is defining a mobile geometric Jacobian. Probably, the most engaging mobile geometric Jacobian is the one corresponding with the Tool "$J^T{}_{ST}$." It describes the relationship between the joint velocity vector and the corresponding velocity of the end-effector tool "$V^T{}_{ST}$" (at each joint configuration). The definition is in the tool coordinate system (see Equation 5.25).

The Tool geometric Jacobian "$J^T{}_{ST}$" has a unique structure by Equation 5.26, where the columns correspond to the joint twists written to the tool frame at the current configuration by Equation 5.27. To build it, we use the inverted adjoint transformation of Equation 5.20.

$$V_{ST}^T = J_{ST}^T(\theta)\dot{\theta} \quad (5.25)$$

$$J_{ST}^T(\theta) = \left[\xi_1^+ \cdots \xi_{n-1}^+ \; \xi_n^+\right] \quad (5.26)$$

$$\xi_i^+ = Ad^{-1}_{\left(e^{\hat{\xi}_i\theta_i}\ldots e^{\hat{\xi}_n\theta_n} H_{ST}(0)\right)} \xi_i \quad (5.27)$$

The adjoint transformation as Equation 5.28 also relates the Spatial and Tool Jacobians.

$$J_{ST}^S(\theta) = Ad_{H_{ST}(\theta)} J_{ST}^T(\theta) = Ad_{\left(e^{\hat{\xi}_1\theta_1}\ldots e^{\hat{\xi}_n\theta_n} H_{ST}(0)\right)} J_{ST}^T(\theta) \quad (5.28)$$

5.3.6 Link Spatial and Link Tool Geometric Jacobian

Another very convenient geometric Jacobian is the one related with some link of the rigid body chain. Remarkably, this is very handy when dealing with the dynamics of each robot link.

We will take the previous definitions for the robot spatial and tool to formulate the link Jacobian concerning the spatial and tool coordinate frames.

- **The Link Spatial Geometric Jacobian "$J^S{}_{SL}$"** of Equation 5.29 maps joint velocities to the link velocities. The structure is the same as the robot spatial geometric Jacobian but going only from the base to the "i^{th}" link instead of reaching the tool. Each column of the link spatial Jacobian corresponds to its joint twist, transformed to the actual robot configuration. To do so, we use the adjoint transformation of Equation 5.20.

Differential Kinematics

Of course, the expressions are in the spatial (stationary) coordinate frame.

$$J_{SL}^{S}(\theta) = \begin{bmatrix} \xi_1' \cdots \xi_i' \end{bmatrix} \quad (5.29)$$

$$\xi_j' = Ad_{\left(e^{\hat{\xi}_1\theta_1}\cdots e^{\hat{\xi}_{j-1}\theta_{j-1}}\right)}\xi_j$$

- **The Link Tool Geometric Jacobian "$J^T{}_{SL}$"** as Equation 5.30 maps joint velocities to the link velocities. It has the same structure already explained for the robot tool geometric Jacobian, but instead of going throughout the complete rigid body chain, it goes only from the tool to the "i^{th}" link instead from the tool to the base. The columns correspond to the joint twists written for the tool frame at the current configuration.

To get a practical implementation for the link tool Geometric Jacobian "$J^T{}_{SL}$," we use two different algorithms for the adjoint transformation to map twist from one frame to another. The first one, "Ad" is already explained in detail in previous sections (see Equation 5.20). The second is a new formulation, "A_{ij}," efficient for this definition of the Jacobian and other developments of screw theory. We make use of this new adjoint transformation in Chapter 6 for inverse dynamics.

Of course, the expression is in the tool (mobile) coordinate frame.

$$J_{SL}^{T}(\theta) = Ad_{H_{SL(0)}}^{-1}\begin{bmatrix} A_{i1}\xi_1 \cdots A_{ii}\xi_i \cdots 0_{i+1} \cdots 0_n \end{bmatrix} \quad (5.30)$$

5.3.7 The New Adjoint Transformation (A_{ij})

In mathematics, the adjoint transformation of a Lie group represents the elements of the group as linear transformations of the group's Lie algebra.

We introduce this new notation for the adjoint transformation of Equation 5.31. It is the mechanism to transform twist and other screw elements from one coordinate system to another. We use this transformation to define the link tool geometric Jacobian and this notation for getting formulas for the Inertia, Coriolis, and Potential matrices in dynamics.

$$A_{ij} = Ad^{-1}_{\left(e^{\hat{\xi}_{j+1}\theta_{j+1}}\cdots e^{\hat{\xi}_i\theta_i}\right)} \forall i > j \quad (5.31)$$

$$A_{ij} = I \forall i = j$$

$$A_{ij} = 0 \forall i < j$$

5.3.8 General Solution to Differential Kinematics

The DK problem formalized with screw theory relates the velocities of the robot joints and tool. The general geometric solution to any manipulator DK problem follows a three-phase algorithm: the kinematics map, the forward DK, and the inverse DK (Pardos-Gotor, 2018).

5.3.8.1 The Kinematics Mapping

- Select the **spatial coordinate system "S"** (usually stationary and the base) and the **mobile coordinate system "T"** (typically the tool). There is no rule and complete flexibility to make this definition, and we can choose the most convenient frames according to the application.
- Define the **axis of each joint**, which are the axes "ω_i" for revolute joints and "v_i" for the prismatic joints, and any point "q_i" on any of those axes, even though for kinematics only the point on the revolute joints is necessary.
- Obtain the **twists "ξ_i"** for the joints, knowing for each revolute joint its axis and a point on that axis, and for each prismatic joint its axis.

$$\xi_i = \begin{bmatrix} -\omega_i \times q_i \\ \omega_i \end{bmatrix} \quad \xi_i = \begin{bmatrix} v_i \\ 0 \end{bmatrix}$$

- Get "$H_{ST}(0)$" **the pose of the tool at the reference** (home) robot position. This configuration of the end-effect happens when all the joint magnitudes are zero. This tool pose comes as a homogeneous matrix, necessary because the POE represents a relative mapping.
- Express the **kinematics mapping "$H_{ST}(\theta)$," with the product of all joint screw exponentials (POE) and "$H_{ST}(0)$."**

$$H_{ST}(\theta) = \prod_{i=1}^{n} e^{\hat{\xi}_i \theta_i} H_{ST}(0) = \begin{bmatrix} n & o & a & p \\ 0 & 0 & 0 & 1 \end{bmatrix}$$

5.3.8.2 The Geometric Forward Differential Kinematics

From the velocities of the robot joints (i.e., θ'_1-θ_n'), it is possible to calculate the tool velocity (i.e., V^S_T) in the spatial or inertial coordinate system. This end-effector pose velocity "V^S_T" has two components, the position velocity for the TCP (i.e., v^S_{TCP} = [x', y', z']) and the rotation velocity for the tool system (i.e., ω^S_{ST} = [α', β', γ']). This solution for the forward DK expression is the same as the more classical analytic Jacobian. Nevertheless, in this case, we use the twist velocity "V^S_{ST}" concept with Equation 5.32, whose component "v^S_{ST}" is not the velocity of the origin of the tool frame "v^S_{TCP}." This concept is the reason to need some transformation between these two different translational velocities.

The direct geometric Jacobian of the screw theory needs no differentiation and preserves the geometrical meaning of the rigid body velocity. This feature is an excellent benefit for practical applications. Of course, this Jacobian must be evaluated for each joint configuration, as it is a matrix-valued function. Finally, **the forward DK map has a total geometric formulation** by Equation 5.33.

$$V^S_{ST} = \begin{bmatrix} v^S_{ST} \\ \omega^S_{ST} \end{bmatrix} = J^S_{ST}(\theta)\dot{\theta} = \begin{bmatrix} \xi'_1 & \xi'_2 & \cdots & \xi'_n \end{bmatrix} \begin{bmatrix} \dot{\theta}_1 \\ \vdots \\ \dot{\theta}_n \end{bmatrix} \quad (5.32)$$

Differential Kinematics

$$V_T^S = \begin{bmatrix} v_{TCP}^S \\ \omega_{ST}^S \end{bmatrix} = \begin{bmatrix} \dot{x} \\ \dot{y} \\ \dot{z} \\ \dot{\alpha} \\ \dot{\beta} \\ \dot{\gamma} \end{bmatrix} = \begin{bmatrix} v_{ST}^S + \left[\widehat{\omega_{ST}^S}\right] TCP^S(\theta) \\ \omega_{ST}^S \end{bmatrix} \quad (5.33)$$

$$TCP^S(\theta) = \begin{bmatrix} TCP_X \\ TCP_Y \\ TCP_Z \end{bmatrix} \quad (5.34)$$

$$\left[\widehat{\omega_{ST}^S}\right] = \begin{bmatrix} 0 & -\omega_z & \omega_y \\ \omega_z & 0 & -\omega_x \\ -\omega_y & \omega_x & 0 \end{bmatrix} \quad (5.35)$$

To complete the formulation of the forward DK, we keep the rotation velocity for the tool system (i.e., $\omega^S{}_{ST}$) but modify the translation component (i.e., "$v^S{}_{ST}$") of the screw twist velocity. We add the effect of the rotation velocity on the TCP corresponding to the joint configuration. Mathematically, this effect corresponds with the rotation velocity vector's cross product and the TCP in the spatial frame (i.e., Equation 5.34). Instead of the cross product, we can use a matrix product with the skew-symmetric transformation of Equation 5.35.

5.3.8.3 The Geometric Inverse Differential Kinematics

It computes the robot joints' instantaneous velocity (i.e., θ'_1-θ_n') from the desired tool system velocity (i.e., $V^S{}_T$). The inverse DK expression solution is the same as the one presented with the more classical analytic Jacobian. The input to this problem is the end-effector pose velocity "$V^S{}_T$," which has two components, the position velocity for the TCP (i.e., $v^S{}_{TCP} = [x', y', z']$) and the rotation velocity for the tool (i.e., $\omega^S{}_{ST} = [\alpha', \beta', \gamma']$). However, in this case, we use the twist velocity "$V^S{}_{ST}$" concept, whose component "$v^S{}_{ST}$" is not the velocity of the origin of the tool frame "$v^S{}_{TCP}$." This new concept is the reason to need some transformation between these two different translational velocities.

The inverse geometric Jacobian of the screw theory needs no differentiation and preserves the geometrical meaning of the rigid body velocity. This property is a great benefit for real implementations. The inverse geometric Jacobian can be obtained from the expression of the direct geometric Jacobian by different algorithms. For instance, with the Moore-Penrose generalized inverse formula. The inverse geometric Jacobian has a different value for each joint configuration. The inverse DK map of Equation 5.36 has a geometric formulation to obtain the joint velocities. The expression for the detailed solution in Equation 5.37 considers the six-element vector (i.e., $[x', y', z', \alpha', \beta', \gamma']$) of the tool system velocity (i.e., $V^S{}_T$).

$$\dot{\theta} = \left[J_{ST}^S(\theta) \right]^{-1} V_{ST}^S = \left[J_{ST}^S(\theta) \right]^{-1} \begin{bmatrix} v_{ST}^S \\ \omega_{ST}^S \end{bmatrix} = \begin{bmatrix} \xi_1' & \xi_2' \cdots \xi_n' \end{bmatrix}^{-1} \begin{bmatrix} v_{TCP}^S - \omega_{ST}^S \times TCP^S(\theta) \\ \omega_{ST}^S \end{bmatrix} \quad (5.36)$$

$$\dot{\theta} = \begin{bmatrix} \dot{\theta}_1 \\ \vdots \\ \dot{\theta}_n \end{bmatrix} = \begin{bmatrix} \xi_1' & \xi_2' \cdots \xi_n' \end{bmatrix}^{-1} \begin{bmatrix} \begin{bmatrix} \dot{x} \\ \dot{y} \\ \dot{z} \end{bmatrix} - \begin{bmatrix} \hat{\dot{\alpha}} \\ \hat{\dot{\beta}} \\ \hat{\dot{\gamma}} \end{bmatrix} TCP^S(\theta) \\ \begin{bmatrix} \dot{\alpha} \\ \dot{\beta} \\ \dot{\gamma} \end{bmatrix} \end{bmatrix} \quad (5.37)$$

$$\begin{bmatrix} \hat{\dot{\alpha}} \\ \hat{\dot{\beta}} \\ \hat{\dot{\gamma}} \end{bmatrix} = \begin{bmatrix} 0 & -\dot{\gamma} & \dot{\beta} \\ \dot{\gamma} & 0 & -\dot{\alpha} \\ -\dot{\beta} & \dot{\alpha} & 0 \end{bmatrix} \quad (5.38)$$

To complete the inverse DK formulation, we keep the rotation velocity for the tool system (i.e., $\omega^S{}_{ST}$) but modify the translation component "$v^S{}_{TCP}$" velocity. We subtract the effect of the rotation velocity on the TCP. Mathematically, this effect corresponds with the rotation velocity vector's cross product and the TCP in the spatial frame. Instead of the cross product, we can use a matrix product with the skew-symmetric transformation. Therefore, if the solution's expression is Equation 5.37, the concept is the same, but considering the skew-symmetric matrix of Equation 5.38 for the tool system rotation.

The POE leads to a very natural and explicit description of the robot Jacobian, respecting the geometry of the mechanism and has none of the drawbacks of a local analytic Jacobian. We give the geometric Jacobian of the DK in terms of twists, which are easy to define for any robot. We can **solve the expression for the robot velocities systematically and without the need for differentiation**, leading to computationally better implementations (Sipser, 2021).

To learn how to solve DK problems is advisable to practice the general procedure. In this chapter, there are several exercises with typical robot architectures in the following sections. They will permit us to consolidate the skills to apply the forward and inverse DK. These examples have different joints and configurations. The solved robot architectures are Puma (e.g., ABB IRB120), Bending-Backwards (ABB IRB1600), Gantry (e.g., ABB 6620LX), Scara (e.g., ABB IRB910SC), Collaborative (e.g., UNIVERSAL UR16e), and a Redundant manipulator (e.g., KUKA IIWA).

To implement the exercises there are very well-known libraries, like the Robotics Toolbox (RTB) (Corke, 2017). Nevertheless, it is advisable to review the Screw Theory Toolbox for Robotics "ST24R" (Pardos-Gotor, 2021a) to understand better the concepts working in the examples of the following sections. Besides, "ST24R" includes the useful new formulations to forward and inverse DK just presented.

Differential Kinematics

5.3.9 Puma Robots (e.g., ABB IRB120)

In this example, we develop some exercises for the ABB IRB120. Remember that we did not even finish the analytic Jacobian for this robot in a previous section because of its difficulty in differentiating its FK expression. Here, we focus on the screw geometric Jacobian, which is much easier to solve for this robot's forward and inverse differential problems (Figure 5.5).

With screw theory, it is even possible to define the spatial manipulator Jacobian by inspection. However, this approach is somehow not trivial with a robot of six DoF like this one (Paden and Sastry, 1988). To do that, we would need to define each twist's axis and the "mobile points" for the rotation axes in any pose, which is the most challenging part of that approach. Therefore, it is more practical to solve geometric Jacobian with the screw theory definition.

We appreciate the power of the screw theory while working with velocities, as this approach is much more meaningful and more comfortable to define for robots with many DoF. Of course, this approach can be extended to other mechanisms whenever we are working with velocities.

5.3.9.1 Geometric Jacobian by Definition

The great advantage of a representation for the geometric Jacobian is the systematic implementation. Each column of the Jacobian is the joint twist transformed to the current configuration with the adjoint transformation of Equation 5.20.

We can define the mobile twists $(\xi'_1 \text{-} \xi'_6)$ algebraically. The contribution of each joint velocity to the tool velocity is independent of the configuration of later joints in the chain (see Equation 5.39).

$$\xi'_1 = \xi_1$$

$$\xi'_2 = Ad_{\left(e^{\hat{\xi}_1 \theta_1}\right)} \xi_2$$

$$\xi'_3 = Ad_{\left(e^{\hat{\xi}_1 \theta_1} e^{\hat{\xi}_2 \theta_2}\right)} \xi_3 \quad (5.39)$$

$$\xi'_4 = Ad_{\left(e^{\hat{\xi}_1 \theta_1} e^{\hat{\xi}_2 \theta_2} e^{\hat{\xi}_3 \theta_3}\right)} \xi_4$$

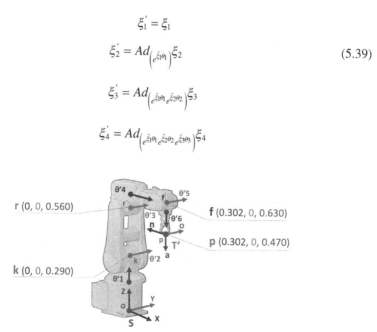

FIGURE 5.5 Puma ABB IRB120 differential kinematics POE parameters.

$$\xi'_5 = Ad_{\left(e^{\hat{\xi}_1\theta_1}e^{\hat{\xi}_2\theta_2}e^{\hat{\xi}_3\theta_3}e^{\hat{\xi}_4\theta_4}\right)}\xi_5$$

$$\xi'_6 = Ad_{\left(e^{\hat{\xi}_1\theta_1}e^{\hat{\xi}_2\theta_2}e^{\hat{\xi}_3\theta_3}e^{\hat{\xi}_4\theta_4}e^{\hat{\xi}_5\theta_5}\right)}\xi_6$$

Now, it is possible to get the geometric Jacobian according to Equation 5.40.

$$J_{ST}^S(\theta) = \begin{bmatrix} \xi'_1 & \xi'_2 & \xi'_3 & \xi'_4 & \xi'_5 & \xi'_6 \end{bmatrix} \quad (5.40)$$

5.3.9.2 Forward Differential Kinematics with Geometric Jacobian

We have chosen for this exercise a tool trajectory parallel to the floor. It is the same path used for the exercises of the Scara robot in this chapter, so it is possible to compare the behavior of these two architectures to develop the same task.

Because the Jacobian is a valued matrix, we need to obtain the joint magnitudes at the target pose "t_1" (see Figure 5.6). Therefore, we must solve the IK of the robot for the tool pose "t_1." We can solve this problem with any of the geometric algorithms presented in the IK chapter.

The forward DK solves the velocity of the tool by Equation 5.42, formed by the TCP velocity (i.e., $v_{TCP}^S = [x', y', z']$) and the rotation velocity of the tool (i.e., $\omega_{ST}^S = [\alpha', \beta', \gamma']$). This velocity is a function of the geometric Jacobian and the joint velocities (i.e., $[\theta_1'\text{-}\theta_6']$) at the target "$t_1$," to follow the expected target in the trajectory. The intermediate step in this geometric approach is to solve the twist velocity by Equation 5.41.

$$V_{ST}^S = \begin{bmatrix} v_{ST}^S \\ \omega_{ST}^S \end{bmatrix} = J_{ST}^S(\theta)\dot{\theta} = \begin{bmatrix} \xi'_1 & \xi'_2 & \xi'_3 & \xi'_4 & \xi'_5 & \xi'_6 \end{bmatrix} \begin{bmatrix} \dot{\theta}_1 \\ \dot{\theta}_2 \\ \dot{\theta}_3 \\ \dot{\theta}_4 \\ \dot{\theta}_5 \\ \dot{\theta}_6 \end{bmatrix} \quad (5.41)$$

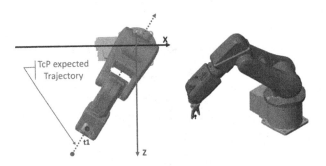

FIGURE 5.6 Puma ABB IRB120 forward differential kinematics.

Differential Kinematics

$$V_T^S = \begin{bmatrix} v_{TCP}^S \\ \omega_{ST}^S \end{bmatrix} = \begin{bmatrix} \dot{x} \\ \dot{y} \\ \dot{z} \\ \dot{\alpha} \\ \dot{\beta} \\ \dot{\gamma} \end{bmatrix} = \begin{bmatrix} v_{ST}^S + \widehat{\omega_{ST}^S} TCP^S(\theta) \\ \omega_{ST}^S \end{bmatrix} \quad (5.42)$$

There is an example of practicing this problem (Exercise 5.3.7.2a). The inputs are the joint velocities to follow the expected trajectory, and the output is the tool velocity (i.e., linear and angular) in the spatial coordinate system.

The code for Exercise 5.3.9a is in the internet hosting for the software of this book[3].

5.3.9.3 Inverse Differential Kinematics with Geometric Jacobian

The joints velocities are a function of the inverse geometric spatial Jacobian and the tool velocities by Equation 5.43. Given the translation velocity for the TCP (i.e., $v^S{}_{TCP}$ = [x', y', z']) and the rotation velocity for the tool system (i.e., $\omega^S{}_{ST}$=[α,' β,' γ']), we obtain the robot joint velocities (i.e., [θ_1'-θ_6']), always in the spatial frame by Equation 5.44.

$$\dot{\theta} = \left[J_{ST}^S(\theta)\right]^{-1} V_{ST}^S = \left[J_{ST}^S(\theta)\right]^{-1} \begin{bmatrix} v_{ST}^S \\ \omega_{ST}^S \end{bmatrix} = \left[J_{ST}^S(\theta)\right]^{-1} \begin{bmatrix} v_{TCP}^S - \omega_{ST}^S \times TCP^S(\theta) \\ \omega_{ST}^S \end{bmatrix} \quad (5.43)$$

$$\begin{bmatrix} \dot{\theta}_1 \\ \dot{\theta}_2 \\ \dot{\theta}_3 \\ \dot{\theta}_4 \\ \dot{\theta}_5 \\ \dot{\theta}_6 \end{bmatrix} = \begin{bmatrix} \xi'_1 & \xi'_2 & \xi'_3 & \xi'_4 & \xi'_5 & \xi'_6 \end{bmatrix}^{-1} \begin{bmatrix} \begin{bmatrix} \dot{x} \\ \dot{y} \\ \dot{z} \end{bmatrix} - \begin{bmatrix} \dot{\alpha} \\ \dot{\beta} \\ \dot{\gamma} \end{bmatrix} TCP^S(\theta) \\ \begin{bmatrix} \dot{\alpha} \\ \dot{\beta} \\ \dot{\gamma} \end{bmatrix} \end{bmatrix} \quad (5.44)$$

The inverse geometric Jacobian is a matrix-valued function that gets a different value for each joint configuration, and the definitions are always in the spatial coordinate system.

We present an example (Exercise 5.3.7.2b), which illustrates this inverse DK problem. There are four poses (t_1-t_2-t_3-t_4) for the puma robot in a particular desired linear trajectory for the tool (see details in Figure 5.7). We solve Equation 5.44 to get the robot joint velocities to make the TCP follow a linear trajectory on a plane without any differentiation and with a precise geometric Jacobian valuable formulation for the complete workspace. We can implement a systematic algorithm to solve the inverse DK. Besides, the geometric Jacobian facilitates easy translation between different coordinate systems.

The tool must develop the expected trajectory with a constant velocity. The joint velocities are inside acceptable limits for three of the targets (i.e., t_1-t_2-t_4). The tool must go through the robot's base just by the spatial origin to follow the desired

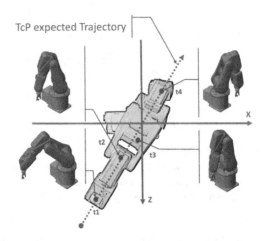

FIGURE 5.7 Puma ABB IRB120 inverse differential kinematics.

trajectory, which is not possible. The robot can develop a turnaround to avoid collision with its base to perform this trajectory in a quasi-optimal way. Then, the robot can retake the expected linear trajectory. We can analytically identify this detour with a point "t_3" located in the path between "t_2" and "t_4." This configuration "t_3" is the same used in the Scara example. However, there is no singularity in this case, and all joint velocities are inside the mechanical limits of the robot.

This exercise demonstrates that we have better possibilities for developing a task along a specific trajectory when we count on a robot with more DoF. The Puma robot with six DoF generates the same linear trajectory fine. Conversely, for the same path, the Scara robot finds unsolvable singularities.

We contribute with another example (Exercise 5.3.7.2c) to study together the forward and inverse DK with random targets.

The code for Exercise 5.3.9b[4] and Exercise 5.3.9c[5] are in the internet hosting for the software of this book.

5.3.10 PUMA ROBOTS (E.G., ABB IRB120) "TOOL-UP"

This DK exercise for the same Puma robot has a different home configuration to illustrate the freedom that screw theory provides to the kinematics analysis. Here, we define the spatial coordinate system with a different orientation. We choose the robot reference pose with a tool-up direction (see the parameters in Figure 5.8). Besides, we do not assume to know the actual sign of rotation for the joints of the commercial robot to build a more hypothetical case. This example checks the geometric inverse and forward DK for a random target in the workspace.

The formulations are the same presented in the previous section.

- The forward DK solves the velocity of the tool by Equation 5.42, formed by the TCP velocity (i.e., [x', y', z']) and the rotation velocity of the mobile tool system (i.e., [α', β', γ']), both expressed in the spatial system. The formula gets

Differential Kinematics

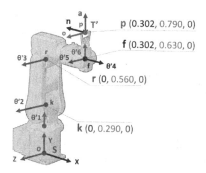

FIGURE 5.8 Puma ABB IRB120 "Tool-Up" differential kinematics POE parameters.

first the twist velocity by Equation 5.41 with the geometric Jacobian and the joint's velocity (i.e., $[\theta_1'\text{-}\theta_6']$).
- The inverse DK gets the joints' velocity, knowing the inverse geometric spatial Jacobian and the tool velocities by Equation 5.43. Given the translation velocity for the TCP (i.e., [x', y', z']) and the rotation velocity for the mobile tool system (i.e., [$\alpha,' \beta,' \gamma'$]), we get the robot joint's velocity (i.e., $[\theta_1'\text{-}\theta_6']$), always expressing these magnitudes in the spatial frame by Equation 5.44.

The code for Exercise 5.3.10 is in the internet hosting for the software of this book[6].

5.3.11 Bending Backwards Robots (e.g., ABB IRB1600)

This DK exercise sees a slightly different robot architecture, a Bending Backwards manipulator (Figure 5.9). This design does not need to rotate the robot to reach for things behind it. Swinging the arm backwards extends the working range of the robot. This example checks the geometric inverse and forward DK for a random target in the workspace.

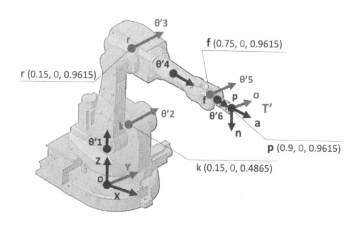

FIGURE 5.9 Bending Backwards ABB IRB1600 differential kinematics POE parameters.

The formulations are the same presented in the previous sections because all these architectures have six DoF. The differences are in the twists' definitions.

- The forward DK solves the velocity of the tool by Equation 5.32, formed by the TCP velocity (i.e., $[x', y', z']$) and the rotation velocity of the mobile tool system (i.e., $[\alpha', \beta', \gamma']$), both expressed in the spatial coordinate system. The formula gets first the twist velocity by Equation 5.33 with the geometric Jacobian and the joint's velocity (i.e., $[\theta_1'-\theta_6']$).
- The inverse DK gets the joints' velocity as a function of the inverse geometric spatial Jacobian and the tool velocities by Equation 5.36. Given the translation velocity for the TCP (i.e., $[x', y', z']$) and the rotation velocity for the mobile tool system (i.e., $[\alpha', \beta', \gamma']$), we get the robot joint's velocity (i.e., $[\theta_1'-\theta_6']$), always expressing these magnitudes in the spatial frame by Equation 5.37.

The code for Exercise 5.3.11 is in the internet hosting for the software of this book[7].

5.3.12 GANTRY ROBOTS (E.G., ABB IRB6620LX)

This DK exercise deals with this Gantry robotic architecture, which combines the advantages of both a long linear axis for the first joint plus five revolute joints (see Figure 5.10). We know how the treatment of the prismatic joint is the same with screw theory, and the resulting exercise mathematics is the same as in previous examples.

The formulations of the previous examples are valid here because all these architectures have six DoF, even though this robot has one prismatic joint, which of course has its adequate twist definition.

- The forward DK solves the velocity of the tool by Equation 5.32, formed by the TCP velocity (i.e., $[x', y', z']$) and the rotation velocity of the mobile tool system (i.e., $[\alpha', \beta', \gamma']$), both expressed in the spatial coordinate system. The formula gets first the twist velocity by Equation 5.33 with the geometric Jacobian and the joint's velocity (i.e., $[\theta_1'-\theta_6']$).

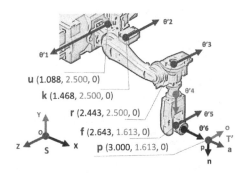

FIGURE 5.10 Gantry ABB IRB6620LX differential kinematics POE parameters.

Differential Kinematics

- The inverse DK gets the joints' velocity as a function of the inverse geometric spatial Jacobian and the tool velocities by Equation 5.36. Given the translation velocity for the TCP (i.e., [x', y', z']) and the rotation velocity for the mobile tool system (i.e., [α', β', γ']), we get the robot joint's velocity (i.e., [θ_1'-θ_6']), always expressing these magnitudes in the spatial frame by Equation 5.37.

The code for Exercise 5.3.12 is in the internet hosting for the software of this book[8].

5.3.13 SCARA ROBOTS (E.G., ABB IRB910SC)

This DK example of a Scara type robot is enlightening for the differential topic. It allows us to compare the analytic Jacobian with the geometric Jacobian. We solved the DK of the Scara robot with the analytic approach of the previous section. We take the same exercise, but we apply the geometric Jacobian formulation for both the forward and the inverse DK.

When working with velocities in screw theory, we get the advantage of getting the spatial geometric Jacobian by inspection or systematically, but always without any differentiation. This particularity is the first of the many benefits provided by this approach.

5.3.13.1 Geometric Jacobian by Inspection

We evaluate the geometric spatial Jacobian by inserting the twists associated with each joint in its current configuration. This approach looks like the twists' definition for the FK or IK, but **instead of getting a fixed point on the axis of rotation twists, we need moving the point on those axes**. These moving points arise because the Jacobian is a matrix-valued representation and has a different value for each robot pose.

It is not enough to know the values for the reference position points (i.e., o, r, f), instead, **we need a formulation for the mobile points (i.e., q_1', q_2', q_4')** (see Figure 5.11). It is necessary to imagine the geometrical evolution of some point belonging to the joint axis, as moved by the previous joints, to get these points by inspection. In this case, it is somehow easy to formulate the evolution of points "r" and "f," which become the mobile points "q_2'" and "q_4'" when affected by the rotations "ω_1" and

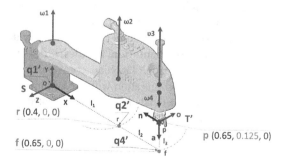

FIGURE 5.11 Scara ABB IRB910SC parameters for Geometric Jacobian.

"ω_2" respectively. We will immediately see these expressions when defining the geometric Jacobian by inspection.

- First, define **the axis of each joint**, which are the axes "ω_i" for revolute joints and "v_i" for the prismatic joint.

$$\omega_1 = \begin{bmatrix} 0 \\ 1 \\ 0 \end{bmatrix} \quad \omega_2 = \begin{bmatrix} 0 \\ 1 \\ 0 \end{bmatrix} \quad v_3 = \begin{bmatrix} 0 \\ 1 \\ 0 \end{bmatrix} \quad \omega_4 = \begin{bmatrix} 0 \\ -1 \\ 0 \end{bmatrix}$$

- Second, we define **the mobile points for each rotation** (q_1', q_2', q_4'), which is relatively easy by inspecting this mechanism (see Figure 5.11).

$$q_1' = \begin{bmatrix} 0 \\ 0 \\ 0 \end{bmatrix} \quad q_2' = \begin{bmatrix} l_1 C_1 \\ 0 \\ -l_1 S_1 \end{bmatrix} \quad q_4' = \begin{bmatrix} l_2 C_{12} + l_1 C_1 \\ 0 \\ -l_2 S_{12} - l_1 S_1 \end{bmatrix}$$

- Third, we can get **the mobile twists** (ξ_1'... ξ_4').

$$\xi_1' = \begin{bmatrix} -\omega_1 \times q_1' \\ \omega_1 \end{bmatrix} \quad \xi_2' = \begin{bmatrix} -\omega_2 \times q_2' \\ \omega_2 \end{bmatrix} \quad \xi_3' = \begin{bmatrix} v_3 \\ 0 \end{bmatrix} \quad \xi_4' = \begin{bmatrix} -\omega_4 \times q_4' \\ \omega_4 \end{bmatrix}$$

- Finally, it is possible to get **the geometric Jacobian** with Equation 5.45.

$$J_{ST}^S(\theta) = \begin{bmatrix} \xi_1' & \xi_2' & \xi_3' & \xi_4' \end{bmatrix} = \begin{bmatrix} 0 & l_1 S_1 & 0 & -l_2 S_{12} - l_1 S_1 \\ 0 & 0 & 1 & 0 \\ 0 & l_1 C_1 & 0 & -l_2 C_{12} - l_1 C_1 \\ 0 & 0 & 0 & 0 \\ 1 & 1 & 0 & -1 \\ 0 & 0 & 0 & 0 \end{bmatrix} \quad (5.45)$$

5.3.13.2 Geometric Jacobian by Definition

Having the possibility to get the manipulator (geometric or spatial) Jacobian by inspection is friendly but challenging when the robot has many DoF or a complicated mechanical configuration. However, we can count on the screw theory, which gives **the great advantage of representing the geometric Jacobian by definition, which is very suitable for systematic implementation. Besides, it does not need any differentiation.** Each column of the spatial Jacobian corresponds to its joint twist, transformed into the current robot configuration with the adjoint transformation.

We can directly define the mobile twists (ξ_1'-ξ_4'). The contribution of each joint velocity to the tool velocity is independent of the configuration of later joints in the chain (see Equation 5.46).

$$\xi_1' = \xi_1 \quad (5.46)$$

Differential Kinematics

$$\xi_2' = Ad_{\left(e^{\hat{\xi}_1\theta_1}\right)}\xi_2$$

$$\xi_3' = Ad_{\left(e^{\hat{\xi}_1\theta_1}e^{\hat{\xi}_2\theta_2}\right)}\xi_3$$

$$\xi_4' = Ad_{\left(e^{\hat{\xi}_1\theta_1}e^{\hat{\xi}_2\theta_2}e^{\hat{\xi}_3\theta_3}\right)}\xi_4$$

Finally, the resulting geometric Jacobian by definition is Equation 5.47. The aftermath is the same as the geometric Jacobian obtained by inspection Equation 5.45.

$$J_{ST}^S(\theta) = \begin{bmatrix} \xi_1' & \xi_2' & \xi_3' & \xi_4' \end{bmatrix} = \begin{bmatrix} 0 & l_1 S_1 & 0 & -l_2 S_{12} - l_1 S_1 \\ 0 & 0 & 1 & 0 \\ 0 & l_1 C_1 & 0 & -l_2 C_{12} - l_1 C_1 \\ 0 & 0 & 0 & 0 \\ 1 & 1 & 0 & -1 \\ 0 & 0 & 0 & 0 \end{bmatrix} \quad (5.47)$$

For the rest of this section, we have two exercises for the Scara with the forward and inverse DK. They reveal the clear advantages of working with the geometric screw theory approach. We define the geometric Jacobian of the DK map in terms of twists, which are easy to determine for any robot manipulator. The POE leads to a very natural and explicit description of the robot Jacobian, which highlights the geometry of the mechanism and has none of the drawbacks of a local analytic Jacobian.

5.3.13.3 Forward Differential Kinematics with Geometric Jacobian

We aim for the velocity of the tool with Equation 5.49, formed by the TCP velocity (i.e., $v_{TCP}^S = [x', y', z']$) and the rotation velocity of the tool system (i.e., $\omega_{ST}^S = [\alpha,' \beta,' \gamma']$). The intermediate step in this approach solves the twist velocity by Equation 5.48 as a function of the geometric Jacobian and the joint velocities (i.e., $[\theta_1'\text{-}\theta_4']$).

$$V_{ST}^S = \begin{bmatrix} v_{ST}^S \\ \omega_{ST}^S \end{bmatrix} = J_{ST}^S(\theta)\dot{\theta} = \begin{bmatrix} \xi_1' & \xi_2' & \xi_3' & \xi_4' \end{bmatrix} \begin{bmatrix} \dot{\theta}_1 \\ \dot{\theta}_2 \\ \dot{\theta}_3 \\ \dot{\theta}_4 \end{bmatrix} \quad (5.48)$$

$$V_T^S = \begin{bmatrix} v_{TCP}^S \\ \omega_{ST}^S \end{bmatrix} = \begin{bmatrix} \dot{x} \\ \dot{y} \\ \dot{z} \\ \dot{\alpha} \\ \dot{\beta} \\ \dot{\gamma} \end{bmatrix} = \begin{bmatrix} v_{ST}^S + \widehat{[\omega_{ST}^S]} TCP^S(\theta) \\ \omega_{ST}^S \end{bmatrix} \quad (5.49)$$

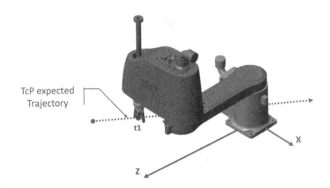

FIGURE 5.12 Scara ABB IRB910SC geometric forward differential kinematics.

The geometric Jacobian is a matrix-valued function, and therefore it gets a different value for each joint configuration, always defined in the spatial coordinate system.

To illustrate this forward DK, we propose an exercise for the Scara robot in a specific desired trajectory for the manipulator tool (see Figure 5.12). We select a particular pose for the robot "t_1." Then, we obtain the joint magnitudes for this pose from solving the IK. We choose the joint velocities to follow the expected trajectory. We can check with Exercise 5.3.7.6a how the output velocities are the same as the exercise solved with the classical analytic Jacobian (see Exercise 5.2.3.1a).

The code for Exercise 5.3.13a is in the internet hosting for the software of this book[9].

5.3.13.4 Inverse Differential Kinematics with Geometric Jacobian

We get the joints' velocity as a function of the inverse geometric spatial Jacobian and the tool velocities by Equation 5.50. Given the translation velocity for the TCP (i.e., $v^S_{TCP} = [x', y', z']$) and the rotation velocity for the tool system (i.e., $\omega^S_{ST} = [\alpha', \beta', \gamma']$), we obtain the robot joint velocities (i.e., $[\theta_1' \text{-} \theta_4']$), always in the spatial frame by Equation 5.51.

$$\dot{\theta} = \left[J^S_{ST}(\theta) \right]^{-1} V^S_{ST} = \left[J^S_{ST}(\theta) \right]^{-1} \begin{bmatrix} v^S_{ST} \\ \omega^S_{ST} \end{bmatrix} = \left[J^S_{ST}(\theta) \right]^{-1} \begin{bmatrix} v^S_{TCP} - \omega^S_{ST} \times TCP^S(\theta) \\ \omega^S_{ST} \end{bmatrix} \quad (5.50)$$

$$\begin{bmatrix} \dot{\theta}_1 \\ \dot{\theta}_2 \\ \dot{\theta}_3 \\ \dot{\theta}_4 \end{bmatrix} = \begin{bmatrix} \xi_1' & \xi_2' & \xi_3' & \xi_4' \end{bmatrix}^{-1} \begin{bmatrix} \begin{bmatrix} \dot{x} \\ \dot{y} \\ \dot{z} \end{bmatrix} - \begin{bmatrix} \widehat{\dot{\alpha}} \\ \dot{\beta} \\ \dot{\gamma} \end{bmatrix} TCP^S(\theta) \\ \begin{bmatrix} \dot{\alpha} \\ \dot{\beta} \\ \dot{\gamma} \end{bmatrix} \end{bmatrix} \quad (5.51)$$

We remind that the inverse geometric Jacobian is a matrix-valued function that gets a different quantity for each joint configuration in the spatial coordinate system.

Differential Kinematics

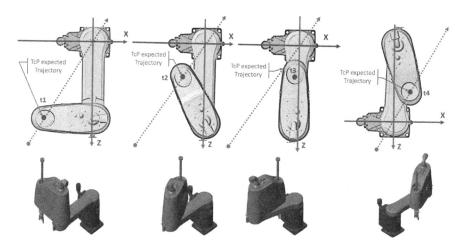

FIGURE 5.13 Scara ABB IRB910SC geometric inverse differential kinematics.

To illustrate this inverse DK, we present an example (Exercise 5.3.7.6b) with four poses (t_1-t_2-t_3-t_4) for the Scara robot in a particular desired linear trajectory for the robot tool (see details in Figure 5.13). We solve Equation 5.51 to get the robot joint velocities to make the TCP follow a linear trajectory on a plane parallel to the floor. The tool must develop the expected trajectory with a constant velocity. The joint velocities are inside acceptable limits for three of the targets (i.e., t_1-t_2-t_4). The velocities around the target "t_3" acquire unacceptably high values. It happens because we are close to a singularity.

The only way to follow the desired trajectory is to make the tool go through the robot's base just by the origin of the spatial system, which is not possible. The robot could approximate the linear trajectory avoiding the mechanical restriction that means a collision with the base. The robot must make a very abrupt change of configuration between targets "t_2" and "t_4" to perform this motion reasonably.

We can even get a hint of the trajectory development from the graphic (Figure 5.13) since the first joint must perform a rotation of value "π" in very little time. Intuitively, we see that this need would require a formidable acceleration for this joint to comply with the motion.

Analytically, this situation translates into considering a target "t_3" located in the path between "t_2" and "t_4." For instance, we take a point "t_3" which makes "θ_2" close to "π." Then, the inverse of the Jacobian becomes almost infinite, and therefore it results in tremendous speeds for the two first joints. These calculation results give us velocities for the joints so big that the robot mechanics' limitation prevents us from reaching such values. In sum, we are just in one configuration of singularity.

The results are the same as those for the inverse DK solved with the analytic Jacobian (see Exercise 5.2.3.1b), but in this case, having the great advantage of avoiding the differentiation. Besides, the geometric Jacobian facilitates a systematic algorithm implementation and the easy translation between different coordinate systems.

The code for Exercise 5.3.13b is in the internet hosting for the software of this book[10].

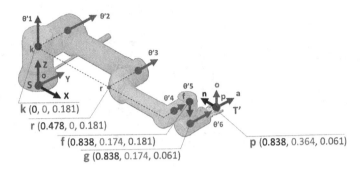

FIGURE 5.14 Collaborative UNIVERSAL UR16e differential kinematics POE parameters.

5.3.14 COLLABORATIVE ROBOTS (E.G., UNIVERSAL UR16E)

This DK exercise deals with a typical collaborative manipulator, such as the robot UR16e of UNIVERSAL, made for human-robot cooperation in the workspace (see Figure 5.14).

The formulations are the same as the previous exercises in this chapter, with also six DoF.

- The forward DK solves the velocity of the tool by Equation 5.32, formed by the TCP velocity (i.e., [x', y', z']) and the rotation velocity of the mobile tool system (i.e., [α', β', γ']), both expressed in the spatial coordinate system. The formula gets first the twist velocity by Equation 5.33 with the geometric Jacobian and the joint's velocity (i.e., [θ_1'-θ_6']).
- The inverse DK gets the joints' velocity as a function of the inverse geometric spatial Jacobian and the tool velocities by Equation 5.36. Given the translation velocity for the TCP (i.e., [x', y', z']) and the rotation velocity for the mobile tool system (i.e., [α', β', γ']), we get the robot joint's velocity (i.e., [θ_1'-θ_6']), always expressing these magnitudes in the spatial frame by Equation 5.37.

The code for Exercise 5.3.14 is in the internet hosting for the software of this book[11].

5.3.15 REDUNDANT ROBOTS (E.G., KUKA IIWA)

This DK exercise is for the lightweight robot KUKA IIWA, made for human-robot collaboration in the workspace (Figure 5.15). This manipulator is redundant as it counts with seven joints. Nonetheless, the general approach of screw theory to solve DK works without problem.

The formulations are quite like those of the previous examples but adjusted for the seven DoF of this robot.

- The forward DK solves the velocity of the tool by Equation 5.52, formed by the TCP velocity (i.e., [x', y', z']) and the rotation velocity of the mobile tool system (i.e., [α', β', γ']), both expressed in the spatial coordinate system.

Differential Kinematics

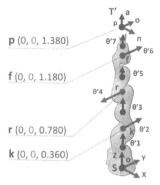

p (0, 0, 1.380)

f (0, 0, 1.180)

r (0, 0, 0.780)

k (0, 0, 0.360)

FIGURE 5.15 Redundant KUKA IIWA differential kinematics POE parameters.

The formula gets first the twist velocity by Equation 5.53 with the geometric Jacobian and the joint's velocity (i.e., $[\theta_1'\text{-}\theta_7']$).

$$V_T^S = \begin{bmatrix} v_{TCP}^S \\ \omega_{ST}^S \end{bmatrix} = \begin{bmatrix} \dot{x} \\ \dot{y} \\ \dot{z} \\ \dot{\alpha} \\ \dot{\beta} \\ \dot{\gamma} \end{bmatrix} = \begin{bmatrix} v_{ST}^S + \widehat{\begin{bmatrix} \omega_{ST}^S \end{bmatrix}} TCP^S(\theta) \\ \omega_{ST}^S \end{bmatrix} \tag{5.52}$$

$$V_{ST}^S = \begin{bmatrix} v_{ST}^S \\ \omega_{ST}^S \end{bmatrix} = J_{ST}^S(\theta)\dot{\theta} = \begin{bmatrix} \xi'_1 & \xi'_2 & \xi'_3 & \xi'_4 & \xi'_5 & \xi'_6 & \xi'_7 \end{bmatrix} \begin{bmatrix} \dot{\theta}_1 \\ \dot{\theta}_2 \\ \dot{\theta}_3 \\ \dot{\theta}_4 \\ \dot{\theta}_5 \\ \dot{\theta}_6 \\ \dot{\theta}_7 \end{bmatrix} \tag{5.53}$$

- The inverse DK gets the joints' velocity (i.e., $[\theta_1'\text{-}\theta_7']$), as a function of the inverse geometric spatial Jacobian and the tool velocities (Equation 5.54), given by the translation velocity for the TCP (i.e., $[x', y', z']$) and the rotation velocity for the mobile tool system (i.e., $[\alpha', \beta', \gamma']$).

$$\begin{bmatrix} \dot{\theta}_1 \\ \dot{\theta}_2 \\ \dot{\theta}_3 \\ \dot{\theta}_4 \\ \dot{\theta}_5 \\ \dot{\theta}_6 \\ \dot{\theta}_7 \end{bmatrix} = \begin{bmatrix} \xi'_1 & \xi'_2 & \xi'_3 & \xi'_4 & \xi'_5 & \xi'_6 & \xi'_7 \end{bmatrix}^{-1} \begin{bmatrix} \begin{bmatrix} \dot{x} \\ \dot{y} \\ \dot{z} \end{bmatrix} - \widehat{\begin{bmatrix} \dot{\alpha} \\ \dot{\beta} \\ \dot{\gamma} \end{bmatrix}} TCP^S(\theta) \\ \begin{bmatrix} \dot{\alpha} \\ \dot{\beta} \\ \dot{\gamma} \end{bmatrix} \end{bmatrix} \tag{5.54}$$

The code for Exercise 5.3.15 is in the internet hosting for the software of this book[12].

5.4 SUMMARY

Differential Kinematics deals with the robot velocity. It defines the relationship between the joint velocities and the tool's linear and angular velocities. The Jacobian concept is crucial to DK, as this operator relates the velocities for both the forward and inverse DK.

The key idea presented in this chapter is the difference between the concepts of analytic and geometric Jacobian. With the first approach, it is imperative to differentiate the kinematics map. Alternatively, with the geometric Jacobian, it is possible to obtain the inverse (Equation 5.55) and forward (Equation 5.56) DK without any differentiation, using the joints' twist. We introduce more screw theory terminology, as the "Twist Velocity" defined by Equation 5.57, inherently linked to the concept of geometric Jacobian and robot velocity.

$$\dot{\theta} = \begin{bmatrix} \dot{\theta}_1 \\ \vdots \\ \dot{\theta}_n \end{bmatrix} = \begin{bmatrix} \xi_1' & \xi_2' & \cdots & \xi_n' \end{bmatrix}^{-1} \begin{bmatrix} \begin{bmatrix} \dot{x} \\ \dot{y} \\ \dot{z} \end{bmatrix} - \begin{bmatrix} \dot{\alpha} \\ \dot{\beta} \\ \dot{\gamma} \end{bmatrix} \widehat{TCP^S}(\theta) \\ \begin{bmatrix} \dot{\alpha} \\ \dot{\beta} \\ \dot{\gamma} \end{bmatrix} \end{bmatrix} \quad (5.55)$$

$$V_T^S = \begin{bmatrix} v_{TCP}^S \\ \omega_{ST}^S \end{bmatrix} = \begin{bmatrix} \dot{x} \\ \dot{y} \\ \dot{z} \\ \dot{\alpha} \\ \dot{\beta} \\ \dot{\gamma} \end{bmatrix} = \begin{bmatrix} v_{ST}^S + \widehat{\left[\omega_{ST}^S\right]} TCP^S(\theta) \\ \omega_{ST}^S \end{bmatrix} \quad (5.56)$$

$$V_{ST}^S = \begin{bmatrix} v_{ST}^S \\ \omega_{ST}^S \end{bmatrix} = J_{ST}^S(\theta)\dot{\theta} = \begin{bmatrix} \xi_1' & \xi_2' & \cdots & \xi_n' \end{bmatrix} \begin{bmatrix} \dot{\theta}_1 \\ \vdots \\ \dot{\theta}_n \end{bmatrix} \quad (5.57)$$

We develop with detail an exercise with a Scara robot to illustrate the differences between the analytic and geometric DK. Besides, this example is apparent to introduce the concept of robot singularity.

Besides, there are some examples to prove the advantages of the geometric Jacobian: a Puma robot (i.e., ABB IRB120), a Bending Backwards robot (i.e., ABB IRB1600), a Gantry robot (i.e., ABB IRB6620LX), a Collaborative robot (e.g., UNIVERSAL UR16e), and a Redundant manipulator (i.e., KUKA IIWA). Of course, extending the geometric Jacobian concept to other robot architectures and more complex mechanical configurations is possible.

Differential Kinematics

We can employ the geometric Jacobian to solve a trajectory generation when there is no suitable IK direct geometric solution, as we will see in more detail throughout Chapter 7.

NOTES

1 Pardos-Gotor, J.M. (2021). *Screw Theory in Robotics*. Github. https://github.com/DrPardosGotor/Screw-Theory-in-Robotics/blob/master/Exercises/Exercise_5_2_4a.m
2 Pardos-Gotor, J.M. (2021). *Screw Theory in Robotics*. Github. https://github.com/DrPardosGotor/Screw-Theory-in-Robotics/blob/master/Exercises/Exercise_5_2_4b.m
3 Pardos-Gotor, J.M. (2021). *Screw Theory in Robotics*. Github. https://github.com/DrPardosGotor/Screw-Theory-in-Robotics/blob/master/Exercises/Exercise_5_3_9a.m
4 Pardos-Gotor, J.M. (2021). *Screw Theory in Robotics*. Github. https://github.com/DrPardosGotor/Screw-Theory-in-Robotics/blob/master/Exercises/Exercise_5_3_9b.m
5 Pardos-Gotor, J.M. (2021). *Screw Theory in Robotics*. Github. https://github.com/DrPardosGotor/Screw-Theory-in-Robotics/blob/master/Exercises/Exercise_5_3_9c.m
6 Pardos-Gotor, J.M. (2021). *Screw Theory in Robotics*. Github. https://github.com/DrPardosGotor/Screw-Theory-in-Robotics/blob/master/Exercises/Exercise_5_3_10.m
7 Pardos-Gotor, J.M. (2021). *Screw Theory in Robotics*. Github. https://github.com/DrPardosGotor/Screw-Theory-in-Robotics/blob/master/Exercises/Exercise_5_3_11.m
8 Pardos-Gotor, J.M. (2021). *Screw Theory in Robotics*. Github. https://github.com/DrPardosGotor/Screw-Theory-in-Robotics/blob/master/Exercises/Exercise_5_3_12.m
9 Pardos-Gotor, J.M. (2021). *Screw Theory in Robotics*. Github. https://github.com/DrPardosGotor/Screw-Theory-in-Robotics/blob/master/Exercises/Exercise_5_3_13a.m
10 Pardos-Gotor, J.M. (2021). *Screw Theory in Robotics*. Github. https://github.com/DrPardosGotor/Screw-Theory-in-Robotics/blob/master/Exercises/Exercise_5_3_13b.m
11 Pardos-Gotor, J.M. (2021). *Screw Theory in Robotics*. Github. https://github.com/DrPardosGotor/Screw-Theory-in-Robotics/blob/master/Exercises/Exercise_5_3_14.m
12 Pardos-Gotor, J.M. (2021). *Screw Theory in Robotics*. Github. https://github.com/DrPardosGotor/Screw-Theory-in-Robotics/blob/master/Exercises/Exercise_5_3_15.m

6 Inverse Dynamics

"I have explained the phenomena of the heavens and our sea by the force of gravity, but I have not yet assigned a cause to gravity."

—Isaac Newton

6.1 PROBLEM STATEMENT IN ROBOTICS

Dynamics is the branch of physics developed in classical mechanics concerned with studying forces and their effects on motion (Abraham and Marsden, 1999). In the realm of robot dynamics today, we can see incredible developments, such as the "Atlas" humanoid developed by Boston Dynamics. However, to reach this cutting-edge application, we must begin by understanding the fundamentals of robot screw theory dynamics and some of its fundamental tools, like the Product of Exponentials (POE) (Brockett, 1983), and this is the scope of this chapter.

We present the dynamics of a robot manipulator using the mathematical terms of the screw theory, Lie algebras, and its extensions (Davidson and Hunt, 2004) (e.g., Screws, Twists, Wrenches, POE, geometric Jacobian, Adjoint Transformations, Spatial Vector Algebra). We start by studying the role that screw kinematics plays in the equations of rigid body motion for then extending these methodologies to robot dynamics (Lynch and Park, 2017).

The dynamics describe how the manipulator moves in response to applied forces, including those of the joint actuators. This description of the dynamics of robot motion uses a nonlinear set of second-order ordinary differential equations, which depend on the mechanism's kinematics and inertial properties (Marsden and Ratiu, 1999). It is a formidable difficulty because there are very few methods for solving nonlinear differential equations exactly. Besides, the proper robot dynamics formulation is the necessary previous step to get any reasonable robotics control, which is the system that contributes to producing the designed movement of the manipulator or robot mechanism.

Isaac Newton was the first to formulate the fundamental physical laws that govern dynamics in classical non-relativistic physics, especially his second law of motion. Later, Euler introduced reaction forces between bodies. It is possible to generate the equations of motion by using the Newton–Euler equations to each of the links in the robot multibody system. Dynamics work with magnitudes such as displacement, velocity, acceleration, and inertia regarding the forces acting on the mechanical system.

We need a dynamic model of the robot for many applications: design of control laws, trajectory generation algorithms, simulation, design optimization, or motion planning (Choset and Lynch, 2005).

Rigid body dynamics is a classical subject, and there is a tendency to regard robot dynamics as a solved problem. However, robot dynamics is still a challenging open area of research because robots have high-dimensional, very complex structures and constraints. All are factors that make the formulation of the robot dynamics less than straightforward.

Our focus for dynamics formulations is limited to algorithms that rely on differential geometry methodologies (Millman and Parker, 1997), particularly the theory of Lie groups and its extensions (Ohwovoriole and Roth, 1981). The advantages of this geometric approach have been recognized for some time by practitioners of classical screw theory. Featherstone showed how the dynamics of rigid multibody systems could be formulated by using and extending Lie algebra constructions (Featherstone, 2016). This chapter includes practical examples of robot dynamics solved by both approaches, the first with the classical screw theory and the second with Featherstone's spatial vector algebra.

Usually, the problem to solve is Inverse Dynamics (ID), consists of finding the forces applied to the robot's actuators to generate the mechanism's desired movement (Mason, 2001).

The geometric methods for open chains have extensions to closed chains and robots subject to other constraints, modeling rigid body contact and robots with soft elements such as variable stiffness actuators. On top of that, there are algorithms for motion optimization, optimal design, and other applications (Greenwood, 2006). However, these implementations are out of this book's scope, and we will focus on the ID problems.

We get started with the classical screw theory dynamics (Selig, 2005), founded on the Lagrange approach. This method is equivalent to the Newton–Euler equations and has fewer equations to describe the motion of a system (Lee et al., 2005). The Lagrange equations only need to compute the potential and kinetic energies of the system. Besides, Lagrange formulations allow us to determine and exploit robot dynamics' structural properties, which is helpful for high-level analysis and control in robotics.

6.2 THE LAGRANGE CHARACTERIZATION

Once a suitable set of generalized coordinates has been chosen (e.g., the angles of the revolute joints and the linear displacement for the prismatic joints). According to the generalized coordinates, we can generate the equations of motion using the Lagrange equations, expressing the forces applied to the system. These components are the generalized forces "Γ." We define the Lagrangian "L" as the difference between the kinetic "K" and potential "V" energy as expressed by Equation 6.1 to describe the equations of motion. The Lagrange equations for the motion dynamics of a mechanical system with generalized coordinates "θ" and Lagrangian "L" are given by Equation 6.2, where "Γ" is the generalized forces that act on the mechanism.

$$L(\theta,\dot{\theta}) = K(\theta,\dot{\theta}) - V(\theta) \tag{6.1}$$

$$\frac{d}{dt}\frac{\partial L}{\partial \dot{\theta}} - \frac{\partial L}{\partial \theta} = \Gamma \tag{6.2}$$

Inverse Dynamics

The ID problem consists of finding the forces that we will apply to the actuators of the robot joints to generate the mechanism's desired movement, mainly the tool. Generally, these joint magnitudes are the torques for those of revolution and the forces for those prismatic. Once we know the robot's motion equations, if the dynamic model were perfect, we could find the necessary torques and forces directly using the Lagrange equations. In practice, it is evident that due to errors in the model, noise, or slip-ups in the initial conditions, it is necessary to apply a control system, as we will briefly see later.

We use the theory of screws and Lie mathematics to define the Lagrangian in terms of the positions and speeds of a robot's joints. This global characterization of the dynamics of a rigid body subject to external forces and torques examines the dynamics in terms of twists and wrenches. Then, the Lagrange expression for an open-chain manipulator becomes Equation 6.3. We can use a second-order vector differential equation for the robot's motion as a function of the applied joint torques by Equation 6.4. In this expression, the first term "M" collects the robot mass or inertia matrix, the second includes the matrix "C" that considers the centrifugal and Coriolis terms, and the third matrix "N" which consists of the potential and external forces (e.g., gravity and friction).

$$L(\theta,\dot{\theta}) = \frac{1}{2}\dot{\theta}^T M(\theta)\dot{\theta} - V(\theta) \tag{6.3}$$

$$M(\theta)\ddot{\theta} + C(\theta,\dot{\theta})\dot{\theta} + N(\theta,\dot{\theta}) = \Gamma \tag{6.4}$$

This formulation of Equation 6.4 is the classical second-order vector differential equation for the robot screw theory ID (Park et al., 1995). It solves the torques' value and forces "Γ" that we must apply to the robot actuators for achieving the positions, speeds, and accelerations that we want in the joints of the robot to comply with the necessary tool or link trajectory.

The Inertia "M" and Coriolis "C" matrices summarize the manipulator's inertial properties and have some essential properties. The passivity property (i.e., $M'-2C$ is a skew-symmetric matrix) implies that the robot system's net energy is conserved for the cases without friction. The passivity property is crucial in the proof of many control laws for robot manipulators.

The expression (6.4) includes new concepts which can be defined in terms of the Lie algebras.

The screw theory offers the formidable possibility of solving this formulation of robot dynamics employing linear algebra techniques (Strang, 2009). Furthermore, the tools of Lie algebra allow expressing this dynamic problem practically by geometry. Utilizing a set of POE operations (Brockett et al., 1993), we can represent the mass or inertia, Coriolis, and potential or gravity matrices. We will not need to differentiate the expressions for the robot dynamics, and we can solve the second-order nonlinear differential equations precisely with closed-form implementations (Murray et al., 2017).

- **The Inertia Matrix of the Mechanism "$M_{ij}(\theta)$"**: this includes the mass and inertial forces related to the acceleration of the joints. We can express the inertia with different formulations, and we present two of them just next.
- **The Link Jacobian Inertia Matrix of the Robot "$M_{Jsl}(\theta)$"** of Equation 6.5: this formulation employs the link tool geometric Jacobian "J^T_{SL}" of Equation 5.29. We can formulate the link tool Jacobian with both definitions for the classical (i.e., Equation 5.20) and new (i.e., Equation 5.31) adjoint transformations. Besides, another new concept comes to play, the inertia matrix for each link "$\mathcal{M}_i$" as Equation 6.6. It is a tensor with a general diagonal form because we aligned its coordinate axes with the link's principal axes. This matrix's structure relies on attaching the body coordinate frame at the link center of mass. This link inertia matrix needs for its definition the mass of the link "m_i" and the moments of inertia about the X, Y, and Z axes of the link frame "I_{xi}," "I_{yi}," and "I_{zi}."

$$M_{Jsl}(\theta) = M_{ij}(\theta) = \sum_{i=1}^{n} \left[J^T_{SL}(\theta) \right]^T \mathcal{M}_i \, J^T_{SL}(\theta) \tag{6.5}$$

$$\mathcal{M}_i = \begin{bmatrix} m_i I_3 & 0 \\ 0 & \Psi_i \end{bmatrix} = \begin{bmatrix} m_i & 0 & 0 & 0 & 0 & 0 \\ 0 & m_i & 0 & 0 & 0 & 0 \\ 0 & 0 & m_i & 0 & 0 & 0 \\ 0 & 0 & 0 & I_{xi} & 0 & 0 \\ 0 & 0 & 0 & 0 & I_{yi} & 0 \\ 0 & 0 & 0 & 0 & 0 & I_{zi} \end{bmatrix} \tag{6.6}$$

- **The New Adjoint Inertia Matrix of the Robot "$M_{Aij}(\theta)$"** of Equation 6.7: it is an algebraic representation based on the new adjoint representation of Equation 5.31, the twists of the links, and a new definition for the inertias. This description is the transformed inertia matrix for each link "$\mathcal{M}'_i$" expressed as Equation 6.8. It represents the inertia matrix for each link given by Equation 6.6 but reflected in the base frame of the robot. For doing that, it applies the classical adjoint transformation (5.20) to the coordinate frame associated with the center of mass for each link "$H_{sli}(0)$" at the home reference pose of the robot.

$$M_{Aij}(\theta) = M_{ij}(\theta) = \sum_{l=\max(i,j)}^{n} \xi_i^T A_{li}^T \mathcal{M}'_l A_{lj} \xi_j \tag{6.7}$$

$$\mathcal{M}'_i = \left(Ad^{-1}_{H_{sli}(0)} \right)^T \mathcal{M}_i \, Ad^{-1}_{H_{sli}(0)} \tag{6.8}$$

Both definitions for the Inertia Matrix of the Robot "$M_{ij}(\theta)$" are equivalent and can be used indistinctly in the formulation of the classical screw theory robot dynamics of Equation 6.4, represented as "$M(\theta)$" in relation with the robot accelerations.

Inverse Dynamics

- **The Coriolis Matrix of the Robot "$C(\theta,\theta')$" of Equation 6.9:** this gives the information for the centrifugal and Coriolis terms non-inertia frames, which are implicit of generalized coordinates. Note that there are other ways to define this matrix. However, this choice has important properties that we shall later exploit (e.g., passivity property). To determine the Coriolis matrix in terms of Lie algebra, we will formulate it using the so-called Christoffel symbols "Γ_{ijk}" which correspond with partial derivatives of the inertia matrix "$M(\theta)$," which makes it difficult to calculate the dynamics of the robot with efficiency. Fortunately, a great advantage of Lie algebra is the possibility of working with direct expressions of matrices, making feasible to solve these Christoffel terms with an algebraic expansion of matrices defined according to Equation 6.10.

$$C_{ij}\left(\theta,\dot{\theta}\right) = \sum_{k=1}^{n}\Gamma_{ijk}\dot{\theta}_k = \frac{1}{2}\sum_{k=1}^{n}\left(\frac{\partial M_{ij}}{\partial \theta_k} + \frac{\partial M_{ik}}{\partial \theta_j} - \frac{\partial M_{kj}}{\partial \theta_i}\right)\dot{\theta}_l \tag{6.9}$$

$$\frac{\partial M_{ij}}{\partial \theta_k} = \sum_{l=\max(i,j)}^{n}\left(\left[A_{ki}\xi_i,\xi_k\right]^T A_{lk}^T \mathcal{M}_l' A_{lj}\xi_j + \xi_i^T A_{li}^T \mathcal{M}_l' A_{lk}\left[A_{kj}\xi_j,\xi_k\right]\right) \tag{6.10}$$

So far, we have calculated the most complex dynamic attributes (i.e., the "M" and "C" matrices). We can determine them by knowing the joints' twists and the mass, center of mass, and inertial tensors of the links. The expressions of the adjoint transformation "A_{ij}" are the only expressions that depend on the robot's configuration (i.e., θ). We need the desired joint velocities to obtain the Coriolis matrix, but we know the generalized velocities because they are inputs the robot ID problem. Now we only must solve the potential matrix "N."

- **The Potential Matrix of the Robot "$N(\theta,\theta')$":** if we neglect friction, which is often a reasonable assumption, the potential matrix is a function only of gravity. In many texts, the gravity forces are obtained by inspection or by the differentiation of ad hoc potential expressions. This approach is possible for robots with few Degrees of Freedom (DoF) and, in any case, is not a systematic approach. Conversely, in this book, we present three different systematic formulations for the gravity matrix (Pardos-Gotor, 2019).

  The first approach takes the symbolic derivative of all potential energies for the robot's links, which we will call the Gravity Symbolic Matrix "$\mathcal{N}_{sym}$." The second is a new formulation that employs the adjoint transformation, and we call it the Gravity Twist Matrix "$\mathcal{N}_{\xi}$." The third is an approach that works with the Lie generalized forces expression (i.e., wrench), which we will name as the Gravity Wrench Matrix "$\mathcal{N}_{\mathcal{F}}$." The first and third formulations have antecedents in the literature, while the second is new, but for all three, we will introduce novel implementations. These developments can be studied in detail with the associate material to this book, viewing the functions programmed in the Screw Theory Toolbox for Robotics "ST24R" (Pardos-Gotor, 2021a) and the code in the numerous dynamics' exercises.

- **The Gravity Symbolic Matrix "$\mathcal{N}_{SYM}(\theta)$":** it is obtained with the derivative of the potential energy for all robot links and is a solution obtained with the differentiation of the potential function to each generalized coordinate. The partial derivatives of the potential energy's function due to gravity, concern to each magnitude (i.e., θ_1-θ_n) of the robot joints. This new formulation of Equation 6.11 needs symbolic calculation tools to derive the expressions of potential energy. In our case, we solve it with the new function in the "ST24R" toolbox. The function uses the center of gravity of the link "CM_i" for the robot configuration, which we calculate with the POE of Equation 6.12 to solve the Forward Kinematics (FK) of the link in the spatial frame. The formulation works for any direction of gravity application, which can be practical and useful for mobile robots. This symbolic solution is more operative than it might seem since the partial derivatives only have to be calculated once for any robot. We can store the expression and evaluate it for each set of values of the generalized coordinates from the first execution. In this way, the calculation for the gravity matrix comes as a closed and direct function (i.e., without any iteration), so it is computational efficiency can be good.

$$N_{SYM}(\theta) = N_i(\theta) = -\sum_{xyz}\left(\frac{\partial}{\partial \theta_i}\sum_{i=1}^{n}\begin{bmatrix} m_i g_x & 0 & 0 \\ 0 & m_i g_y & 0 \\ 0 & 0 & m_i g_z \end{bmatrix} CM_i(\theta)\right) \quad (6.11)$$

$$H_{SLi}^{S}(\theta) = \prod_{j=1}^{i} e^{\hat{\xi}_j \theta_j} \cdot H_{SLi}^{S}(0) = \begin{bmatrix} n & o & a & CM_i(\theta) \\ 0 & 0 & 0 & 1 \end{bmatrix} \quad (6.12)$$

- **The Gravity Twist Matrix "$\mathcal{N}_\xi(\theta)$":** we introduce this new definition that uses the new adjoint transformation of Equation 5.31. We present the gravity twist concept "ξ_g," which is given by the pure translation corresponding to the direction of application of gravity. This formulation works for the most typical cases with manipulative robots, for which gravity applies in the negative direction of the spatial reference system (e.g., Z-axis, but it can be another). With this expression of Equation 6.13, we avoid making any derivative, and the gravity matrix is defined algebraically with a clear geometric sense.

$$N_\xi(\theta) = N_i(\theta) = -\sum_{l=i}^{n} \xi_i^T A_{li}^T \mathcal{M}_l' A_{l1}(\xi_g g) \quad (6.13)$$

- **The Gravity Wrench Matrix "$\mathcal{N}_\mathcal{F}(\theta)$":** this is a new formulation that uses the screw theory generalized force concept (i.e., wrench). Gravity is a translational force (i.e., only linear wrench) applied to the robot's links. The potential of gravity on the mechanism is the sum of all gravity wrenches. They apply to the center of the mass of each link of the robot and then transform it into the joint space with Equation 6.14. The wrenches are passed to the stationary space through the link spatial geometric Jacobian by Equation 6.15 considered for each link. To obtain the gravity wrench for each link with Equation 6.16, we use the gravity "g" and its axis of application "ω_g," the mass of the link "m_i" and the actual position of its center of mass "CM_i." Each CM is obtained with the POE to solve the link's FK for the current configuration with Equation 6.17. The Jacobian and the gravity wrench are a function of the robot's position at each moment. They depend on the configuration given by the different positions of the joints. We defined these expressions with the inverse kinematics (IK) problem's solution as a function of the tool's pose.

This gravity wrench solution has many advantages:

- It globally characterizes the effect of gravity.
- It does not use symbolic tools.
- It does not need differential treatment.
- It is entirely algebraic and has a straightforward implementation with great computational performance.
- It works for any direction of application of gravity, which is practical and useful for mobile or space robots.

$$N_{\mathcal{F}}(\theta) = N_i(\theta) = -\sum_{i=1}^{n} J_{SL}^S(\theta) \mathcal{F}_{gi}^S(\theta) \tag{6.14}$$

$$J_{SL}^S(\theta) = \begin{bmatrix} \xi_1' \cdots \xi_i' \end{bmatrix}; \xi_j' = Ad_{\left(e^{\hat{\xi}_1 \theta_1} \cdots e^{\hat{\xi}_{j-1} \theta_{j-1}}\right)} \xi_j \tag{6.15}$$

$$\mathcal{F}_{gi}^S(\theta) = m_i g \begin{bmatrix} \omega_g \\ -\omega_g \times q_i \end{bmatrix} = m_i g \begin{bmatrix} \omega_g \\ -\omega_g \times CM_i(\theta) \end{bmatrix} \tag{6.16}$$

$$H_{SLi}^S(\theta) = \prod_{j=1}^{i} e^{\hat{\xi}_j \theta_j} \cdot H_{SLi}^S(0) = \begin{bmatrix} n & o & a & CM_i(\theta) \\ 0 & 0 & 0 & 1 \end{bmatrix} \tag{6.17}$$

We have presented a totally algebraic and almost entirely geometric formulation for solving the dynamics problem of a manipulative robot. The motion equations are expressed as matrices (i.e., sets of geometric POE), with the tremendous computational and control advantage that this formulation implies.

Besides, to improve the calculation efficiency, it is possible to obtain the inertia "M," Coriolis "C," and potential "N" matrices algebraically according to the expressions presented. The main contribution of this classical screw dynamics approach is that it allows abandoning methods limited by the need to use local parameters, moving on to a genuinely global characterization of the dynamics problem. Therefore, we can use these concepts to formulate the equations of motion of rigid multibody systems efficiently and reference frame-invariant.

There are remarkable achievements in the field of robot dynamics in the last few years (Siciliano and Khatib, 2016). For instance, the locomotion experimentation with the robot's "Atlas" or the "Spot," both of "Boston Dynamics." Other examples are the droids "Cassie" and "Digit" of "Agility Robotics." These works are defining the state of the art of dynamics control for robotic systems. It is truly inspiring for everyone involved in this discipline.

Nonetheless, the most significant challenge is in managing the complexity of the new robotics models. Techniques for dimension reduction of the high-dimensional models and reliable methods for generating dynamically feasible trajectories in real time are all issues that we need to address before robots can perform valuable tasks in human environments. These advanced challenges are out of the scope of this book, but the basic dynamics knowledge is available within the manipulator examples of this chapter.

6.2.1 General Non-Recursive Solution to Inverse Dynamics

The ID formalized with screw theory and the Lagrange characterization relate the robot position, velocity, and acceleration to the generalized forces of the joints. We can develop the non-recursive geometric solution to any manipulator ID problem following this algorithm for the joint space.

- Select the **Spatial coordinate system "S"** (usually stationary, such as the robot base) and the **Mobile coordinate system "T"** (typically the tool). There is no rule and complete flexibility to make this definition, and we can choose the most convenient frames according to the application.
- Define the **axis of each joint**, which are the axes "ω_i" for revolute joints and "v_i" for the prismatic joints, **and a point** "q_i" on any of those axes.
- Obtain the **Twists "ξ_i"** for the joints, knowing for each revolute joint its axis and a point on that axis, and for each prismatic joint its axis.

$$\xi_i = \begin{bmatrix} -\omega_i \times q_i \\ \omega_i \end{bmatrix} \quad \xi_i = \begin{bmatrix} v_i \\ 0 \end{bmatrix}$$

- Get "$H_{ST}(0)$" the **pose of the tool at the reference** (home) robot position. This configuration of the end-effect happens when all the joint magnitudes are zero. This tool pose comes as a homogeneous matrix, necessary because the POE represents a relative mapping.
- Express the **kinematics mapping "$H_{ST}(\theta)$,"** with the product of all joint screw exponentials (POE) and "$H_{ST}(0)$."

$$H_{ST}(\theta) = \prod_{i=1}^{n} e^{\hat{\xi}_i \theta_i} H_{ST}(0) = \begin{bmatrix} n & o & a & p \\ 0 & 0 & 0 & 1 \end{bmatrix}$$

- Obtain **the robot dynamics parameters for each link**: the mass "m_i," the center of mass location "CM_{iX}-CM_{iY}-CM_{iZ}" defined in the spatial coordinate system, and the moments of inertia "I_{iX}-I_{iY}-I_{iZ}" described in the link coordinate

Inverse Dynamics

system. **The link coordinate system is associated with the center of mass of each link. It has the exact orientation of the spatial coordinate system** (e.g., the stationary system related to the manipulator base). When there are no products of inertia, we propose a compact representation (i.e., LM^S) for all robot dynamic parameters in a unique matrix, which contains by columns the information for each link. If there were products of inertia, we include them in the complete inertia tensor "Ψi" of Equation 6.18.

$$LM^S = \begin{bmatrix} CM_{1x}^S & CM_{2x}^S & \cdots & CM_{nx}^S \\ CM_{1y}^S & CM_{2y}^S & \cdots & CM_{ny}^S \\ CM_{1z}^S & CM_{2z}^S & \cdots & CM_{nz}^S \\ I_{1x}^L & I_{2x}^L & \cdots & I_{nx}^L \\ I_{1y}^L & I_{2y}^L & \cdots & I_{ny}^L \\ I_{1z}^L & I_{2z}^L & \cdots & I_{nz}^L \\ m_1 & m_2 & \cdots & m_n \end{bmatrix}$$

- Calculate **the Inertia Matrix of the mechanism "$M_{ij}(\theta)$."** We propose using the Link Jacobian formulation, even though there is another alternative with the new adjoint inertia presented before.

$$M_{ij}(\theta) = \sum_{i=1}^{n} \left[J_{SL}^T(\theta) \right]^T \mathcal{M}_i J_{SL}^T(\theta)$$

$$\mathcal{M}_i = \begin{bmatrix} m_i I_3 & 0 \\ 0 & \Psi_i \end{bmatrix} \quad (6.18)$$

- Calculate **the Coriolis Matrix of the Robot "$C(\theta,\theta')$."**

$$C_{ij}(\theta,\dot\theta) = \sum_{k=1}^{n} \Gamma_{ijk}\dot\theta_k = \frac{1}{2}\sum_{k=1}^{n}\left(\frac{\partial M_{ij}}{\partial \theta_k} + \frac{\partial M_{ik}}{\partial \theta_j} - \frac{\partial M_{kj}}{\partial \theta_i} \right)\dot\theta_k$$

$$\frac{\partial M_{ij}}{\partial \theta_k} = \sum_{l=\max(i,j)}^{n} \left(\left[A_{ki}\xi_i,\xi_k \right]^T A_{lk}^T \mathcal{M}_l' A_{lj} \xi_j + \xi_i^T A_{li}^T \mathcal{M}_l' A_{lk} \left[A_{kj}\xi_j,\xi_k \right] \right)$$

- Define **the Potential Action vector "PoAcc"** (e.g., the gravity action vector).

$$PoAcc = \begin{bmatrix} g_x \\ g_y \\ g_z \end{bmatrix}$$

- Calculate **the Potential Matrix of the Robot "$N(\theta,\theta')$."** We propose to use the new **Gravity Wrench Matrix formulation**, even though there are also other alternatives as presented before in this chapter.

$$N_i(\theta) = -\sum_{i=1}^{n} J_{SL}^{S}(\theta) \mathcal{F}_{gi}^{S}(\theta)$$

- Define **the target with joints' position, velocity, and acceleration,** for the desired task trajectory.

$$\theta = \begin{bmatrix} \theta_1 & \theta_2 & \cdots & \theta_n \end{bmatrix}$$

$$\dot{\theta} = \begin{bmatrix} \dot{\theta}_1 & \dot{\theta}_2 & \cdots & \dot{\theta}_n \end{bmatrix}$$

$$\ddot{\theta} = \begin{bmatrix} \ddot{\theta}_1 & \ddot{\theta}_2 & \cdots & \ddot{\theta}_n \end{bmatrix}$$

- **Solve the classical Lagrange expression for the ID problem** to obtain the generalized forces, which once applied to the joint actuator, will produce the robot mechanism to comply with the dynamics function, irrespective of the manipulator number DoF.

$$M(\theta)\ddot{\theta} + C(\theta,\dot{\theta})\dot{\theta} + N(\theta,\dot{\theta}) = \Gamma$$

Find the dynamic parameters for real robots is a tricky task. For instance, the dynamic model of the commercial robot is not published by many manufacturers because this is key and relevant information for industrial applications. However, this knowledge is indispensable for any simulation and control based on the system dynamic model. Identifying the dynamic parameters for a commercial manipulator is a field of research out of the scope of this book. We assume that the robots' dynamic parameters are already known for our developments and exercises, either because the manufacturer provides them or identified by experimentation.

The definition of the joints' position, velocity, and acceleration has to do with the Trajectory Generation problem, which we will address in more detail in the next chapter. It suffices to understand that any robot task needs to define path planning, trajectory planning, and trajectory control. If the trajectory is in the task space, we need some IK or inverse differential kinematics algorithm to solve the corresponding joint trajectories. In our case, we can count on all geometric screw theory algorithms presented in Chapters 4 and 5. After getting a joint position trajectory, we must produce trajectories for the joints' velocity and acceleration, which must be inside the mechanism's limits. Ideally, the three trajectories (i.e., position, velocity, and acceleration) described over time might be differentiable for continuously stitching together diverse trajectory segments. In the end, we discretize the trajectories in a set of points, and this is a collection of targets. For this chapter's aim, we focus on the ID

Inverse Dynamics

for only one of these targets. We see the complete trajectory dynamics control later with the simulations of Chapter 8.

From this moment on, we can solve the Lagrange classical screw theory ID problems with effectiveness, based on pure algebraic and geometric methods for all types of manipulators and many robotic and mechanical systems (Stokes and Brockett, 1996).

In the next section, we present an example of applying the general non-recursive ID solution to an ABB IRB120 Puma-type manipulator. In further sections, there are similar exercises for other architectures but presented more briefly, as the implementations follow the thorough presentation for this Puma robot.

There are some toolboxes useful to implement next exercises, such as the MATLAB® Robotics Toolbox (RTB) (Corke, 2017). However, it is advisable to review the Screw Theory Toolbox for Robotics "ST24R" (Pardos-Gotor, 2021a) to understand better the concepts working in the examples of the following sections, as because it supports some new functions necessary to run the code of the exercises.

6.2.2 PUMA ROBOTS (E.G., ABB IRB120)

To consolidate the understanding of the formulations presented in the previous section, we carried out an illustrative exercise for a commercial robot with a Puma-type configuration. We illustrate the ID problem in Figure 6.1.

This ID exercise consists of determining the torques to be applied by the robot's motors to make the manipulator follow the desired tool trajectory point. The target expresses the position, velocity, and acceleration (Figure 6.1). The joints' magnitudes are a function of the tool's configuration, and we get them with the solution of the IK problem of the robot for each point of the trajectory. In Chapter 4, we have already seen that we can solve it geometrically with the screw theory POE.

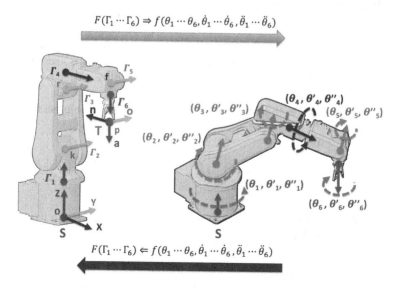

FIGURE 6.1 Puma ABB IRB120 inverse dynamics problem definition.

162   Screw Theory in Robotics

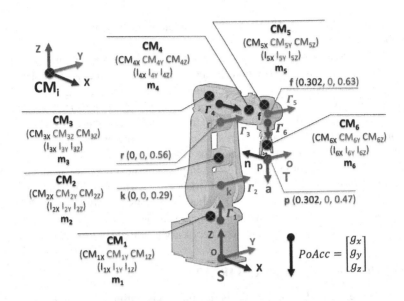

FIGURE 6.2   Puma ABB IRB120 Lagrange inverse dynamics parameters.

The graphical representation of the robot dynamic scheme (e.g., Figure 6.2) has no explicit dynamics parameters. These values are only a good approximation for this exercise but perfectly valid for our didactical purposes. For real applications, it will be necessary to find out the actual values before a commercial implementation.

In this exercise, we solve ID with three different algorithms to check the performance of several formulations presented along with this chapter. It is possible to review all details for the expressions in the exercise code. Hereafter we include only the basic formulations for the completeness of this example.

- Select the **Spatial coordinate system "S"** (usually stationary and the base) and the **Mobile coordinate system "T"** (typically the tool). There is no rule and complete flexibility to make this definition, and we can choose the most convenient frames.
- Define the **axis of each joint**, which are the axes "ω_i" and **a point on those axes**.

$$\omega_1 = \begin{bmatrix} 0 \\ 0 \\ 1 \end{bmatrix} \quad \omega_2 = \begin{bmatrix} 0 \\ 1 \\ 0 \end{bmatrix} \quad \omega_3 = \begin{bmatrix} 0 \\ 1 \\ 0 \end{bmatrix} \quad \omega_4 = \begin{bmatrix} 1 \\ 0 \\ 0 \end{bmatrix} \quad \omega_5 = \begin{bmatrix} 0 \\ 1 \\ 0 \end{bmatrix} \quad \omega_6 = \begin{bmatrix} 0 \\ 0 \\ -1 \end{bmatrix}$$

- Obtain the **Twists "ξ_i"** for the joints, knowing for each revolute joint its axis and a point on that axis.

$$\xi_1 = \begin{bmatrix} -\omega_1 \times o \\ \omega_1 \end{bmatrix} \quad \xi_2 = \begin{bmatrix} -\omega_2 \times k \\ \omega_2 \end{bmatrix} \quad \xi_3 = \begin{bmatrix} -\omega_3 \times r \\ \omega_3 \end{bmatrix}$$

Inverse Dynamics

$$\xi_4 = \begin{bmatrix} -\omega_4 \times f \\ \omega_4 \end{bmatrix} \quad \xi_5 = \begin{bmatrix} -\omega_5 \times f \\ \omega_5 \end{bmatrix} \quad \xi_6 = \begin{bmatrix} -\omega_6 \times f \\ \omega_6 \end{bmatrix}$$

- Get "$H_{ST}(0)$" the pose of the tool at the reference (home) robot position. This configuration of the end-effect happens when all the joint magnitudes are zero. This tool pose comes as a homogeneous matrix, necessary because the POE represents a relative mapping.

$$H_{ST}(0) = T_{xyz}\begin{bmatrix} p_x \\ p_y \\ p_z \end{bmatrix} R_Y(\pi) = \begin{bmatrix} -1 & 0 & 0 & p_x \\ 0 & 1 & 0 & 0 \\ 0 & 0 & -1 & p_z \\ 0 & 0 & 0 & 1 \end{bmatrix}$$

- Express the **kinematics mapping "$H_{ST}(\theta)$," with the product of all joint screw exponentials (POE) and "$H_{ST}(0)$."**

$$H_{ST}(\theta) = e^{\hat{\xi}_1\theta_1} e^{\hat{\xi}_2\theta_2} e^{\hat{\xi}_3\theta_3} e^{\hat{\xi}_4\theta_4} e^{\hat{\xi}_5\theta_5} e^{\hat{\xi}_6\theta_6} H_{ST}(0) = \begin{bmatrix} n & o & a & p \\ 0 & 0 & 0 & 1 \end{bmatrix}$$

- Obtain **the robot dynamics parameters for each link**: the mass "m_i," the center of mass location "CM_{ix}-CM_{iy}-CM_{iz}," and the moments of inertia "I_{ix}-I_{iy}-I_{iz}." **The coordinate frame associated with the center of mass of each link, for simplicity, has the same orientation as the spatial coordinate system** (e.g., the stationary system related to the manipulator base). When there are no products of inertia, we propose a compact representation (i.e., LM^S) for all robot dynamic parameters in a unique matrix, which contains by columns the information for each link.

$$LM^S = \begin{bmatrix} CM^S_{1x} & CM^S_{2x} & CM^S_{3x} & CM^S_{4x} & CM^S_{5x} & CM^S_{6x} \\ CM^S_{1y} & CM^S_{2y} & CM^S_{3y} & CM^S_{4y} & CM^S_{5y} & CM^S_{6y} \\ CM^S_{1z} & CM^S_{2z} & CM^S_{3z} & CM^S_{4z} & CM^S_{5z} & CM^S_{6z} \\ I^L_{1x} & I^L_{2x} & I^L_{3x} & I^L_{4x} & I^L_{5x} & I^L_{6x} \\ I^L_{1y} & I^L_{2y} & I^L_{3y} & I^L_{4y} & I^L_{5y} & I^L_{6y} \\ I^L_{1z} & I^L_{2z} & I^L_{3z} & I^L_{4z} & I^L_{5z} & I^L_{6z} \\ m_1 & m_2 & m_3 & m_4 & m_5 & m_6 \end{bmatrix}$$

- Define **the Potential Action vector "$PoAcc$"** (e.g., the gravity action vector).

$$PoAcc = \begin{bmatrix} g_x \\ g_y \\ g_z \end{bmatrix} = \begin{bmatrix} 0 \\ 0 \\ -9.81 \end{bmatrix}$$

- **Define the Target with joints' position, velocity, and acceleration,** for the desired task trajectory.

$$\theta = \begin{bmatrix} \theta_1 & \theta_2 & \theta_3 & \theta_4 & \theta_5 & \theta_6 \end{bmatrix}$$

$$\dot{\theta} = \begin{bmatrix} \dot{\theta}_1 & \dot{\theta}_2 & \dot{\theta}_3 & \dot{\theta}_4 & \dot{\theta}_5 & \dot{\theta}_6 \end{bmatrix}$$

$$\ddot{\theta} = \begin{bmatrix} \ddot{\theta}_1 & \ddot{\theta}_2 & \ddot{\theta}_3 & \ddot{\theta}_4 & \ddot{\theta}_5 & \ddot{\theta}_6 \end{bmatrix}$$

- **The First Algorithm – Symbolic:** where "M" employs the link Jacobian inertia matrix formulation with Equation 6.5, the Coriolis matrix "C" according to Equation 6.9 and the Potential "N" with the new Gravity Symbolic Matrix expression of Equation 6.9.

$$M_{Jsl}(\theta) = M_{ij}(\theta) = \sum_{i=1}^{6} \left[J_{SL}^T(\theta) \right]^T \mathcal{M}_i J_{SL}^T(\theta)$$

$$C_{ij}(\theta,\dot{\theta}) = \sum_{k=1}^{6} \Gamma_{ijk}\dot{\theta}_k = \frac{1}{2}\sum_{k=1}^{6}\left(\frac{\partial M_{ij}}{\partial \theta_k} + \frac{\partial M_{ik}}{\partial \theta_j} - \frac{\partial M_{kj}}{\partial \theta_i} \right)\dot{\theta}_k$$

$$N_{SYM}(\theta) = N_i(\theta) = -\sum_{xyz}\left(\frac{\partial}{\partial \theta_i} \sum_{i=1}^{6} \begin{bmatrix} m_i g_x & 0 & 0 \\ 0 & m_i g_y & 0 \\ 0 & 0 & m_i g_z \end{bmatrix} CM_i(\theta) \right)$$

- **The Second Algorithm – Twist:** whose formulation is based mainly on the new adjoint transformation, we get "M" with Equation 6.7, the Coriolis matrix "C" with Equation 6.9 and the Potential "N" with the new gravity twist matrix formulation of Equation 6.13.

$$M_{Aij}(\theta) = M_{ij}(\theta) = \sum_{l=\max(i,j)}^{6} \xi_i^T A_{li}^T \mathcal{M}_l' A_{lj} \xi_j$$

$$C_{ij}(\theta,\dot{\theta}) = \sum_{k=1}^{6} \Gamma_{ijk}\dot{\theta}_k = \frac{1}{2}\sum_{k=1}^{6}\left(\frac{\partial M_{ij}}{\partial \theta_k} + \frac{\partial M_{ik}}{\partial \theta_j} - \frac{\partial M_{kj}}{\partial \theta_i} \right)\dot{\theta}_k$$

$$N_\xi(\theta) = N_i(\theta) = -\sum_{l=i}^{6} \xi_i^T A_{li}^T \mathcal{M}_l' A_{l1} (\xi_g g)$$

Inverse Dynamics

- **The Third Algorithm – Wrench** is a complete algebraic solution. We get "*M*" with to the link Jacobian inertia matrix formulation with Equation 6.5, the Coriolis matrix "*C*" according to Equation 6.9, and the Potential "*N*" with the new formulation according to the gravity wrench matrix expression of Equation 6.14.

$$M_{Jsl}(\theta) = M_{ij}(\theta) = \sum_{i=1}^{6} \left[J_{SL}^T(\theta) \right]^T \mathcal{M}_i J_{SL}^T(\theta)$$

$$C_{ij}(\theta,\dot{\theta}) = \sum_{k=1}^{6} \Gamma_{ijk}\dot{\theta}_k = \frac{1}{2}\sum_{k=1}^{6}\left(\frac{\partial M_{ij}}{\partial \theta_k} + \frac{\partial M_{ik}}{\partial \theta_j} - \frac{\partial M_{kj}}{\partial \theta_i}\right)\dot{\theta}_k$$

$$N_{\mathcal{F}}(\theta) = N_i(\theta) = -\sum_{i=1}^{6} J_{SL}^S(\theta)\mathcal{F}_{gi}^S(\theta)$$

- Solve **the classical Lagrange expression for the ID problem** to obtain the generalized forces (i.e., torques in this case), which applied to the actuators will produce the robot mechanism to comply with the dynamics function, reaching the target position, velocity, and acceleration for the joints.

$$M_{ij}(\theta)\begin{bmatrix}\ddot{\theta}_1\\\ddot{\theta}_2\\\ddot{\theta}_3\\\ddot{\theta}_4\\\ddot{\theta}_5\\\ddot{\theta}_6\end{bmatrix} + C_{ij}(\theta,\dot{\theta})\begin{bmatrix}\dot{\theta}_1\\\dot{\theta}_2\\\dot{\theta}_3\\\dot{\theta}_4\\\dot{\theta}_5\\\dot{\theta}_6\end{bmatrix} + N_i(\theta,\dot{\theta}) = \begin{bmatrix}\Gamma_1\\\Gamma_2\\\Gamma_3\\\Gamma_4\\\Gamma_5\\\Gamma_6\end{bmatrix}$$

The code for Exercise 6.2.2 is in the internet hosting for the software of this book[1].

To abound in the knowledge, we present some more examples with commercial manipulators: another example with the same ABB IRB120 with other reference configuration, a Bending Backwards robot with six DoF (i.e., ABB IRB1600), a Gantry type with six DoF (i.e., ABB IRB6620LX), a Scara type robot with four DoF (i.e., ABB IRB920SC), a Collaborative robot with six DoF (i.e., UNIVERSAL UR16e) and a Redundant manipulator with seven DoF (i.e., KUKA IIWA). However, we will show these exercises more briefly, as the details are the same as those presented in this example.

6.2.3 Puma Robots (e.g., ABB IRB120) "Tool-Up"

This ID exercise for the same Puma robot has a different home configuration to illustrate the freedom that screw theory provides to the kinematics analysis (Figure 6.3).

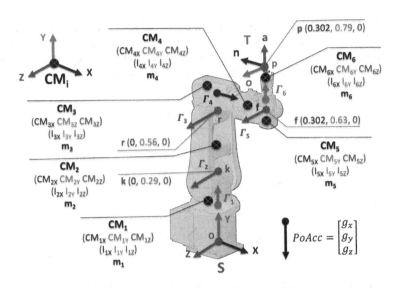

FIGURE 6.3 Puma ABB IRB120 "Tool-Up" Lagrange inverse dynamics parameters.

Here, we define the spatial coordinate system with a different orientation. We choose the robot reference pose with a tool-up orientation. Besides, we do not assume to know the actual sign of rotation for the joints of the commercial robot to build a more hypothetical case.

This example checks the Lagrange ID formulation for a random target in the workspace, implementing three different algorithms to check the performance of several formulations (for more details, review Exercise 6.2.2). The only difference comes from the other kinematics maps as the last twist changes because of the various configurations.

The code for Exercise 6.2.3 is in the internet hosting for the software of this book[2].

6.2.4 Bending Backwards Robots (e.g., ABB IRB1600)

This ID exercise sees a slightly different robot architecture, a Bending Backwards manipulator (Figure 6.4). This design does not need to rotate the robot to reach for things behind it. Swinging the arm backwards extends the working range of the robot.

This example checks the Lagrange ID formulation for a random target in the workspace, implementing three different algorithms to check the performance of several formulations (for more details, review Exercise 6.2.2). The only differences come from the other kinematics map because the twists are proper to each manipulator.

The code for Exercise 6.2.4 is in the internet hosting for the software of this book[3].

6.2.5 Gantry Robots (e.g., ABB IRB6620LX)

This ID exercise deals with this Gantry robotic architecture (see Figure 6.5), which combines the advantages of both a long linear axis for the first joint plus five revolute

Inverse Dynamics

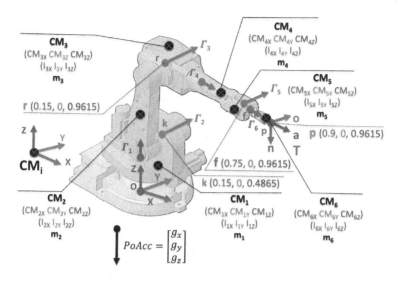

FIGURE 6.4 Bending Backwards ABB IRB1600 Lagrange inverse dynamics parameters.

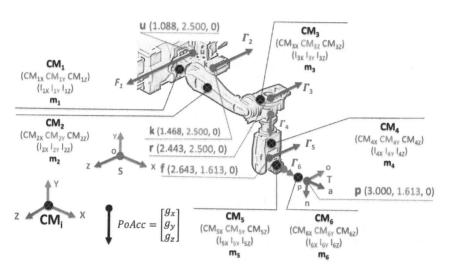

FIGURE 6.5 Gantry ABB IRB6620LX Lagrange inverse dynamics parameters.

joints. We know how the treatment of the prismatic joint is the same with screw theory, and the resulting exercise mathematics is like in previous examples.

This example checks the Lagrange ID formulation for a random target in the workspace, implementing three different algorithms to check the performance of several formulations (for more details, review Exercise 6.2.2). The only differences come from the other kinematics map because the twists are proper to each manipulator.

The code for Exercise 6.2.5 is in the internet hosting for the software of this book[4].

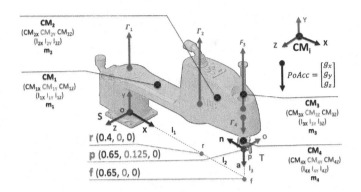

FIGURE 6.6 Scara ABB IRB910SC Lagrange inverse dynamics parameters.

6.2.6 Scara Robots (e.g., ABB IRB910SC)

This ID exercise has a two parallel joint axes mechanism, advantageous for many assembly operations. For the sake of showing the freedom given by the screw theory in the selection of the coordinate systems, we have chosen the typical virtual reality spatial frame orientation (see Figure 6.6). This example has only four joints, but obviously, the general approach applies.

This example checks the Lagrange ID formulation for a random target in the workspace, implementing three different algorithms to check the performance of several formulations (for more details, review Exercise 6.2.2). The differences come from the different number of DoF, as the previous exercises were for a manipulator with six joints, and this one has only four. Nevertheless, the mathematics work equally fine. For any doubt on the kinematics map definition, you can go to any previous chapters (particularly the third) for the twist's definition.

The code for Exercise 6.2.6 is in the internet hosting for the software of this book[5].

6.2.7 Collaborative Robots (e.g., UNIVERSAL UR16e)

This ID exercise deals with a typical collaborative manipulator made for human-robot cooperation in the workspace. The necessary parameters for the mathematics formulation are in Figure 6.7.

This example checks the Lagrange ID formulation for a random target in the workspace, implementing three different algorithms to check the performance of several formulations (for more details, review Exercise 6.2.2). The only differences come from the other kinematics map because the twists are proper to each manipulator.

The code for Exercise 6.2.7 is in the internet hosting for the software of this book[6].

6.2.8 Redundant Robots (e.g., KUKA IIWA)

This ID exercise has the novelty of solving a robot with seven DoF. We can check the excellent performance of the formulations because they work for any number of joints. The dynamics parameters are in Figure 6.8.

Inverse Dynamics 169

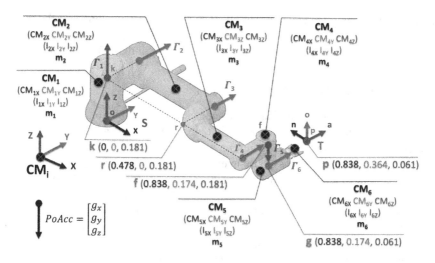

FIGURE 6.7 Collaborative UNIVERSAL UR16e Lagrange inverse dynamics parameters.

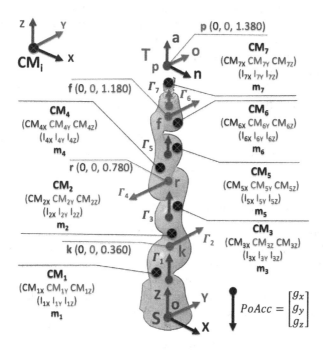

FIGURE 6.8 Redundant KUKA IIWA Lagrange inverse dynamics parameters.

This example checks the Lagrange ID formulation for a random target in the workspace, implementing three different algorithms to check the performance of several formulations (for more details, review Exercise 6.2.2). The differences come from the different number of DoF, as this robot has seven joints. Nonetheless, the mathematical approach works equally fine. For any doubt on the kinematics map definition, you can go to any previous chapters (particularly the third) for the twist's definition.

The code for Exercise 6.2.8 is in the internet hosting for the software of this book[7].

6.3 ROBOT DYNAMICS CONTROL

Robots need a control system that enables the automatic regulation of the actuators' forces, which generate the joints' movement, necessary for the mechanism to follow a reference path (Craig, 2004). Therefore, we need to design a robust control law to correct the forces applied in response to errors from the planned trajectory.

There are many techniques to design stable controls for robotics and some of them are geometric (Jurdjevic, 1997). Nevertheless, the stability of some algorithms requires that the equations of motion comply with the passivity property (i.e., as presented before in this text). We must adjust control for each robot, so the laws given in this section start to design a controller for every manipulator.

There are two basic ways to solve the control problem. First, we will refer to **"Control in the Joint Space,"** which turns a given task into a desired path for the robot's joints so that the control law defines the necessary forces or torques for the joints to follow the trajectories. Second, we define **"Control in the Task Space,"** which converts the dynamics and control in terms of the workspace to regulate the robot tool's positions and orientations.

A well-known control procedure is that of feedback linearization, which produces linear and decoupled joint space dynamics. However, for applications such as tool control, it is also helpful to develop the same type of feedback linearization but in the workspace.

If the tool is in physical contact with the environment, the dynamics problem is more complicated, and we must control the position, and the force exerted. The introduction of any restrictions complicates the model.

6.3.1 ROBOTICS CONTROL IN THE JOINT SPACE

The description of this control problem begins with the formulation presented in this chapter for the classical screw theory ID (Equation 6.4). The inputs come from the desired targets for the joints' generalized coordinates (i.e., the position, velocity, and acceleration). The targets for the joints are a function of the end-effector configuration, and we get them by solving the IK problem for each point of the task trajectory. In the fourth chapter of this book, we have already seen that we can solve it in a geometric, effective, and efficient way with the screw theory POE and the canonical subproblems (Paden, 1986).

Once we get the joints' position trajectory, we can define their velocity and acceleration with multiple trajectory generation techniques. We will see some of them in the next chapter. The three joint trajectories (i.e., position, velocity, and acceleration) described over time might be differentiable. Besides, we discretize the trajectories along with a collection of targets. Then, we define the set of torques or the robot joints that make the mechanism comply with a trajectory target with Equation 6.19.

$$M(\theta_d)\ddot{\theta}_d + C(\theta_d,\dot{\theta}_d)\dot{\theta}_d + N(\theta_d,\dot{\theta}_d) = \Gamma \qquad (6.19)$$

Inverse Dynamics

We need a control law to correct many kinds of errors in the following of an actual trajectory. It is necessary a solution like the "**Computed Torque Control Law**," which adds state feedback to Equation 6.19, to obtain the control Equation 6.20, with the inclusion of two terms to compensate the velocity and position errors. It is compulsory to adjust a constant gain matrix "K_v" for the velocity error (Equation 6.21). It is the same for the gain matrix "K_p" for the position error (Equation 6.22). This computed torque control is an example of feedback linearization, allowing many tools to synthesize linear controls. There are many other types of regulators (e.g., PD, PDI), but the computed torque control has shown high throughput experimental results in many developments. However, it can be a computationally expensive technique.

$$M(\theta)\left(\ddot{\theta}_{ref} - \dot{\theta}_{err} - \theta_{err}\right) + C(\theta,\dot{\theta})\dot{\theta} + N(\theta,\dot{\theta}) = \Gamma \qquad (6.20)$$

$$\dot{\theta}_{err} = K_v\left(\dot{\theta} - \dot{\theta}_{ref}\right) \qquad (6.21)$$

$$\theta_{err} = K_p\left(\theta - \theta_{ref}\right) \qquad (6.22)$$

We will see some examples with computed torque control law in Chapter 8 devoted to Simulations.

6.3.2 Robotics Control in the Task Space

There are several difficulties for joint space control, such as excessive computation time to solve the IK, and the task operates in terms of the tool path. For many applications, we are interested in defining the dynamic equations of motion as a function of the robot's workspace as Equation 6.23, where the valuable parameters of the system are given by Equation 6.24, which is a function of the dynamic parameters (i.e., "M," "C" and "N") and the geometric spatial Jacobian.

$$\tilde{M}(\theta_d)\ddot{x}_d + \tilde{C}(\theta_d,\dot{\theta}_d)\dot{x}_d + \tilde{N}(\theta_d,\dot{\theta}_d) = \tilde{\Gamma} \qquad (6.23)$$

$$\tilde{M} = J_{ST}^{-T} M J_{ST}^{-1};\ \tilde{C} = J_{ST}^{-T}\left[CJ_{ST}^{-1} + M\frac{d}{dt}(J_{ST}^{-1})\right];\ \tilde{N} = J_{ST}^{-T} N;\ \tilde{\Gamma} = J_{ST}^{-T}\Gamma \qquad (6.24)$$

Once we define the system, we can also apply a control law to correct disturbances and errors to obtain a control (Equation 6.25). The advantage of this formulation is that the choice of the gain matrices to correct the task is much more comfortable. It is unnecessary to solve the IK at each step, but we need to solve the Jacobian at any target of the trajectory.

$$\tilde{M}(\theta)\left(\ddot{x}_{ref} - \dot{x}_{err} - x_{err}\right) + \tilde{C}(\theta,\dot{\theta})\dot{x} + \tilde{N}(\theta,\dot{\theta}) = \tilde{\Gamma} \qquad (6.25)$$

$$\dot{x}_{err} = K_v\left(\dot{x} - \dot{x}_{ref}\right)$$

$$x_{err} = K_p \left(x - x_{ref} \right)$$

However, the scope of this book is in the screw theory geometric algorithms for kinematics and dynamics. So, we do not contemplate advancing further in the aspects of control (Ploen, 1997), left for a possible later edition. Nonetheless, in the simulations of Chapter 8, there are many control examples where we implement several control laws.

6.4 SPATIAL VECTOR ALGEBRA

Rigid body dynamics is a classical subject but still an open area of research because of the high-dimensional and intricate robotics mechanical structures. The motion equations' nonlinear nature and the physical constraints make the robot dynamics a challenging computational problem.

Dealing with dynamics is troublesome when using 3D vectors, and therefore it is better to use the notation of 6D vectors instead. This approach combines into a unified set of quantities the linear and angular aspects of motion. The benefit is around a six-fold reduction in the volume of algebra. Featherstone showed how to formulate rigid multibody systems dynamics using and extending constructs from classical screw theory with Plücker coordinates (Featherstone, 2016).

The **Spatial Velocity** of a rigid body motion (Figure 6.9a) is a 6D vector that integrates the linear "v" and angular "ω" velocities. The basis vectors for Plücker coordinates (Figure 6.9b) are defined by Equation 6.26. Do not confuse this concept with the "Twist Velocity" of the classical screw theory used previously.

$$\hat{v} = \omega_x d_{ox} + \omega_y d_{oy} + \omega_z d_{oz} + v_x d_x + v_y d_y + v_z d_z \tag{6.26}$$

The **Spatial Force** of a rigid body motion (Figure 6.10a) is a 6D vector that integrates the linear "f" and angular "n" forces. The basis vectors for Plücker coordinates (Figure 6.10b) are defined by Equation 6.27. Do not confuse this concept with the "Wrench" of the classical screw theory, even though the resemblance exists.

$$\hat{f} = n_x e_x + n_y e_y + n_z e_z + f_x e_{ox} + f_y e_{oy} + f_z e_{oz} \tag{6.27}$$

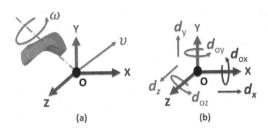

FIGURE 6.9 (a) Rigid body motion diagram. (b) Rigid body Plücker motion coordinates.

Inverse Dynamics

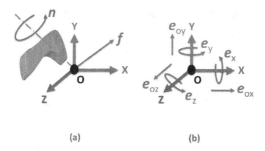

FIGURE 6.10 (a) Rigid body force diagram. (b) Rigid body Plücker force coordinates.

6.4.1 Coordinate Transforms

The formulations of the fundamental Plücker transformations are 6×6 matrices (Figure 6.11). Nevertheless, it is necessary to pay attention to the convention used for these transformations, which change from one text to another. For instance, Featherstone employs a notation of the Plücker transform from a Cartesian coordinate A to another B, different from the classical approach, which we apply throughout this book. If we contrast the matrices, we can realize that Featherstone's convention describes the coordinate transforms associated with explaining how it would have to move frame B to coincide with frame A. Therefore, this formulation translates motion magnitudes expressed in frame A to the B system, which is just the opposite approach to classical, used in this book.

Function for Basic ROTATION on X axis	Function for Basic ROTATION on Y axis
$X_X(\alpha) = \begin{bmatrix} Rot_X(\alpha) & 0 \\ 0 & Rot_X(\alpha) \end{bmatrix}$	$X_Y(\beta) = \begin{bmatrix} Rot_Y(\beta) & 0 \\ 0 & Rot_Y(\beta) \end{bmatrix}$
$Rot_X(\alpha) = \begin{bmatrix} 1 & 0 & 0 \\ 0 & \cos\alpha & -\sin\alpha \\ 0 & \sin\alpha & \cos\alpha \end{bmatrix}$	$Rot_Y(\beta) = \begin{bmatrix} \cos\beta & 0 & \sin\beta \\ 0 & 1 & 0 \\ -\sin\beta & 0 & \cos\beta \end{bmatrix}$
Function for Basic ROTATION on Z axis	**Function for Basic DISPLACEMENT XYZ**
$X_Z(\gamma) = \begin{bmatrix} Rot_Z(\gamma) & 0 \\ 0 & Rot_Z(\gamma) \end{bmatrix}$	$X_{XYZ}(p) = \begin{bmatrix} 0 & 1 \\ \hat{p} & 0 \end{bmatrix}$
$Rot_Z(\gamma) = \begin{bmatrix} \cos\gamma & -\sin\gamma & 0 \\ \sin\gamma & \cos\gamma & 0 \\ 0 & 0 & 1 \end{bmatrix}$	$\hat{p} = \begin{bmatrix} 0 & -p_Z & p_Y \\ p_Z & 0 & -p_X \\ -p_Y & p_X & 0 \end{bmatrix}$

FIGURE 6.11 Four basic Plücker transformations for rotations X-Y-Z and translation.

6.4.2 Mechanics of a Constrained Rigid Body System

Featherstone uses the modified Denavit–Hartenberg (DH) parameters to describe the kinematics for modeling the rigid body chains. Then, we transform the coordinate frames associated with the links with the Plücker formulations, and the joint models are those of the spatial vector algebra. Differently, a nice novelty of this book is the use of the screw theory POE to describe the kinematics system model in union with the spatial vector algebra.

For the mechanics' treatment, the Lagrange approach has been used in the previous section with the classical screw theory, coming out with non-recursive closed-form solutions for the ID problem. Nonetheless, the modern dynamics algorithms which rely on the spatial vector algebra terminology have superior efficiency.

6.5 THE NEWTON–EULER EQUATIONS

For a manipulator, the Newton–Euler equations describe the dynamics for translation and rotation of the whole rigid body chain.

Many robot dynamics algorithms use methodologies from the screw theory, differential geometry, and Lie groups. These methods have many computational advantages, and some of them derive from the use of six-dimensional vectors to express motion and force magnitudes (Park, 1994). As a somehow extension of the screw theory, the remarkable Recursive Newton–Euler Algorithm (RNEA) for ID proposed by Featherstone uses the spatial vector algebra. The computational complexity of RNEA is $O(n)$, whereas the complexity for non-recursive algorithms, typically based on the Lagrange approach, generally is $O(n^4)$, "n" being the number of DoF (Sipser, 2021).

We propose a screw theory innovation to solve the ID problem jointly with Featherstone's achievement of the RNEA. We use the POE formulation to analyze the robot kinematics. The POE is much easier to define and understand compared with using the DH convention. Therefore, the resulting motion transformations from one link to another, necessary for the recursive implementation, will result in a more understandable and straightforward process.

6.5.1 General Recursive Solution to Inverse Dynamics RNEA with POE

The ID formalized with the RNEA relates the robot position, velocity, and acceleration to the generalized forces of the joints. We develop this algorithm to solve the ID problem for a rigid multibody chain mechanism with a two-pass method. It makes an outward pass through the rigid body tree from base to the tool for calculating the link velocities, accelerations, and forces and then, the inward pass from the tool to the base, during which we obtain the joint forces. We can calculate the rigid body (i.e., link) forces in any of the passes, but we usually choose the first. The RNEA finds the joint forces or torques "τ" for the input target given by the joint positions, velocities, and accelerations. The implementation of this RNEA with the POE applied to a general kinematics tree follows these steps (see the details of Figure 6.12 to follow the complete formulation).

Inverse Dynamics

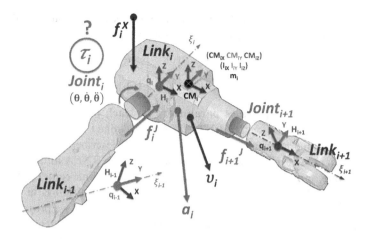

FIGURE 6.12 The RNEA link dynamics with the POE kinematics.

- Select **the Spatial coordinate system "S"** (usually stationary and at the base) and the **Mobile coordinate system "T"** (typically the tool). There is no rule and complete flexibility to make this definition, and we can choose the most convenient frames according to the application. Define the spatial coordinate system "S" with a homogeneous matrix "$H_0(0)$," according to the free selection for its orientation matrix "R_0" and position "q_0."

$$H_0(0) = \begin{bmatrix} R_0 & q_0 \\ 0 & 1 \end{bmatrix}_{4 \times 4}$$

- Define **the axis of each joint**, which are the axes "ω_i" for revolute joints and "v_i" for the prismatic joints.
- Define **the spatial vector motion Subspace "S_i" for each joint**, based on the twist axis. For revolute joints the axis "ω_i" defines the half superior of the subspace vector, and the half inferior is zero. For the prismatic joints, the axis "v_i" defines the half inferior of the subspace vector and half superior is zero.

$$S_i = \begin{bmatrix} \omega_i \\ 0 \end{bmatrix}_{6 \times 1} \quad S_i = \begin{bmatrix} 0 \\ v_i \end{bmatrix}_{6 \times 1}$$

- Choose **the Points "q_i" on the axis of each joint**, defined in the spatial coordinate system. Whatever point on the axis works but we must be very careful to get a different point for each joint. We know it is possible to use the same point for several twists with crossing axes. For the RNEA we need to distinguish them, to position a different coordinate system on each link. **These points define the position for the link coordinate system whose orientation, to make it simple, will be the same as the spatial coordinate system** (e.g., the stationary system associated with the manipulator base).

- Obtain **the Twists "ξ_i" for the joints,** knowing for each revolute joint its axis "ω_i" and a point on that axis "q_i," and for each prismatic joint its axis "v_i."

$$\xi_i = \begin{bmatrix} -\omega_i \times q_i \\ \omega_i \end{bmatrix} \quad \xi_i = \begin{bmatrix} v_i \\ 0 \end{bmatrix}$$

- Obtain **the robot dynamics parameters for each link**: the mass "m_i," the center of mass location "***CMiX-CMiY-CM***$_{iZ}$" defined in the link coordinate system, and the moments of inertia "***IiX-IiY-I***$_{iz}$" described in the link coordinate system. **There is a crucial difference between RNEA and the classical Lagrange approach. For the latter, we defined all link centers of mass to the spatial frame, whereas for this RNEA, the definition is in each link coordinate frame.** When there are no products of inertia, we propose a compact representation (i.e., LM^L) for all robot dynamic parameters in a unique matrix, which contains by columns the information for each link. If there were products of inertia, we include them in the complete inertia tensor "Ψ_i" of Equation 6.31. Usually, we define the inertia tensor at the link center of mass. If that is the case, it is essential to realize **we must translate the moments of inertia or the complete inertia tensor to the link coordinate system, which has its position on the point of the twist,** which does not coincide in general with the center of mass.

$$LM^L = \begin{bmatrix} CM^L_{1x} & CM^L_{2x} & \cdots & CM^L_{nx} \\ CM^L_{1y} & CM^L_{2y} & \cdots & CM^L_{ny} \\ CM^L_{1z} & CM^L_{2z} & \cdots & CM^L_{nz} \\ I^L_{1x} & I^L_{2x} & \cdots & I^L_{nx} \\ I^L_{1y} & I^L_{2y} & \cdots & I^L_{ny} \\ I^L_{1z} & I^L_{2z} & \cdots & I^L_{nz} \\ m_1 & m_2 & \cdots & m_n \end{bmatrix}$$

- Define **the Potential Action vector "*PoAcc*"** (e.g., the gravity action vector).

$$PoAcc = \begin{bmatrix} g_x \\ g_y \\ g_z \end{bmatrix}$$

- Define **the Target with joints' position, velocity, and acceleration,** for the desired task trajectory.

$$\theta = \begin{bmatrix} \theta_1 & \theta_2 & \cdots & \theta_n \end{bmatrix}$$
$$\dot{\theta} = \begin{bmatrix} \dot{\theta}_1 & \dot{\theta}_2 & \cdots & \dot{\theta}_n \end{bmatrix}$$
$$\ddot{\theta} = \begin{bmatrix} \ddot{\theta}_1 & \ddot{\theta}_2 & \cdots & \ddot{\theta}_n \end{bmatrix}$$

Inverse Dynamics

- Define **each link coordinate system (outward recursive Pass "$i=1$ to n")** "$H_i(\theta)$" with the product of the joint exponentials "$POE_i(\theta)$" and the home configuration of the system "$H_i(0)$" (Equation 6.28). The exponential for the spatial system (e.g., base of the robot) is the identity matrix. Each link coordinate system has its position on the related twist point "q_i" and inherits the home configuration "$H_i(0)$" orientation from the spatial coordinate system.

$$H_i(\theta) = POE_i(\theta) H_i(0) = \begin{bmatrix} R_i & p_i \\ 0 & 1 \end{bmatrix}_{4x4} \quad (6.28)$$

$$POE_i(\theta) = POE_{i-1}(\theta) e^{\hat{\xi}_i \theta_i} \ ; \ POE_0 = I_4$$

$$H_i(0) = \begin{bmatrix} I_3 & q_i - q_{i-1} \\ 0 & 1 \end{bmatrix}_{4x4} H_{i-1}(0)$$

- Obtain **each link Plücker transformation (Outward recursive Pass "$i=1$ to n")** "X_i" converting the link coordinate system calculated before with the Equation 6.28.

$$X_i(\theta) = \begin{bmatrix} R_i & 0 \\ p_i \times R_i & R_i \end{bmatrix}_{6x6}$$

- Calculate **each Link Velocity (Outward recursive pass "$i=1$ to n")** "v_i" for the whole rigid multibody (see Equation 6.29). The velocity of the stationary base is zero. The transformation in Plücker coordinates "X_i^{i-1}" is the relative from one link to the previous one, and we must obtain it with the proper composition of the fundamental transformations from "X_i" to "X_{i-1}."

$$v_i = X_i^{i-1} v_{i-1} + S_i \dot{\theta}_i \quad (6.29)$$

$$v_0 = 0$$

- Calculate **each Link Acceleration (Outward recursive pass "$i=1$ to n")** "a_i" for the whole rigid multibody (Equation 6.30). The operator "$\times$" is for the cross product of the spatial vector of motion (Featherstone, 2016). The potential action vector is, in many cases, only the gravity action vector. We model the gravitational field as a fictitious acceleration of the base. To do this, we replace the initial acceleration value with the inverse of gravity. With this trick, the values for the link's accelerations and forces are not actual values, as they include an offset by gravity acceleration and force. Nonetheless, the result of the ID is correct, as we are interested only in the result of the joint torques to comply with the motion.

$$a_i = X_i^{i-1} a_{i-1} + S_i \ddot{\theta}_i + v_i \times S_i \dot{\theta}_i \quad (6.30)$$

$$a_0 = \begin{bmatrix} 0 \\ -PoAcc \end{bmatrix}_{6 \times 1}$$

- Calculate **each total Link Force (Outward recursive pass "$i=1$ to n")** "f_i^L" required to produce the rigid-body acceleration. We obtain the net force acting on each link by Equation 6.32, knowing the velocity and acceleration of this body. We define the link inertia tensor "I_i" by Equation 6.31 in Plücker coordinates, with the mass of the link "m_i" and the rotational inertia "Ψ_i." We define all this information in the link coordinate system. The operator "$\times^*$" is for the cross product of the spatial vector of force (Featherstone, 2016).

$$I_i = \begin{bmatrix} \Psi_i & 0 \\ 0 & m_i I_3 \end{bmatrix} \tag{6.31}$$

$$f_i^L = I_i a_i + v_i \times^* I_i v_i \tag{6.32}$$

- Calculate **each Joint Transmitted Force (Inward recursive pass "$i=n$ to 1")** "f_i^J" between the links (Equation 6.34). The result "f_i^J" is the force transmitted from the predecessor link across the joint, and "f_{i+1}^J" the force transmitted to the successor link. There is no transferred force to a successor for any last link in an open rigid body chain. This "f_{n+1}^J" (Equation 6.33) is null, and then the transformation "$X^n{}_{n+1}$" does not affect, but many times is helpful to define it as corresponding to the robot tool coordinate system. This definition would permit the formulation of Equation 6.34 to work for any "n" value. Besides, we assume to know all external forces (including gravity) acting on the link "f_i^X." The transformation in Plücker coordinates "X_a^b" is the relative from one link to another and must be obtained with the proper composition of the fundamental transformations from "X_a" and "X_b."

$$f_{n+1}^J = 0 \tag{6.33}$$

$$f_i^J = f_i^L - \left(X_0^i\right)^T f_i^X + \left(X_{i+1}^i\right)^T f_{i+1}^J \tag{6.34}$$

- Calculate **each Joint Generalize Force (Inward recursive pass "$i=n$ to 1")** "τ_i" with the motion subspace of each joint by Equation 6.35.

$$\tau_i = S_i^T f_i^J \tag{6.35}$$

We have already experimented with the advantages of dealing with robot kinematics using the screw theory tools. Among them, the liberty to define the coordinate systems of interest is also included. Commonly, we describe only the frames for the spatial and tool systems. Conversely, any DH approach needs to define a coordinate system associated with each link (Denavit and Hartenberg, 1955). In this case, the

RNEA for ID demands defining a coordinate system on each link, so the DH algorithm might be the obvious choice. However, there are different ways to solve a link coordinate systems' recursive definition with screw theory. We propose one easy and transparent with the utilization of the POE. The link coordinate system has the same orientation as the spatial coordinate system, and the position coincides with the point of each joint twist. This definition is convenient and straightforward. Following the screw theory, it is even possible to define the link frames in points of interest, even outside the link (see detail in Figure 6.12).

We perform the computation of each link in its body coordinates. Therefore, all the quantities refer to the body frame. Consequently, the spatial vector motion transformation in Plücker coordinates "X" must be inserted to pass the velocity and acceleration from the predecessor to the successor link in the analysis. Similarly, we introduce the spatial vector force transformation to transport the joint force from the successor to the predecessor link. The Plücker transformation for forces works dually to the transformation for motion and velocities. Therefore, it is entirely feasible to manage them indistinctly, as far as we follow the logic of the Plücker definition of transformations. So, inverting and transposing the motion transformations can be used for the forces transformations as well.

This algorithm approach replaces the initial zero acceleration of the robot base for a fictitious value to balance gravity, which is an external force to the link. This trick is perfectly acceptable when the only result of interest is the joint forces.

Finding the dynamic parameters for real robots is quite tricky. Many manufacturers do not publish the commercial robot's dynamic model because this is key and relevant for industrial applications. However, this knowledge is indispensable for any simulation and control based on the system dynamic model. Identifying the minimal set of base dynamic parameters for a model is a field of research out of the scope of this book. We assume that the robots' dynamic parameters are already known for our developments and exercises in this chapter, either because the manufacturer provides them or identified by experimentation. Consequently, in the next exercises, the values for the dynamic parameters are only approximations, perfectly valid for our theory didactical purposes, but not suitable for industrial implementations.

The RNEA determines the torques to be applied by the robot's actuators to make the manipulator complying with the designed target. The target includes the information for the joint's position, velocity, and acceleration. Frequently, the joints' magnitudes are a function of the tool's pose and obtained by solving the robot's IK problem for each point of the tool trajectory. Chapter 4 showed how to solve the IK in a geometric, effective, and efficient way with the screw theory POE and canonical subproblems. Chapter 5 presented another alternative for getting the joint trajectories with the inverse differential kinematics approach. Chapter 7 covers the trajectory generation problem in detail, and we will learn how to discretize a trajectory in a set of points, which is a collection of targets. For this chapter's exercises, we focus on the ID for only one of these targets.

This RNEA is much more efficient, as we know it has a computational complexity of $O(n)$. In contrast, the classical non-recursive Lagrange alternative has a complexity of $O(n^4)$, "n" being the number of DoF. Therefore, the RNEA results more conveniently for real-time ID control applications. It will be possible to contrast the efficiency of both approaches in the ulterior Chapter 8 with the corresponding simulations. However, the effectiveness of both methods is the same, as they get the exact closed-form solution for the set of joint forces.

We could remark the interest of the novelty introduced in this book for the RNEA, which combines the screw theory POE and the Spatial Vector Algebra. It substitutes with many advantages the DH approach for the kinematics analysis of the rigid multibody tree, which is the convention for the standard RNEA implementation for robotics. However, the POE has many benefits, as we showed along with this text, and as we will apprehend more comprehensibly with the exercises of this chapter. The RNEA with POE solves ID problems very efficiently with pure algebraic and geometric methods for almost any robotic and mechanical system.

In the next section, we present an example of applying the general recursive solution to ID to an ABB IRB120 manipulator. The rest of the sections in this chapter add the same exercise for other architectures but presented more briefly, as the implementation follows the thorough presentation for the Puma robot.

It is advisable to review the Screw Theory Toolbox for Robotics "ST24R" (Pardos-Gotor, 2021a) to understand better the concepts working in the examples of the following sections.

6.5.2 PUMA ROBOTS (E.G., ABB IRB120)

To strengthen the new formulations introduced, we present an illustrative exercise for a commercial robot with a Puma configuration (Figure 6.13). The ID formalized with the RNEA relates the robot target expressed in the joint space in terms of position, velocity, and acceleration to the generalized forces, which we must apply to the joints controllers to comply with the defined motion. The joints' magnitudes are a function of the tool's configuration, and we obtain them by solving the IK problem of the robot for each point of the trajectory.

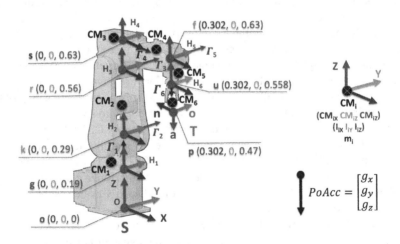

FIGURE 6.13 Puma ABB IRB120 RNEA with POE inverse dynamics parameters.

Inverse Dynamics

The graphical representation of the robot dynamics schematic (e.g., Figure 6.13) has no explicit parameters but represents all the necessary kinematics parameters for developing the algorithm. The dynamics values are only a good approximation for this exercise but perfectly valid for our didactical purposes.

In this exercise, we apply the new ID RNEA with POE general solution. Hereafter we include only the information for the completeness of this example (see Figure 6.13 to follow the formulation).

- Select **the Spatial coordinate system "S"** (at the base) and the **Mobile coordinate system "T"** (the tool). There is no rule and complete flexibility to make this definition, and we can choose the most convenient frames for the application. Define the spatial coordinate system "S" with a homogeneous matrix "$H_0(0)$," according to the free selection for its orientation matrix "R_0" and position vector "q_0."

$$H_0(0) = \begin{bmatrix} R_0 & q_0 \\ 0 & 1 \end{bmatrix} = \begin{bmatrix} 1 & 0 & 0 & 0 \\ 0 & 1 & 0 & 0 \\ 0 & 0 & 1 & 0 \\ 0 & 0 & 0 & 1 \end{bmatrix}$$

- Define **the axis of each joint**, which are the axes "ω_i" for revolute joints.

$$\omega_1 = \begin{bmatrix} 0 \\ 0 \\ 1 \end{bmatrix} \quad \omega_2 = \begin{bmatrix} 0 \\ 1 \\ 0 \end{bmatrix} \quad \omega_3 = \begin{bmatrix} 0 \\ 1 \\ 0 \end{bmatrix} \quad \omega_4 = \begin{bmatrix} 1 \\ 0 \\ 0 \end{bmatrix} \quad \omega_5 = \begin{bmatrix} 0 \\ 1 \\ 0 \end{bmatrix} \quad \omega_6 = \begin{bmatrix} 0 \\ 0 \\ -1 \end{bmatrix}$$

- Define **the spatial vector motion Subspace "S_i" for each joint**, which considers the twist axis. For revolute joints, the axis "ω_i" defines the half superior of the subspace vector, and the half inferior is zero.

$$S_1 = \begin{bmatrix} \omega_1 \\ 0 \end{bmatrix} \quad S_2 = \begin{bmatrix} \omega_2 \\ 0 \end{bmatrix} \quad S_3 = \begin{bmatrix} \omega_3 \\ 0 \end{bmatrix} \quad S_4 = \begin{bmatrix} \omega_4 \\ 0 \end{bmatrix} \quad S_5 = \begin{bmatrix} \omega_5 \\ 0 \end{bmatrix} \quad S_6 = \begin{bmatrix} \omega_6 \\ 0 \end{bmatrix}$$

- Choose **the Points "q_i" on the axis of each joint**, defined in the spatial coordinate system. Whatever point on the axis works but we must be very careful to get a different point for each joint. We know it is possible to use the same point for several twists with crossing axes. For the RNEA we need to distinguish them, to position a different coordinate system on each link. **These points define the position for the link coordinate system whose orientation, to make it simple, will be the same as the spatial coordinate system** (e.g., the stationary system associated with the manipulator base).
- Obtain **the Twists "ξ_i" for the joints**, knowing for each revolute joint its axis "ω_i" and a point on that axis "q_i."

$$\xi_1 = \begin{bmatrix} -\omega_1 \times g \\ \omega_1 \end{bmatrix} \quad \xi_2 = \begin{bmatrix} -\omega_2 \times k \\ \omega_2 \end{bmatrix} \quad \xi_3 = \begin{bmatrix} -\omega_3 \times r \\ \omega_3 \end{bmatrix}$$

$$\xi_4 = \begin{bmatrix} -\omega_4 \times s \\ \omega_4 \end{bmatrix} \quad \xi_5 = \begin{bmatrix} -\omega_5 \times f \\ \omega_5 \end{bmatrix} \quad \xi_6 = \begin{bmatrix} -\omega_6 \times u \\ \omega_6 \end{bmatrix}$$

- Obtain **the robot dynamics parameters for each link**: the mass "m_i," the center of mass location "$CMiX\text{-}CMiY\text{-}CM_{iZ}$" defined in the link coordinate system, and the moments of inertia "$IiX\text{-}IiY\text{-}I_{iZ}$" described in the link coordinate system. **There is a crucial difference between RNEA and the classical Lagrange approach. For the latter, we defined all link centers of mass to the spatial frame, whereas for this RNEA, the definition is in each link coordinate frame.** When there are no products of inertia, we propose a compact representation (i.e., LM^L) for all robot dynamic parameters in a unique matrix, which contains by columns the information for each link. If there were products of inertia, we include them in the complete inertia tensor "Ψ_i." Usually, we define the inertia tensor at the link center of mass. If that is the case, it is essential to realize **we must translate the moments of inertia or the complete inertia tensor to the link coordinate system, which has its position on the point of the twist,** which does not coincide in general with the center of mass.

$$LM^L = \begin{bmatrix} CM^L_{1x} & CM^L_{2x} & CM^L_{3x} & CM^L_{4x} & CM^L_{5x} & CM^L_{6x} \\ CM^L_{1y} & CM^L_{2y} & CM^L_{3y} & CM^L_{4y} & CM^L_{5y} & CM^L_{6y} \\ CM^L_{1z} & CM^L_{2z} & CM^L_{3z} & CM^L_{4z} & CM^L_{5z} & CM^L_{6z} \\ I^L_{1x} & I^L_{2x} & I^L_{3x} & I^L_{4x} & I^L_{5x} & I^L_{6x} \\ I^L_{1y} & I^L_{2y} & I^L_{3y} & I^L_{4y} & I^L_{5y} & I^L_{6y} \\ I^L_{1z} & I^L_{2z} & I^L_{3z} & I^L_{4z} & I^L_{5z} & I^L_{6z} \\ m_1 & m_2 & m_3 & m_4 & m_5 & m_6 \end{bmatrix}$$

- Define **the Potential Action vector "$PoAcc$,"** which for this exercise is only the gravity "a_g."

$$PoAcc = \begin{bmatrix} 0 \\ 0 \\ -a_g \end{bmatrix}$$

- Define **the Target with joints' position, velocity, and acceleration,** for the desired task trajectory.

$$\theta = \begin{bmatrix} \theta_1 & \theta_2 & \theta_3 & \theta_4 & \theta_5 & \theta_6 \end{bmatrix}$$

$$\dot{\theta} = \begin{bmatrix} \dot{\theta}_1 & \dot{\theta}_2 & \dot{\theta}_3 & \dot{\theta}_4 & \dot{\theta}_5 & \dot{\theta}_6 \end{bmatrix}$$

$$\ddot{\theta} = \begin{bmatrix} \ddot{\theta}_1 & \ddot{\theta}_2 & \ddot{\theta}_3 & \ddot{\theta}_4 & \ddot{\theta}_5 & \ddot{\theta}_6 \end{bmatrix}$$

Inverse Dynamics

- Define **each link coordinate system (Outward recursive Pass "$i=1$ to n")** "$H_i(\theta)$" with the product of the joint exponentials "$POE_i(\theta)$" and the home configuration of the system "$H_i(0)$". The exponential for the spatial system (e.g., base of the robot) is the identity matrix. Each link coordinate system has its position on the related twist point "q_i" and inherits the home configuration "$H_i(0)$" orientation from the spatial coordinate system.

$$H_i(\theta) = POE_i(\theta) H_i(0) = \begin{bmatrix} R_i & p_i \\ 0 & 1 \end{bmatrix}_{4 \times 4}$$

$$POE_i(\theta) = POE_{i-1}(\theta) e^{\hat{\xi}_i \theta_i} \; ; \; POE_0 = I_4$$

$$H_i(0) = \begin{bmatrix} I_3 & q_i - q_{i-1} \\ 0 & 1 \end{bmatrix}_{4 \times 4} H_{i-1}(0)$$

- Obtain **each link Plücker transformation (Outward recursive Pass "$i=1$ to n")** "X_i" converting the link coordinate system calculated before with the POE.

$$X_i(\theta) = \begin{bmatrix} R_i & 0 \\ p_i \times R_i & R_i \end{bmatrix}_{6 \times 6}$$

- Calculate **each Link Velocity (Outward recursive pass "$i=1$ to n")** "v_i" for the whole robot body. The velocity of the stationary base is zero. The Plücker transformation "X_i^{i-1}" is the relative from one link to the previous one, and we must obtain it with the proper composition of the total transformations from "X_i" to "X_{i-1}."

$$v_0 = 0$$

$$v_i = X_i^{i-1} v_{i-1} + S_i \dot{\theta}_i$$

- Calculate **each Link Acceleration (Outward recursive pass "$i=1$ to n")** "a_i" for the whole rigid multibody. The operator "×" is for the cross product of the spatial vector of motion (Featherstone, 2016). The Potential Action vector is, in many cases, only the gravity action vector. We model the gravitational field as a fictitious acceleration of the base. To do this, we replace the initial acceleration value with the inverse of gravity. With this trick, the values for the link's accelerations and forces are not actual values, as they include an offset by gravity acceleration and force. Nonetheless, the result of the ID is correct, as we are interested only in the result of the Joint torques to comply with the motion

$$a_0 = \begin{bmatrix} 0 \\ -PoAcc \end{bmatrix} = \begin{bmatrix} 0 \\ 0 \\ 0 \\ 0 \\ 0 \\ a_g \end{bmatrix}$$

$$a_i = X_i^{i-1} a_{i-1} + S_i \ddot{\theta}_i + v_i \times S_i \dot{\theta}_i$$

- Calculate **each total Link Force (Outward recursive pass "$i=1$ to n")** "f_i^L" required to produce the rigid-body acceleration. We obtain the net force acting on each link, knowing the velocity and acceleration of this body. The definition of link inertia tensor "I_i" is in Plücker coordinates, with the mass of the link "m_i" and the rotational inertia "Ψ_i." We define all of them in the link coordinate system. The operator "×*" is for the cross product of the spatial vector of force (Featherstone, 2016).

$$I_i = \begin{bmatrix} \Psi_i & 0 \\ 0 & m_i I_3 \end{bmatrix}$$

$$f_i^L = I_i a_i + v_i \times^* I_i v_i$$

- Calculate **each Joint Transmitted Force (Inward recursive pass "$i=n$ to 1")** "f_i^J" between the links. The result "f_i^J" is the force transmitted from the predecessor link across the joint, and "f_{i+1}^J" the force transmitted to the successor link. There is no transmitted force to a successor for any last link in an open rigid body chain. This "f_{n+1}^J" is null, and then the transformation "X_{n+1}^n" does not affect, but many times is helpful to define it as corresponding to the robot tool. This definition permits the formulation to work for any "n" value. Besides, we assume to know all external forces (including gravity) acting on the link "f_i^X." The transformation in Plücker coordinates "X_a^b" is the relative from one link to another and must be obtained with the proper composition of the total transformations from "X_a" and "X_b."

$$f_{n+1}^J = 0$$

$$f_i^J = f_i^L - \left(X_0^i\right)^T f_i^X + \left(X_{i+1}^i\right)^T f_{i+1}^J$$

- Calculate **each Joint Generalize Force (Inward recursive pass "$i=n$ to 1")** "τ_i" with the motion subspace of each joint.

$$\tau_i = S_i^T f_i^J$$

Inverse Dynamics

- At the end of the inward recursive pass, we finalize **the complete result of the ID problem with the torques for all joints** to comply with the joint target definition in terms of position, velocity, and acceleration.

$$\tau = \begin{bmatrix} \Gamma_1 \\ \Gamma_2 \\ \Gamma_3 \\ \Gamma_4 \\ \Gamma_5 \\ \Gamma_6 \end{bmatrix}$$

We include two exercises, the first (Exercises 6.5.2a), where we can find the practical implementations for the RNEA algorithm with the POE applied to solve this robot's ID according to the procedure presented here before. The second exercise (Exercises 6.5.2b) enlightens very much about the performance of the two ID algorithms presented in this chapter. We address the same problem, and the most efficient for the job is always by RNEA, as we know it has a computational complexity of $O(6)$. In contrast, the classical non-recursive Lagrange alternative has a complexity of $O(6^4)$. The RNEA performs at least two orders of magnitude better in terms of computational cost than the non-recursive option, giving the joint forces' solutions at least some hundred times faster. Of course, the resulting joint torques are equal with both algorithms, as the effectiveness is the same.

The code for Exercise 6.5.2a[8] and Exercise 6.5.2b[9] are in the internet hosting for the software of this book.

To abound in the knowledge, we present some more ID examples with commercial manipulators: another example with the same ABB IRB120 with other reference configuration, a Bending Backwards robot with six DoF (i.e., ABB IRB1600), a Gantry type with six DoF (i.e., ABB IRB6620LX), a Scara type robot with four DoF (i.e., ABB IRB920SC), a Collaborative robot with six DoF (i.e., UNIVERSAL UR16e) and a Redundant manipulator with seven DoF (i.e., KUKA IIWA). However, we will show these exercises more briefly, as the details are the same as those presented in this example.

6.5.3 PUMA ROBOTS (E.G., ABB IRB120) "TOOL-UP"

This ID exercise for the same Puma robot has a different home configuration to illustrate the freedom that screw theory provides to the kinematics analysis (Figure 6.14). Here, we define the spatial coordinate system with a different orientation. We choose the robot reference pose with a tool-up orientation. Besides, we do not assume to know the actual sign of rotation for the joints of the commercial robot to build a more hypothetical case.

This example checks the RNEA with POE formulation to solve the ID for a random target in the workspace (for more details, review Exercise 6.5.2). The only difference comes from the kinematics map because the last twist changes orientation.

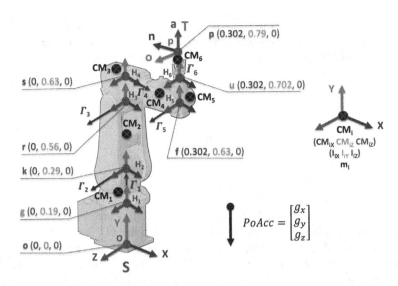

FIGURE 6.14 Puma ABB IRB120 "Tool-Up" RNEA with POE inverse dynamics parameters.

There are two exercises, the first to test the RNEA alone and the second to compare the performance of both algorithms, the RNEA and the non-recursive of Lagrange.

The code for Exercise 6.5.3a[10] and Exercise 6.5.3b[11] are in the internet hosting for the software of this book.

6.5.4 Bending Backwards Robots (e.g., ABB IRB1600)

This ID exercise sees a slightly different robot architecture, a Bending Backwards manipulator (Figure 6.15). This design does not need to rotate the robot to reach for

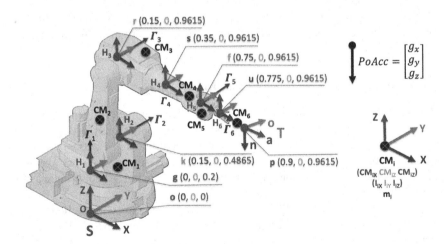

FIGURE 6.15 Bending Backwards ABB IRB1600 RNEA with POE ID parameters.

Inverse Dynamics

things behind it. Swinging the arm backwards extends the working range of the robot.

This example checks the RNEA with POE formulation to solve the ID for a random target in the workspace (for more details, review Exercise 6.5.2). The only differences come from the different kinematics maps because the twists are proper to each manipulator.

There are two exercises, the first to test the RNEA alone and the second to compare the performance of both algorithms, the RNEA and the non-recursive of Lagrange.

The code for Exercise 6.5.4a[12] and Exercise 6.5.4b[13] are in the internet hosting for the software of this book.

6.5.5 GANTRY ROBOTS (E.G., ABB IRB6620LX)

This ID exercise deals with this Gantry robotic architecture (see Figure 6.16), which combines the advantages of both a long linear axis for the first joint plus five revolute joints. We know how the treatment of the prismatic joint is the same with screw theory, and the resulting exercise mathematics is like in previous examples.

This example checks the RNEA with POE formulation to solve the ID for a random target in the workspace (for more details, review Exercise 6.5.2). The only differences come from the different kinematics maps because the twists are proper to each manipulator.

There are two exercises, the first to test the RNEA alone and the second to compare the performance of both algorithms, the RNEA and the non-recursive of Lagrange.

The code for Exercise 6.5.5a[14] and Exercise 6.5.5b[15] are in the internet hosting for the software of this book.

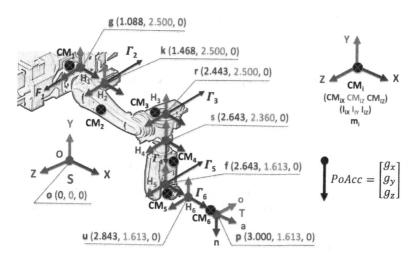

FIGURE 6.16 Gantry ABB IRB6620LX RNEA with POE inverse dynamics parameters.

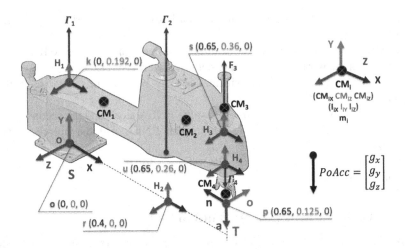

FIGURE 6.17 Scara ABB IRB910SC RNEA with POE inverse dynamics parameters.

6.5.6 Scara Robots (e.g., ABB IRB910SC)

This ID exercise has a two parallel joint axes mechanism, advantageous for many assembly operations. For the sake of showing the freedom given by the screw theory in the selection of the coordinate systems, we have chosen the typical virtual reality spatial frame orientation (see Figure 6.17). This example has only four joints, but the general approach applies. Even with a second link coordinate system out of the mechanism, the algorithm performs well.

This example checks the RNEA with POE formulation to solve the ID for a random target in the workspace (for more details, review the general formulation of Section 6.5.1). The differences come from the different number of DoF, as the previous exercises were for manipulators with six joints, and this one has only four. Nevertheless, the mathematics work equally fine. For any doubt on the kinematics map definition, you can go to the previous chapters (particularly the third) for the twist's definition.

There are two exercises, the first to test the RNEA alone and the second to compare the performance of both algorithms, the RNEA and the non-recursive of Lagrange.

The code for Exercise 6.5.6a[16] and Exercise 6.5.6b[17] are in the internet hosting for the software of this book.

6.5.7 Collaborative Robots (e.g., Universal UR16e)

This ID exercise deals with a typical collaborative manipulator made for human-robot cooperation in the workspace. The necessary parameters for the mathematics formulation are in Figure 6.18.

Inverse Dynamics

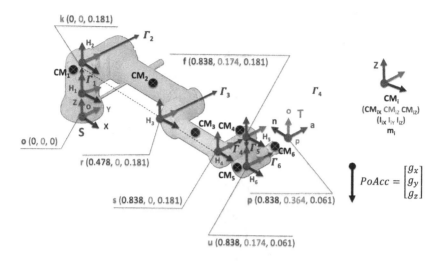

FIGURE 6.18 Collaborative UNIVERSAL UR16e RNEA with POE ID parameters.

This example checks the RNEA with POE formulation to solve the ID for a random target in the workspace (for more details, review Exercise 6.5.2). The only differences come from the different kinematics maps because the twists are proper to each manipulator.

There are two exercises, the first to test the RNEA alone and the second to compare the performance of both algorithms, the RNEA and the non-recursive of Lagrange.

The code for Exercise 6.5.7a[18] and Exercise 6.5.7b[19] are in the internet hosting for the software of this book.

6.5.8 REDUNDANT ROBOTS (E.G., KUKA IIWA)

This ID exercise has the novelty of solving a robot with seven DoF. We can check the excellent performance of the formulations because they work for any number of joints. The information for the dynamics parameters of the manipulator is in Figure 6.19.

This example checks the RNEA with POE formulation to solve the ID for a random target in the workspace (for more details, review the general formulation of Section 6.5.1). The differences come from the number of DoF, as this robot has seven joints. Nonetheless, the mathematical approach works equally fine. For any doubt on the kinematics map definition, you can go to any previous chapters for the twist's definition.

There are two exercises, the first to test the RNEA alone and the second to compare the performance of both algorithms, the RNEA and the non-recursive of Lagrange.

The code for Exercise 6.5.8a[20] and Exercise 6.5.8b[21] are in the internet hosting for the software of this book.

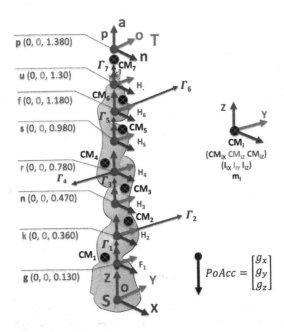

FIGURE 6.19 Redundant KUKA IIWA RNEA with POE inverse dynamics parameters.

6.6 SUMMARY

In this chapter, we have moved on from the realm of kinematics to one of screw theory dynamics. We have introduced some of the most modern geometric dynamics implementations. We presented how the screw-theoretic concepts can be expressed in a reference frame-invariant way to create algorithms for robots forward and ID.

The subject of robot dynamics is still an open area of research because they can have very complex structures with a lot of DoF. The nonlinear equations of motion with the existence of constraints make the formulation of robot dynamics tricky. In this chapter, our goal was to provide robot dynamics algorithms that rely on methods from the screw theory and some elaborated extensions from it.

These tools were mainly inaccessible to engineers because screw theory dynamics requires a new language and notations (e.g., Twist, Wrench, Robot Inertia Matrix, Link Inertia Matrix, Link Jacobian, Robot Coriolis Matrix, Robot Potential Matrix, Spatial Vector Algebra). For this reason, our first target was to address these concepts adequately.

We introduce a geometric expression to solve the ID problem based on the classical Lagrange screw theory. This chapter presents **three novel expressions for the Potential or Gravity matrix. Two of which are genuinely new complete geometric formulas (i.e., Gravity Twist and Gravity Wrench matrices)**. They allow to complete the screw theory forward and ID approach with a geometric formulation, and their convergence is assured.

Featherstone introduced the Spatial Vector Algebra and showed how to formulate rigid multibody systems' dynamics by extending constructs from classical screw

Inverse Dynamics

theory. There is an introduction to the RNEA machinery to solve the ID problem. **We introduce innovation to use the screw theory POE for the kinematics analysis of the RNEA** instead of the standard DH. This idea provides flexibility and clarity to the recursive algorithm. The efficiency of this RNEA algorithm is better than the Lagrange one. We get the joint forces and torques two or three orders of magnitude faster. Therefore, this approach is more convenient for real-time applications.

We exercise all these ID results with some robotics architectures: Puma robot (e.g., ABB IRB120), Bending Backwards robot (e.g., ABB IRB1600), Gantry robot (e.g., ABB IRB6620LX), Scara robot (e.g., ABB IRB910SC), Collaborative robot (e.g., UR16e) and even a Redundant robot (e.g., KUKA IIWA).

To take advantage of the beauty of these mathematical tools applied to robot dynamics, we will have to use the formulations for other essential concepts: dynamics in the workspace or interaction with the environment (e.g., integration of constraints in dynamics). However, all these progressive ideas are out of this book's scope and left for future editions.

The robot's dynamic models are necessary for a wide range of applications, from design and optimization to trajectory generation algorithms or simulation. In the following chapters, we will see some of these applications.

NOTES

1 Pardos-Gotor, J.M. (2021). *Screw Theory in Robotics*. Github. https://github.com/DrPardosGotor/Screw-Theory-in-Robotics/blob/master/Exercises/Exercise_6_2_2.m
2 Pardos-Gotor, J.M. (2021). *Screw Theory in Robotics*. Github. https://github.com/DrPardosGotor/Screw-Theory-in-Robotics/blob/master/Exercises/Exercise_6_2_3.m
3 Pardos-Gotor, J.M. (2021). *Screw Theory in Robotics*. Github. https://github.com/DrPardosGotor/Screw-Theory-in-Robotics/blob/master/Exercises/Exercise_6_2_4.m
4 Pardos-Gotor, J.M. (2021). *Screw Theory in Robotics*. Github. https://github.com/DrPardosGotor/Screw-Theory-in-Robotics/blob/master/Exercises/Exercise_6_2_5.m
5 Pardos-Gotor, J.M. (2021). *Screw Theory in Robotics*. Github. https://github.com/DrPardosGotor/Screw-Theory-in-Robotics/blob/master/Exercises/Exercise_6_2_6.m
6 Pardos-Gotor, J.M. (2021). *Screw Theory in Robotics*. Github. https://github.com/DrPardosGotor/Screw-Theory-in-Robotics/blob/master/Exercises/Exercise_6_2_7.m
7 Pardos-Gotor, J.M. (2021). *Screw Theory in Robotics*. Github. https://github.com/DrPardosGotor/Screw-Theory-in-Robotics/blob/master/Exercises/Exercise_6_2_8.m
8 Pardos-Gotor, J.M. (2021). *Screw Theory in Robotics*. Github. https://github.com/DrPardosGotor/Screw-Theory-in-Robotics/blob/master/Exercises/Exercise_6_5_2a.m
9 Pardos-Gotor, J.M. (2021). *Screw Theory in Robotics*. Github. https://github.com/DrPardosGotor/Screw-Theory-in-Robotics/blob/master/Exercises/Exercise_6_5_2b.m
10 Pardos-Gotor, J.M. (2021). *Screw Theory in Robotics*. Github. https://github.com/DrPardosGotor/Screw-Theory-in-Robotics/blob/master/Exercises/Exercise_6_5_3a.m
11 Pardos-Gotor, J.M. (2021). *Screw Theory in Robotics*. Github. https://github.com/DrPardosGotor/Screw-Theory-in-Robotics/blob/master/Exercises/Exercise_6_5_3b.m
12 Pardos-Gotor, J.M. (2021). *Screw Theory in Robotics*. Github. https://github.com/DrPardosGotor/Screw-Theory-in-Robotics/blob/master/Exercises/Exercise_6_5_4a.m
13 Pardos-Gotor, J.M. (2021). *Screw Theory in Robotics*. Github. https://github.com/DrPardosGotor/Screw-Theory-in-Robotics/blob/master/Exercises/Exercise_6_5_4b.m
14 Pardos-Gotor, J.M. (2021). *Screw Theory in Robotics*. Github. https://github.com/DrPardosGotor/Screw-Theory-in-Robotics/blob/master/Exercises/Exercise_6_5_5a.m

15 Pardos-Gotor, J.M. (2021). *Screw Theory in Robotics*. Github. https://github.com/DrPardosGotor/Screw-Theory-in-Robotics/blob/master/Exercises/Exercise_6_5_5b.m
16 Pardos-Gotor, J.M. (2021). *Screw Theory in Robotics*. Github. https://github.com/DrPardosGotor/Screw-Theory-in-Robotics/blob/master/Exercises/Exercise_6_5_6a.m
17 Pardos-Gotor, J.M. (2021). *Screw Theory in Robotics*. Github. https://github.com/DrPardosGotor/Screw-Theory-in-Robotics/blob/master/Exercises/Exercise_6_5_6b.m
18 Pardos-Gotor, J.M. (2021). *Screw Theory in Robotics*. Github. https://github.com/DrPardosGotor/Screw-Theory-in-Robotics/blob/master/Exercises/Exercise_6_5_7a.m
19 Pardos-Gotor, J.M. (2021). *Screw Theory in Robotics*. Github. https://github.com/DrPardosGotor/Screw-Theory-in-Robotics/blob/master/Exercises/Exercise_6_5_7b.m
20 Pardos-Gotor, J.M. (2021). *Screw Theory in Robotics*. Github. https://github.com/DrPardosGotor/Screw-Theory-in-Robotics/blob/master/Exercises/Exercise_6_5_8a.m
21 Pardos-Gotor, J.M. (2021). *Screw Theory in Robotics*. Github. https://github.com/DrPardosGotor/Screw-Theory-in-Robotics/blob/master/Exercises/Exercise_6_5_8b.m

7 Trajectory Generation

"Every word or concept, clear as it may seem to be, has only a limited range of applicability."

—Werner Heisenberg

7.1 CONCEPTS AND DEFINITIONS

Regarding the scope of this book, we limit our explanations to the idea of Trajectory Generation, even though this is a subset of other more comprehensive problems, such as navigation and motion planning (Siciliano and Khatib, 2016).

The specification of the robot position as a function of space is called a path. Differently, identifying the robot position as a function of time is called a trajectory. A trajectory is the combination of a path, a purely geometric description of the configuration sequence, and a time scaling, which defines when the robot reaches those configurations. A trajectory is a description of how to follow the path over time.

The common hierarchy of the trajectory generation problem includes the following topics:

- **Task planning**: it is the definition of a set of high-level goals, such as "go and weld those two metallic pieces in front of you."
- **Path planning**: it is the generation of a feasible path from a "start point" to a "goal point," going through a set of connected "via points" or "waypoints" (see Figure 7.1). To ensure that the robot dynamics are well defined, the path must be twice differentiable to get a feasible acceleration of the mechanism.
- **Trajectory planning:** it builds a schedule for following a path planning, given constraints such as position, velocity, and acceleration.
- **Trajectory tracking**: it is the control of the mechanical system to execute the trajectory as accurately as possible.

We can consider both concepts, paths and trajectories, in different coordinate reference frames, generally for the task space (i.e., tool space) and the joint space (Lynch and Park, 2017).

- **Task space**: it means the construction of the trajectory, including points and interpolation, is made on the Cartesian pose (i.e., position and orientation) of a specific location on the robot, which is typically the end-effector. The same applies to velocities and accelerations of the tool.
- **Joint space**: it means the building of the trajectory, including points and interpolation, is made on the joint position (angle for revolute joints or displacements for prismatic joints). The velocities and accelerations are also in this joint space.

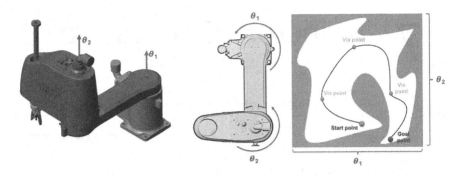

FIGURE 7.1 Path planning valid in the joint configuration space of two Degrees of Freedom.

The trajectories tend to look more natural in the task space than in the joint space because we consider the motion of the tool inside the typical human environment. The drawback is that following a task-space trajectory involves solving Inverse Kinematics (IK) for getting the joint magnitudes, which means a lot more computation if the IK solver applies a numeric optimization. Here again, we appreciate the benefits of the screw theory, as it permits us to solve IK with closed-form geometric solutions (Murray et al., 2017).

There is a choice to design the application with the generation of trajectories in any convenient space. However, the typical problem begins with a task to be executed, which implies a task-space trajectory. From there, we must solve a feasible joint-space trajectory to comply with that task motion. There are several approaches to get the joint-space trajectory, as we have shown in the previous chapters. We use inverse kinematics (IK) and inverse differential kinematics (DK), which we explained thoroughly in the last chapters.

Irrespective of selecting the task-space or joint-space trajectory, various methods create trajectories interpolating the Cartesian pose or joint path configurations over time.

7.1.1 Point-to-Point Position Straight-line Trajectories

The simplest type of path is from rest at one configuration to rest at another. We call this a point-to-point motion. Nonetheless, this is the first classical definition, but there are other concepts for point to point, which considerer an absolute velocity and acceleration for both the start and goal point. This approach is our election for the simulations of the next chapter.

The first type of path we can imagine is for a motion along a straight line. Unfortunately, a straight line in joint space generally does not yield a straight-line movement of the tool in task space. If there is a need to develop a task-space straight-line motion, consider:

- If the path passes near a robot kinematic singularity, the joint velocities could become unreasonably big.
- For some points of the task-space trajectory, the joint-space configurations (or some of them) might lie outside the robot's dexterous workspace, making the motion impossible with due accuracy.

Trajectory Generation

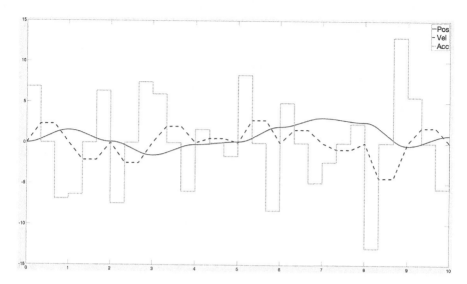

FIGURE 7.2 Trapezoidal motion trajectory example.

The time scaling of the path must ensure that the motion is smooth enough and that any constraints on robot velocity and acceleration are satisfied.

7.1.2 Trapezoidal Position Trajectory

Trapezoidal motion time scaling is quite common in control, and it gets its name from its velocity profiles (see Figure 7.2). These trajectories have a typical form with constant acceleration, zero, and constant deceleration. The velocity has a trapezoidal shape, and the position a linear segment with a parabolic blend. This parameterization makes it relatively easy for implementation to comply with the robot's mechanical limits.

The trapezoidal time scaling has the advantage that if there are constant limits on the joint velocities and accelerations, it is the fastest straight-line motion possible at the cost of discontinuous jumps in acceleration.

7.1.3 Polynomial Position Trajectory

We can interpolate between points using polynomials of various orders. The most common orders used are:

- Cubic (3rd order) needs four boundary conditions of position and velocity at both ends (see Figure 7.3).
- Quintic (5th order) requires six boundary conditions of the position, velocity, and acceleration at both ends (see Figure 7.4).
- Higher-order might be helpful if we need to match at the trajectory points the derivatives of positions.

Polynomial trajectories are useful for stitching segments with zero or nonzero velocity and acceleration. The accelerations are smooth, unlike with trapezoidal motion. The drawback is that the validation is complex, particularly for accelerations

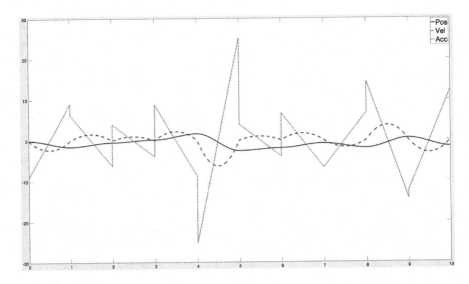

FIGURE 7.3 Polynomial cubic motion trajectory example.

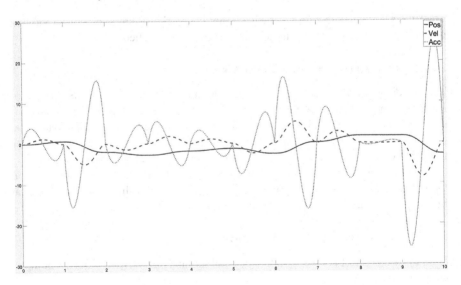

FIGURE 7.4 Polynomial quintic motion trajectory example.

and velocities, because the set boundary conditions may overshoot between trajectory segments.

7.1.4 Spline Position Trajectory

Another way to build trajectories is through splines, which are combinations of polynomials. Splines are polynomials in space that can create complex shapes. The resulting splines can develop a uniform speed. There are many types of splines, but

Trajectory Generation

one commonly used type for motion planning is B-Splines (or basis splines). It permits the refinement of the control points to meet motion requirements regarding the limits of the physical actuators.

7.1.5 Rotation Motion Trajectory

The interpolation of orientation is kind of special since angles are continuously wrapping. In this chapter, we have already discussed mathematical tools for some orientation representations (e.g., Euler angles), which have multiple representations for the same configuration. There are several ways to overcome this pitfall, such as the use of quaternions or twists from screw theory. With these tools, the representation of orientation is unambiguous.

One technique available to design rotation motion trajectories is the SLERP (Spherical Linear Interpolation). It solves the shortest path between two orientations with constant angular velocity about a fixed axis. We can add a time scaling to change the behavior instead of sampling the trajectory at a uniform time spacing.

7.1.6 Trajectory Tracking and Control

It is advisable to verify the magnitudes of the trajectories (i.e., position, velocity, and acceleration) against limits. Velocity data are also significant for low-level control of the manipulator. For example, the velocity trajectory can serve as input to the derivative branch of PID controllers. It is useful also to calculate forward dynamics for model controllers. We do not focus this book on control. However, there are some controllers in the next chapter's simulations, where there are trajectory tracking examples with different approaches (i.e., Inverse Kinematics, Inverse Differential Kinematics, and Inverse Dynamics). The regulators will be PID and Computed Torque controllers.

All reviewed concepts are applied to manipulators in this book, but other robotics systems such as self-driving cars or drones can benefit from these ideas for their applications.

7.2 TRAJECTORY PLANNING

Trajectory generation is not the main topic of this book. However, to help understand the theoretical screw theory core subjects of this text, which are kinematics, differential kinematics, and dynamics for robotics, we need to build some trajectories for testing the algorithms and developments with the simulations of the next chapter.

We do not design high-level goals as task planning for the exercises, such as specific manipulator industrial applications. On the contrary, the tasks will become random generic paths for the robot tool end-effector in the task space to test the screw theory algorithms under very different conditions.

It is necessary to obtain the correct path planning in the joint space to generate the robot's motion. The path goes from a "start point" to a "goal point," crossing through a set of "via points." All these points ought to be possible configurations. The joint path planning must be twice differentiable to get viable velocities and accelerations of the mechanism. In so doing, we ensure well-defined dynamics for the robot.

The joint path planning will have two approaches already seen in this book. The first is to use some of the screw theory IK algorithms presented in chapter four to obtain the joint-space path points corresponding to task-space path points. In contrast, the second technique gets the joints' paths, with the inverse DK algorithms introduced in Chapter 5.

Once we have a path planning in the joint space, the trajectory planning consists of defining a schedule for such joint path planning, given position, velocity, and acceleration. For this section's exercises, we will apply some of the interpolation methods mentioned before (i.e., trapezoidal, cubic polynomial, and quintic polynomial). All the results are a foundation for further exercises of different simulations. However, we must realize that this whole approach is valid for obtaining only offline path planning, entirely suitable for many applications, like repetitive industrial tasks. However, when the applications need real-time results, such as adaptive robot behaviors, we need different online joint path planning algorithms.

The next chapter's simulations aim to facilitate the learning and testing of the central screw theory concepts. In this sense, for many simulations (i.e., IK, DK, and ID), there are two possibilities to generate paths for the robot tool end-effector task space, either in real time with the devices of the dashboard or offline with some predefined file. However, path planning in joint space is always online. Therefore, only a point-to-point trajectory is considered, with trapezoidal motion interpolation in real time. The velocity constraints are within limits, and accelerations keep within limits for the robot's torques and forces.

The trajectory tracking is not in the exercises of this chapter. In contrast, it is put into practice in the simulations of the next chapter, with some control techniques.

7.2.1 General Solution to Trajectory Generation

We propose two general solutions for trajectory generation. There are other approaches, but we limit the exercises of this chapter to these two methods. The first technique uses the screw theory IK algorithms (Selig, 2005). The second methodology employs the DK algorithms. Both ways obtain the joint-space path points corresponding with the tool task-space path points. We apply both alternatives for trajectory generation to produce the position, velocity, and acceleration for the robot joints. The solution follows these steps:

- Create **the end-effector path planning** as the set of targets in task space for the robot's tool inside the dexterous workspace. There are six paths for the Cartesian position and Euler orientation of the end-effector in X-Y-Z. These paths have a set of start point, via points, and goal point.
- Fix **the end-effector trajectory planning** with a **timeline**, choosing an acceptable task-space timestamp. Employ some interpolation method to complete the trajectory between the end-effector path planning points until the timeline's complete discretization. This discretization must be small enough to guarantee limited changes in robot magnitudes between consecutive points.
- Obtain **the joint path planning** in joint space. For this step, we have two alternatives, which we take according to the task characteristics.

Trajectory Generation

- We are solving **IK** for all end-effector path planning points. We have at our disposal all screw theory geometric algorithms developed for different robotics architectures. We use this approach when we have a closed-form formulation and the target is inside the workspace.
- We are using the **inverse DK** tools. We get the new joint magnitudes with the joint coordinates for the initial tool pose and the joint velocities' integration. We get the tool velocities from the difference between the configurations of the point of interest in the tool reference trajectory and the actual tool pose. By the integration of the geometric DK, we obtain the joint coordinates increments. The process goes on for the following points of the expected reference tool trajectory.
- Determine **the joint trajectory planning** by choosing the proper timestamp for the robot and applying it to the whole timeline. We employ some interpolation methods (e.g., trapezoidal, cubic, or quintic polynomial) to complete the trajectory between the points of the joint path planning for the complete discretization of the timeline. The result gives three trajectories for the position, velocity, and acceleration of the joints.
- Check **the actual end-effector trajectory** using the joint trajectories as input for the robot forward kinematics (FK). There is a comparison with the original target end-effector trajectory planning for testing the quality of the algorithm.

The following section presents an example of applying the trajectory generation general solution to an ABB IRB120 manipulator. The rest of the sections in this chapter add a similar exercise for other architectures more briefly, as the implementations follow the same procedure.

It is advisable to review the Screw Theory Toolbox for Robotics "ST24R" (Pardos-Gotor, 2021a) to understand better the concepts working in the examples of the following sections.

7.2.2 PUMA ROBOTS (E.G., ABB IRB120)

To consolidate the understanding of the trajectory generation process, we carried out two illustrative exercises for a Puma-type configuration of six DoF (Paden and Sastry, 1988).

In the first example (Exercise 7.2.2a), we generate a random **end-effector path planning** with one start point, nine via points, and one goal point. The **timeline** has 10 seconds, and the motion of the tool is in the 3D space. Consequently, the **end-effector trajectory planning** includes six trajectories corresponding to the tool's position and orientation in "X-Y-Z." The task has only as points of interest those where we want to obtain the tool's exact position and orientation. These eleven points formed by the start, nine vias, and goal timestamped exactly each second. The rest of the tool trajectory presents a linear interpolation of time. Still, we do not care what the actual evolution of the tool is in the 3D space between two consecutive relevant points. All this design responds to didactical reasons. For an actual application, we can define as many via points as needed. Besides, the timestamp is adjustable to any robot and application. In any case, the meaning of the theory stands still.

To get the **joint path planning** in joint space through the IK for all tool path planning points, we take advantage of the Chapter 4 algorithms to solve the IK of the robot with a geometric solution. In this case, the algorithm is the "**PG5+PG4+PG6+PK1**". It uses some Pardos-Gotor canonical subproblems (i.e., PG5, PG4, and PG6), plus the classical Paden–Kahan subproblem one. All other IK algorithms presented in Chapter 4 for the same robot are equal alternatives by changing the function included in this exercise. The IK algorithm applies to all points (in this case, eleven) of the end-effector path planning. The result is eleven configurations for each joint. The exact geometric nature of the screw theory IK algorithm guarantees that, with these joint configurations applied to the robot actuators, the tool will precisely reach the desired configuration (i.e., position and orientation) for the points of the end-effector path planning.

To determine the **joint trajectory planning** (see Figure 7.5), we must interpolate between the already known eleven points of the joint path, according to the chosen timestamp, along with all the timeline in joint space. The interpolation must ensure that the joint position trajectory is twice differentiable to get feasible velocity and acceleration that comply with the robot's dynamics. In this first exercise, we apply a **trapezoidal interpolation** to complete the trajectory between the joint path planning configurations. The result has a smooth trajectory for joint positions and trapezoidal velocities, which gives the name to the method. Then the joint accelerations have the advantage of being at least limited, even if squared, which in practice might provoke some jerk motion.

Trapezoidal motion is quite common in control. These trajectories have constant acceleration, zero, and constant deceleration (Figure 7.5 dotted line). This characteristic leads to a trapezoidal velocity profile (Figure 7.5 dash-dot line) and a linear segment with a parabolic blend position profile (Figure 7.5 solid line). This parameterization

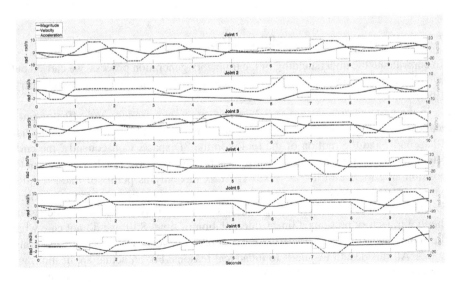

FIGURE 7.5 ABB IRB120 IK joint path planning with trapezoidal interpolation.

Trajectory Generation

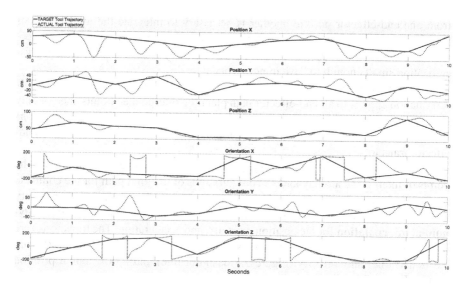

FIGURE 7.6 ABB IRB120 IK end-effector target and actual values.

makes the implementation relatively straightforward with compliance with the robot's position, speed, and acceleration limits.

The last part of the exercise tests the results using the obtained complete joint trajectories as input for the robot FK to get the actual end-effector trajectory's values. Theoretically, before any control application, a contrast is available between **the target end-effector trajectory** (Figure 7.6 solid line) and **the actual end-effector trajectory** (Figure 7.6 dotted line). It is possible to check how well the tool 3D motion's value complies with the target, in terms of timing, position, and orientation (i.e., Cartesian and Euler X-Y-Z), at the relevant points of the end-effector trajectory planning, which are marked every second.

Arrived at this point, we reached the plan for this trajectory exercise. For the rest of the points in the timeline the end-effector trajectory is irrelevant, according to the exercise's definition, as far as it complies with the key target points. Do not be misled by the orientation picture, as sometimes the jump from minus to plus "π" could seem farther than it is. Nonetheless, for the whole timeline, the motions comply with their joint mechanical limits.

The code for Exercise 7.2.2a is in the internet hosting for the software of this book[1].

For the second example (Exercise 7.2.2b), we check the second approach to solve the trajectory generation, in this case, based on the screw theory **DK** (Pardos-Gotor, 2018). In this case, the **end-effector path planning** is the same as the previous example (Exercise 7.2.2a) for comparing the performance of both approaches. The task has only those eleven points of interest, where we need to solve the exact position and orientation of the tool as a function of time. One difference is that here the rest of the end-effector trajectory has a cubic polynomial interpolation. For this DK method, the interpolation of the end-effector path planning is an absolute necessity, whereas it was not for the previous IK approach. The reason is that to move the robot from one end-effector pose to another is necessary to integrate the joint velocities based on the end-effector velocities, which must be discretized for the whole trajectory. This formulation approximates the joint trajectory, which requires a target discretization small enough to have minor errors.

We obtain a **joint path planning** implementing the DK for this robot. The DK formulation applies to all end-effector path planning interpolated points and not only to the eleven relevant points. This joint path planning has the position configurations for each joint corresponding with the task points of the complete discretized end-effector trajectory.

To determine the **joint trajectory planning**, we must interpolate between the points of interest the already known configurations of the joint path for the complete timeline in joint space. The interpolation must ensure that the joint position trajectory is twice differentiable to comply with the robot's dynamics. We apply a **cubic polynomial interpolation** for the position, velocity, and acceleration. The trajectories are useful for stitching contiguous segments with zero or nonzero acceleration (Figure 7.7 dotted line) and velocity (Figure 7.7 dash-dot line) and a polynomial position profile (Figure 7.7 solid line).

The last part of the exercise tests the results using the obtained complete joint trajectories as input for the robot FK to get the actual end-effector trajectory's values. Theoretically, before any control application, a contrast is available between **the target end-effector trajectory** (Figure 7.8 solid line) and **the actual end-effector trajectory** (Figure 7.8 dotted line). It is possible to check how the result is good in terms of timing, position, and orientation (i.e., Cartesian and Euler X-Y-Z) at the end-effector's relevant points in the trajectory planning. Nonetheless, we must not forget that the DK method approximates the exact IK, and then the performance depends on the trajectory discretization and integration method. Here we use a straightforward Euler integration as it suffices the theory of this chapter, but there are more sophisticated

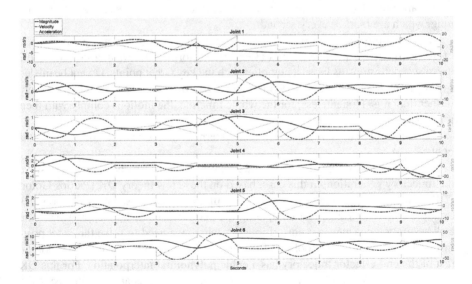

FIGURE 7.7 ABB IRB120 DK joint path planning with cubic interpolation.

Trajectory Generation 203

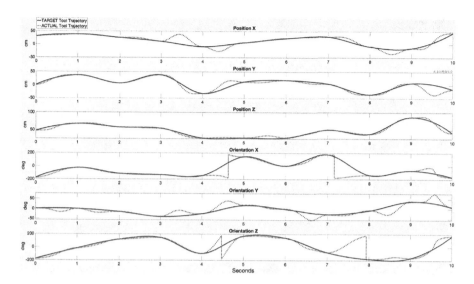

FIGURE 7.8 ABB IRB120 DK end-effector target and actual values.

alternatives indeed. The orientations can be a little bit misleading in the picture because the jump from minus to plus "π" could seem farther away than it is in physical terms.

The code for Exercise 7.2.2b is in the internet hosting for the software of this book[2]. It is advisable to review the details of the Screw Theory Toolbox for Robotics (ST24R) to understand better the concepts that worked in this example (Pardos-Gotor, 2021a).

7.2.3 PUMA ROBOTS (E.G., ABB IRB120) "TOOL-UP"

These trajectory generation exercises for the same Puma robot have a different home configuration to illustrate the freedom that screw theory provides to the kinematics analysis. Here, we define the spatial coordinate system with a different orientation. We choose the robot reference pose with a tool-up direction. Besides, we do not assume to know the actual sign of rotation for the joints of the commercial robot to build a more hypothetical case.

The first example checks the trajectory generation with the IK formulation (Exercise 7.2.3a). The path planning is solved using the screw theory IK "PG5+PG4+PG6+PK1" algorithm presented in Chapter 4. One difference with the previous exercise comes from the kinematics map as the last twist changes. Other differences are the interpolation method for the trajectory planning, which is linear for the end-effector and cubic polynomial for the joints.

The last part of this exercise tests the results using the obtained complete joint trajectories as input for the robot FK to get the actual end-effector trajectory's values. Theoretically, before any control application, a contrast is available between the target end-effector trajectory (Figure 7.9 solid line) and the actual end-effector trajectory (Figure 7.9 dotted line). It is possible to check how the result is exact. The tool 3D motion's value complies with the target in terms of timing, position, and

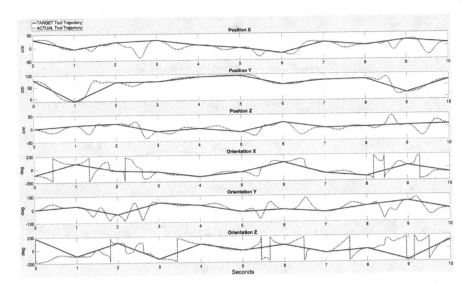

FIGURE 7.9 ABB IRB120 "Tool-Up" IK end-effector target and actual values.

orientation (i.e., Cartesian and Euler) at the relevant points of the end-effector trajectory planning. For the rest of the points in the trajectory, we do not care about how the tool motion evolves.

The second example (Exercise 7.2.3b) tests the same trajectory generation with the DK formulation. To determine the joint trajectory planning, we must interpolate between the points of interest of the already known configurations of the joint path along the timeline. We apply a cubic polynomial interpolation. The trajectories are useful for stitching contiguous segments with zero or nonzero acceleration (Figure 7.10 dotted line) and velocity (Figure 7.10 dash-dot line) and a polynomial position profile (Figure 7.10 solid line).

The last part of this exercise tests the results using the obtained complete joint trajectories as input for the robot FK to get the actual end-effector trajectory's values. Theoretically, before any control application, a contrast is available between the target end-effector trajectory (Figure 7.11 solid line) and the actual end-effector trajectory (Figure 7.11 dotted line). The orientations can be a little bit misleading in the picture because the jump from minus to plus "π" could seem farther away than it is in physical terms.

The code for Exercise 7.2.3a[3] and Exercise 7.2.3b[4] are in the internet hosting for the software of this book.

7.2.4 BENDING BACKWARDS ROBOTS (E.G., ABB IRB1600)

These trajectory generation exercises see a slightly different robot architecture, a Bending Backwards manipulator. This design does not need to rotate the robot to reach for things behind it. Swinging the arm backwards extends the working range of the robot.

Trajectory Generation

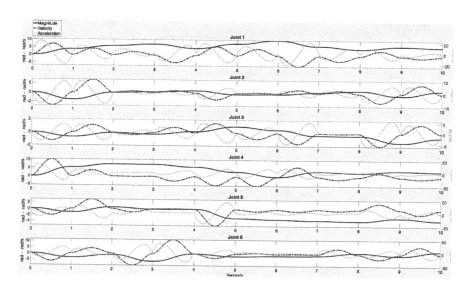

FIGURE 7.10 ABB IRB120 "Tool-Up" DK joint path planning with quintic interpolation.

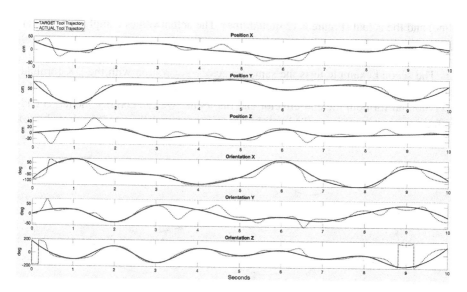

FIGURE 7.11 ABB IRB120 "Tool-Up" DK end-effector target and actual values.

The first example checks the trajectory generation with the IK formulation (Exercise 7.2.4a). The path planning is solved using the screw theory IK "PG7+PG6+PK1" algorithm presented in Chapter 4. One difference with other robots comes from the kinematics map. The interpolation method for the trajectory planning is linear for the end-effector and trapezoidal for the joints. It is possible to check the results, contrasting the target end-effector trajectory (Figure 7.12 solid

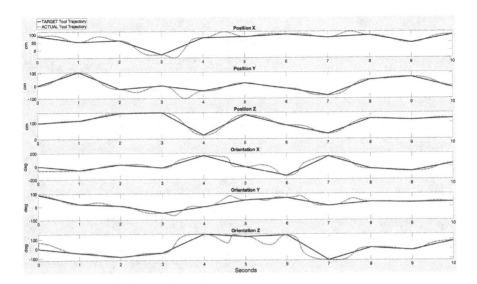

FIGURE 7.12 ABB IRB1600 IK end-effector target and actual values.

line) and the actual (Figure 7.12 dotted line). The actual values comply with the target in terms of timing, position, and orientation (i.e., Cartesian and Euler) at the relevant points.

The second example tests the same trajectory generation with the DK formulation (Exercise 7.2.4b). For the trajectory planning interpolations, we apply a cubic polynomial for both the end-effector and the joints. We can contrast the target end-effector trajectory (Figure 7.13 solid line) and the actual (Figure 7.13 dotted line).

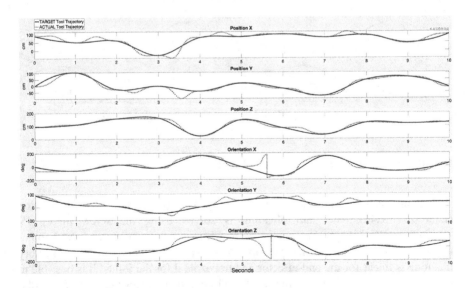

FIGURE 7.13 ABB IRB1600 DK end-effector target and actual values.

Trajectory Generation

The orientations can be a little bit misleading in the picture because the jump from minus to plus "π" could seem farther away than it is in physical terms. Nonetheless, the target tracking is reasonable enough.

The code for Exercise 7.2.4a[5] and Exercise 7.2.4b[6] are in the internet hosting for the software of this book.

7.2.5 Gantry Robots (e.g., ABB IRB6620LX)

These trajectory generation exercises deal with this Gantry robotic architecture, which combines the advantages of both a long linear axis for the first joint plus five revolute joints. We know how the treatment of the prismatic joint is the same with screw theory.

The first example checks the trajectory generation with the IK formulation (Exercise 7.2.5a). The path planning is solved using the screw theory IK "PG1+PG4+PG6+PK1" algorithm presented in Chapter 4. The PG1 is necessary to solve the problem because it has a prismatic joint. One difference with other robots comes from the kinematics map. The interpolation methods for the trajectory planning are linear for the end-effector and trapezoidal for the joints. It is possible to check the results, contrasting the target end-effector trajectory (Figure 7.14 solid line) and the actual (Figure 7.14 dotted line). The actual values comply with the target in terms of timing, position, and orientation (i.e., Cartesian and Euler) at the relevant points.

The second example tests the same trajectory generation with the DK formulation (Exercise 7.2.5b). We apply a cubic polynomial interpolation for the end-effector trajectory and a quintic for the joints. We can contrast the target end-effector trajectory (Figure 7.15 solid line) and the actual (Figure 7.15 dotted line). The target tracking is satisfactory.

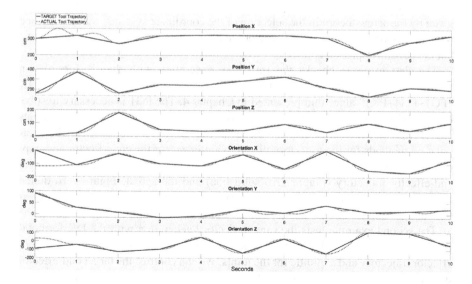

FIGURE 7.14 ABB IRB6620LX IK end-effector target and actual values.

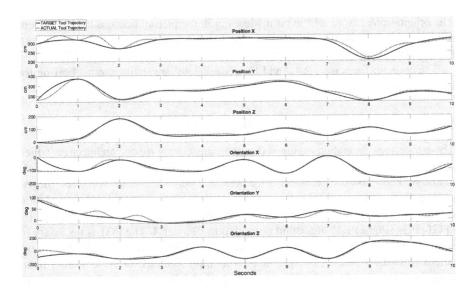

FIGURE 7.15 ABB IRB6620LX DK end-effector target and actual values.

The code for Exercise 7.2.5a[7] and Exercise 7.2.5b[8] are in the internet hosting for the software of this book.

7.2.6 SCARA ROBOTS (E.G., ABB IRB910SC)

These trajectory generation exercises have a two parallel joint axes mechanism, advantageous for many assembly operations. For the sake of showing the freedom given by the screw theory in the selection of the coordinate systems, we have chosen the typical virtual reality spatial frame orientation. This example has only four joints, but the general approach applies.

The first example checks the trajectory generation with the IK formulation (Exercise 7.2.6a). The path planning is solved using the screw theory IK "PG1+PG4+PK1" algorithm presented in Chapter 4. The PG1 is necessary to solve the IK problem because the third joint is prismatic. One difference with other robots comes from the kinematics map and the fewer Degrees of Freedom (DoF). The interpolation methods for the trajectory planning are linear for the end-effector and trapezoidal for the joints. It is possible to check the results, contrasting the target end-effector trajectory (Figure 7.16 solid line) and the actual (Figure 7.16 dotted line). The actual values comply with precision to the target in terms of timing, position, and orientation (i.e., Cartesian and Euler) at the relevant points.

The second example tests the same trajectory generation with the DK formulation (Exercise 7.2.6b). We apply a cubic polynomial interpolation for the end-effector trajectory and a quintic for the joints. We can contrast the target end-effector trajectory (Figure 7.17 solid line) and the actual end-effector trajectory (Figure 7.17 dotted line). The target tracking is good. Nonetheless, it is essential to be careful

Trajectory Generation

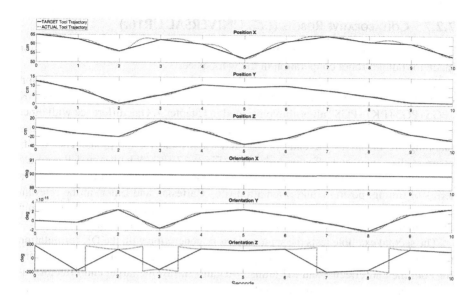

FIGURE 7.16 ABB IRB910SC IK end-effector target and actual values.

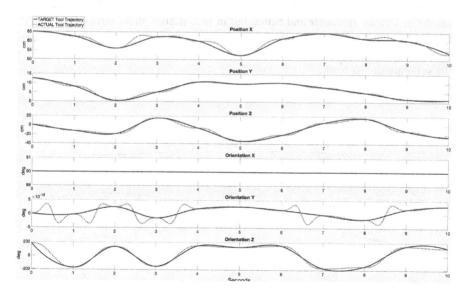

FIGURE 7.17 ABB IRB910SC DK end-effector target and actual values.

with the random paths generated with this exercise because the approximative nature of this approach might give worse results when the amplitude of the path magnitudes is too big.

The code for Exercise 7.2.6a[9] and Exercise 7.2.6b[10] are in the internet hosting for the software of this book.

7.2.7 COLLABORATIVE ROBOTS (E.G., UNIVERSAL UR16E)

These trajectory generation exercises deal with a typical collaborative manipulator made for human-robot cooperation in the workspace.

The first example checks the trajectory generation with the IK formulation (Exercise 7.2.7a). The path planning is solved using the screw theory IK "PG5+PG3+PK1+PG8" algorithm presented in Chapter 4. One difference with other robots comes from the kinematics map. The interpolation methods for the trajectory planning are linear for the end-effector and trapezoidal for the joints. The results contrast the target end-effector trajectory (Figure 7.18 solid line) and the actual (Figure 7.18 dotted line). The actual values comply with precision to the target in terms of timing, position, and orientation (i.e., Cartesian and Euler) at the relevant points. The orientations can be a little bit misleading in the picture because the jump from minus to plus "π" could seem farther away than it is in physical terms.

The second example tests the same trajectory generation with the DK formulation (Exercise 7.2.7b). We apply a cubic polynomial trajectory planning interpolations for both the end-effector and the joints. We can contrast the target end-effector trajectory (Figure 7.19 solid line) and the actual end-effector trajectory (Figure 7.19 dotted line). The target tracking is good. This DK approach's approximative nature is less robust than the previous IK, especially for significant rates of change in the evolution of the target path. The figures could be a little misleading, as it seems the results of DK are smoother and better, but in fact, it gives more error for the points of interest.

The code for Exercise 7.2.7a[11] and Exercise 7.2.7b[12] are in the internet hosting for the software of this book.

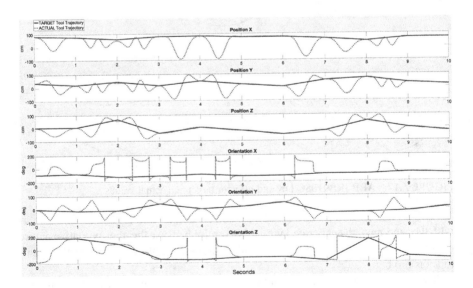

FIGURE 7.18 UNIVERSAL UR16e IK end-effector target and actual values.

Trajectory Generation

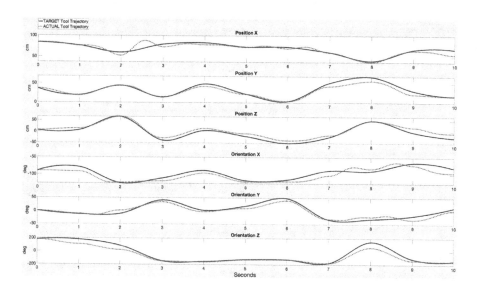

FIGURE 7.19 UNIVERSAL UR16e DK end-effector target and actual values.

7.2.8 REDUNDANT ROBOTS (E.G., KUKA IIWA)

These trajectory generation exercises have the novelty of solving a robot with seven DoF. The performance of the formulations is good working for any number of joints.

The first example checks the trajectory generation with the IK formulation (Exercise 7.2.8a). The path planning is solved using the screw theory IK "PK1+PK3+PK2+PK2+PK2+PK1" algorithm presented in Chapter 4. One difference with other robots comes from the kinematics map for the redundancy of the mechanism. The interpolation methods for the trajectory planning are linear for the end-effector and trapezoidal for the joints. The results contrast the target end-effector trajectory (Figure 7.20 solid line) and the actual (Figure 7.20 dotted line). The actual values comply with precision to the target in terms of timing, position, and orientation (i.e., Cartesian and Euler) at the relevant points. Out of those points, the evolution of the tool is not relevant for the exercise. The orientations can be misleading in the figure because the jump from minus to plus "π" could seem farther away than it is in physical terms.

The second example tests the same trajectory generation with the DK formulation (Exercise 7.2.8b). It is necessary to pay attention to the geometric Jacobian management because it is not symmetric due to the seven DoF. We apply a cubic polynomial for the end-effector and quintic for the joints trajectory planning interpolations. We can contrast the target end-effector trajectory (Figure 7.21 solid line) and the actual end-effector trajectory (Figure 7.21 dotted line). The target tracking is reasonably good, but this DK approach's approximative nature is less robust than the previous IK. The figures could be a little misleading, as it seems the results of DK are smoother and better, but there are some errors between the end-effector target and actual values at the relevant points of interest.

The code for Exercise 7.2.8a[13] and Exercise 7.2.8b[14] are in the internet hosting for the software of this book.

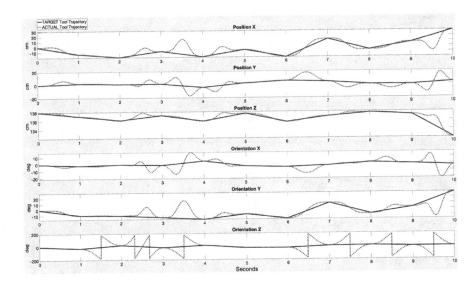

FIGURE 7.20   KUKA IIWA IK end-effector target and actual values.

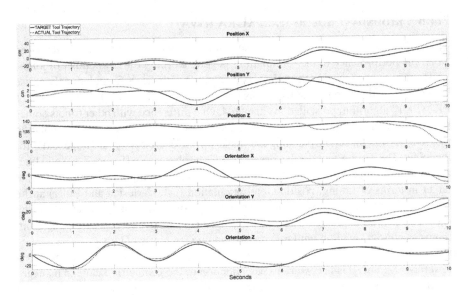

FIGURE 7.21   KUKA IIWA DK end-effector target and actual values.

7.3 SUMMARY

First, we presented the basic concepts and definitions for path planning, trajectory planning, and trajectory tracking.

For the appropriate robot motion, it is necessary to get a good path planning in the joint space, as a feasible path from a "start point" to a "goal point," going through a set of "via points." We create the joint path planning through two methods. The first way uses some of the geometric IK algorithms presented in Chapter 4. The second technique employs the DK introduced in Chapter 5. We obtain the joint-space path planning points corresponding with the tool task-space path planning targets.

The joint trajectory planning builds a time scaling introducing constraints such as position, velocity, and acceleration. An optimal position trajectory planning function must be twice differentiable to facilitate the control of the robot complying with the robot dynamics constraints. For doing that, it is possible to apply several interpolation algorithms (e.g., trapezoidal, cubic polynomial, and quintic polynomial). This approach gives offline path planning. When the applications need real-time results and online joint path planning, other techniques are required.

There are some exercises to put to the test all these trajectory planning ideas: A Puma with six revolute DoF (i.e., ABB IRB120), a Bending Backwards robot with also six revolute DoF (i.e., ABB IRB1600), a Gantry type with one prismatic and five revolute DoF (i.e., ABB IRB6620LX), a Scara with three revolute and one prismatic DoF (i.e., ABB IRB910SC), a Collaborative robot with six revolute DoF (i.e., UNIVERSAL UR16e) and a Redundant manipulator with seven revolute DoF (i.e., KUKA IIWA).

For most of the following chapter simulations, there are two possibilities to generate paths for the robot tool end-effector in the task space, either in real time with the dashboard's commands or offline with some predefined file. However, path planning in joint space works always online. Therefore, only a point-to-point trajectory will be considered, with trapezoidal motion interpolation in real time.

The joint velocity respects the actuator's physical constraints, and accelerations keep the robot joints' safety limits for torques and forces. This trajectory planning's core goal is to make the robot joints pass through the series of targets (i.e., start, goal, via) with a stern specification of the position. The reason is that the target is to make the tool end-effector reach as strictly as possible the shape of the task-space path. In other words, the joint path time scaling does not minimize joint motion neither optimizes the robot energy or productivity. The joint trajectory planning timing is inherited from the task-space path time scaling, always subject to the limits of the actuators.

The sophisticated control techniques for trajectory tracking are not the focus of this book. Nonetheless, some controllers are in the simulations of the next chapter, where there are several trajectory tracking examples, from simple PID control to computed torque algorithms.

NOTES

1 Pardos-Gotor, J.M. (2021). *Screw Theory in Robotics*. Github. https://github.com/DrPardosGotor/Screw-Theory-in-Robotics/blob/master/Exercises/Exercise_7_2_2a.m
2 Pardos-Gotor, J.M. (2021). *Screw Theory in Robotics*. Github. https://github.com/DrPardosGotor/Screw-Theory-in-Robotics/blob/master/Exercises/Exercise_7_2_2b.m
3 Pardos-Gotor, J.M. (2021). *Screw Theory in Robotics*. Github. https://github.com/DrPardosGotor/Screw-Theory-in-Robotics/blob/master/Exercises/Exercise_7_2_3a.m
4 Pardos-Gotor, J.M. (2021). *Screw Theory in Robotics*. Github. https://github.com/DrPardosGotor/Screw-Theory-in-Robotics/blob/master/Exercises/Exercise_7_2_3b.m

5. Pardos-Gotor, J.M. (2021). *Screw Theory in Robotics*. Github. https://github.com/DrPardosGotor/Screw-Theory-in-Robotics/blob/master/Exercises/Exercise_7_2_4a.m
6. Pardos-Gotor, J.M. (2021). *Screw Theory in Robotics*. Github. https://github.com/DrPardosGotor/Screw-Theory-in-Robotics/blob/master/Exercises/Exercise_7_2_4b.m
7. Pardos-Gotor, J.M. (2021). *Screw Theory in Robotics*. Github. https://github.com/DrPardosGotor/Screw-Theory-in-Robotics/blob/master/Exercises/Exercise_7_2_5a.m
8. Pardos-Gotor, J.M. (2021). *Screw Theory in Robotics*. Github. https://github.com/DrPardosGotor/Screw-Theory-in-Robotics/blob/master/Exercises/Exercise_7_2_5b.m
9. Pardos-Gotor, J.M. (2021). *Screw Theory in Robotics*. Github. https://github.com/DrPardosGotor/Screw-Theory-in-Robotics/blob/master/Exercises/Exercise_7_2_6a.m
10. Pardos-Gotor, J.M. (2021). *Screw Theory in Robotics*. Github. https://github.com/DrPardosGotor/Screw-Theory-in-Robotics/blob/master/Exercises/Exercise_7_2_6b.m
11. Pardos-Gotor, J.M. (2021). *Screw Theory in Robotics*. Github. https://github.com/DrPardosGotor/Screw-Theory-in-Robotics/blob/master/Exercises/Exercise_7_2_7a.m
12. Pardos-Gotor, J.M. (2021). *Screw Theory in Robotics*. Github. https://github.com/DrPardosGotor/Screw-Theory-in-Robotics/blob/master/Exercises/Exercise_7_2_7b.m
13. Pardos-Gotor, J.M. (2021). *Screw Theory in Robotics*. Github. https://github.com/DrPardosGotor/Screw-Theory-in-Robotics/blob/master/Exercises/Exercise_7_2_8a.m
14. Pardos-Gotor, J.M. (2021). *Screw Theory in Robotics*. Github. https://github.com/DrPardosGotor/Screw-Theory-in-Robotics/blob/master/Exercises/Exercise_7_2_8b.m

8 Robotics Simulation

"Theoretically, if we could build a machine whose mechanical structure duplicated human physiology, then we could have a machine whose intellectual capacities would duplicate those of human beings."

—Norbert Wiener

8.1 ROBOTICS SIMULATION

For those who have decided to read this book from the beginning and come to this far, congratulations! They have already learned the most important and intellectually daring concepts. This chapter is for the robotics simulation of the manipulators reviewed along with this book. On top of that, it is possible to test the algorithms, developments, and applications with the available simulation tools.

Simulation is a crucial research tool since the beginning of the 20th century. Then, better times for simulation started with the development of computers. The simulation is a powerful visualization instrument, which has a vital role in robotics design. Different tools are ready to analyze robotic manipulators' kinematics and dynamics to design different mechanical configurations and control algorithms (Siciliano and Khatib, 2016).

A robotics simulator is helpful to create applications for a robot without relying on the actual machine, thus saving time and costs (Barrientos et al., 2007). The simulator usually allows us to create objects in a virtual world so that our robot can interact with them, and we can test its behavior. Almost all modern robot simulators have the capability for three-dimensional (3D) environment modeling. Simulation tools must allow to implement the principles of robot motion (Choset and Lynch, 2005) and mimic a variety of mechanisms and control systems (Bloch, 2003).

Most advanced simulation software platforms offer many capabilities, making simulation easier and very close to real life. There are multiple mathematical artifacts to modeling robotic manipulators (Rodriguez et al., 1991). Many tools are compatible with programming languages like C/C++, Python, Java, LabVIEW, or MATLAB® (Corke, 2017). They offer varied feature sets depending on the purpose of the simulation. We mention a couple of simulators from different environments for illustrative purposes: "Robcad" software by SIEMENS for industrial applications and CoppeliaSim more used in research and academy. There are a lot of simulators, and we must find the tool which best suits the requirements of the applications. Creating a complete virtual model of a robot by simulating as many elements as possible can significantly impact the performance of the projects.

Today there is a vibrant and growing community of robotics researchers and practitioners who work with the "Robot Operating System" (ROS). That is why perhaps it is worthwhile to mention the robot simulation "Gazebo" environment, which integrates with ROS. This well-designed simulator makes it possible to rapidly test algorithms, design robots, and train systems using realistic scenarios.

A simulator for developing robotics software and control (Bullo and Lewis, 2004) allows creating programs conveniently written and debugged offline before getting a final version ready to be tested on an actual robot. Consider that one of the keys for the offline simulation programming success depends on how well the simulator recreates the robot's natural environment (Agahi and Kreutz-Delgado, 1994). There is always a trade-off between the simulation precision and the productivity of the application. We must decide what robotics simulation best suits the purpose of research or application.

Characteristics of robot simulation to consider are as follows:

- Reduce costs involved in robot production while simulating various alternatives without affecting physical expenses (i.e., shorter delivery times).
- Diagnose the source code that controls a particular robot or component and test it before implementation.
- Demonstrate a system's feasibility.

Nonetheless, an application can simulate just what it is programmed to do, and it might overlook many internal or external elements. There are more scenarios in the real world than those in the simulator for any robot.

8.1.1 Why Code in MATLAB®?

One of this book's targets is to present screw theory in an integrated way with clear examples to illustrate the robot kinematics and dynamics concepts. We do not relegate the code and exercises to a last chapter or annex because we consider them fundamental for learning.

MATLAB® is the computational foundation for this book. It is an interactive mathematical software environment that makes it easy to work with linear algebra and data analysis. It is a popular package familiar to engineers and researchers. It has a wholly interpreted programming language that permits the creation of complex programs and algorithms.

MATLAB® supports many toolboxes covering many topics and specialties useful in diverse mathematical and robotics fields. There are essential toolboxes available that are very useful for investigating robot problems. The toolboxes are free to use and distribute under the GNU Lesser General Public License. The Robotics System Toolbox™ is well known and provides tools and algorithms for designing, simulating, and testing manipulators, mobile robots, and humanoid robots. MATLAB® allows the design of hardware platforms and analyzes 3D rigid-body mechanics (Abraham and Marsden, 1999). It is possible to work directly with existing CAD (Computer-Aided Design) files and importing Universal Robot Description Format (URDF) files directly into Simulink® software. Once in operation, it is also advantageous to reuse design models as digital twins to optimize real robots' performance.

MATLAB® provides a software environment convenient for robotics researchers and engineers because it works intrinsically integrated with Simulink® and Simscape® packages to design and tune algorithms and generate code.

Robotics Simulation 217

MATLAB® (MathWorks, 2021a) and Simulink® work together, combining textual and graphical programming to design a simulation environment system. There are thousands of algorithms already available for this environment, ready for adding them to the simulation blocks. On top of that, it is possible to analyze and visualize the data with built-in functions. Integrating MATLAB® into the Simulink® environment provides many benefits:

- Connect the control with the algorithms for the robot.
- Develop hardware and connect to the ROS.
- Connect to a wide range of sensors and actuators.
- Generate code for embedded targets like micro-controllers and work with legacy code of existing robotics.
- Connect to low-cost hardware such as "Arduino" and "Raspberry Pi" using pre-built hardware support packages.
- Simplify the design by creating shared code and applications.
- Design and analyze 3D rigid-body mechanics.

We have decided to use Simulink® and Simscape™ as the robotics simulator for this book for all these reasons. It can serve well to test and visualize the critical theoretical concepts presented in this text.

Simscape™ (MathWorks, 2021b) enables the rapid creation of models and simulates multi-domain physical systems in the Simulink® environment. Simscape™ facilitates building component models based on physical connections that directly integrate with block diagrams. We can model systems by assembling components into a schematic. Simscape™ add-on products provide more features and analysis capabilities.

With Simscape™, it is feasible to design and analyze 3D rigid-body mechanics such as a manipulator's arm. We can work with existing CAD files by importing files directly into Simulink® or from software like "SolidWorks." We can build our robot models or use a library of commonly used robots to model applications quickly. Simscape™ is open to importing URDF files to create custom robot models and visual geometries.

With Simscape™, it is beneficial to have the single integration with MATLAB® and Simulink® environment for algorithms development, debugging, and analysis, instead of switching between multiple tools. That integration reduces overall project development time and the chances of introducing errors.

We can simulate the previous chapters' algorithms and test their performance with actual commercial manipulator's hardware. The Screw Theory Toolbox for Robotics (ST24R) (Pardos-Gotor, 2021a) includes all the functions and algorithms to run the Simscape™ simulations and check the screw theory mathematics performance visually. This chapter contains several simulations with commercial manipulators to try all the theoretical developments of previous chapters: Forward Kinematics (FK), Inverse Kinematics (IK), Inverse Differential Kinematics (DK), and Inverse Dynamics (ID).

To follow next simulations the only requisite is a basic knowledge of robotics (Craig, 2004) and the study of the exercises of previous chapters of this book. Using these examples is an excellent kick-off to easily employ them on building other robot applications of different scope (Arbulú et al., 2005).

We use version 2021b of MATLAB® for developing the software, exercises, and simulations of this book.

8.2 SCREW THEORY TOOLBOX FOR ROBOTICS (ST24R)

This toolbox encompasses a collection of functions to analyze the robot kinematics and dynamics through formulas of the theory of Lie groups and the geometry of screws (Murray et al., 2017). Naturally, the toolbox covers all the necessary screw mathematics to follow the examples and simulations. There are other toolboxes with similar functionalities of great quality to cover screw theory (Lynch and Park, 2017).

ST24R includes functions for position and rotation representation, homogeneous and spatial vector algebra transformations, screws, twists, wrenches, geometric Jacobian, adjoint transformations, and conversion between data types.

At this moment, the ST24R is programmed only in MATLAB®. The functions have study documentation, and the details of the software are in an internet repository (Pardos-Gotor, 2021a). This library implements the concepts of this book to make the formulas operational. There is access to the code of the toolbox, exercises, and simulations for real manipulators. They are useful for applications of many robotics structures and configurations.

ST24R is a free software to redistribute or modify under the terms of the GNU Lesser General Public License. It is entirely open to cooperation and improvement.

It is decisive to appreciate that the ST24R software is developed mainly for teaching and learning purposes. Therefore, it is not for industrial use. The code is programmed to enjoy and learn the screw theory concepts, and the execution performance is not a priority. Nevertheless, the mathematical approach's advantages are so significant that they will become the cornerstone of practical implementations.

Beware that the ST24R functions implementing the canonical IK subproblems always give a result. In the case of having a problem without an exact solution, the algorithms will provide an approximation. These functions are beneficial for the simulations and lead to quite natural solutions for many robots' motion. If there is a need for a different approach, it is possible to modify the software code to other utilities.

8.3 FORWARD KINEMATICS SIMULATIONS

This section presents simulations for all exercises presented in Chapter 3 for screw theory FK. The simulation experiments with the actual geometry and dynamics of the manipulators, even though this section only simulates FK formulations (Lilly and Orin, 1994). In the simulations, we apply the Product of Exponentials (POE) (Brockett, 1983) to the mechanical system's kinematics, a digital twin of the real robots. Some dynamics for the used manipulators are approximations to the actual commercial robots, but this does not affect the validation of the FK screw theory. The tools for these simulations are Simulink® environment with Simscape™ and the ST24R library.

Robotics Simulation 219

8.3.1 General Solution to Forward Kinematics Simulation

There are complete models for all FK simulators of the following sections. The structure of each simulator model has seven blocks at the first level (see an example in Figure 8.1). All the necessary files to make any simulator running are in a compressed folder in the internet repository[1].

For anyone with a minimum experience working with MATLAB®, Simulink®, and Simscape™, the understanding of the simulator architecture will be easy to grasp with the following descriptions for the blocks (see Figure 8.1) of the model.

- **Model Basic Configuration Modules** (marked as 1 in the Figure 8.1): any Simscape™ model needs the following modules for the configuration of the simulation: Block Parameters Solver, Mechanism Configuration (where fix the gravity), and the World Frame (considered the spatial and inertial coordinate system for the mechanism). These modules affect the whole model from its kinematics perspective and math approach. Then, be aware of these definitions before applying POE algorithms.
- **Simulation Dashboard** (marked as 2 in the Figure 8.1): it is a command tablet for the robot. The dashboard serves to manage the joints in two ways, manually with six knobs (one for each joint) or with an automatic input trajectory. A selector is precisely to change between these two operational modes, "manual" or "automatic." Besides, a switch simulates the "safety button" functionality to freeze the mechanism whenever this control activates.
- **Robot Target (NOAP)** (marked as 3 in the Figure 8.1): this block contains a cubic object to which a reference system is associated and two displays to indicate the position and orientation (X-Y-Z) of that object concerning the spatial reference system. We use this NOAP object to define the POE FK map result. This block's inputs come out of the dashboard block and are also the command positions for the robot joints. Then, the application of screw mathematics defines the pose of this cube (i.e., NOAP) in the simulator 3D space.

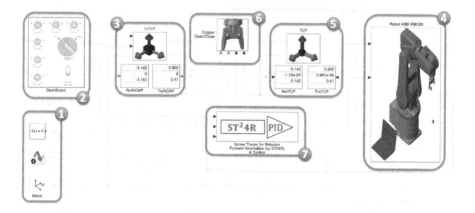

FIGURE 8.1 ABB IRB120 FK simulator complete model.

In building the NOAP object, we use several modules: cartesian joint, gimbal joint, rigid transform, solid, or transform sensor.
- **Manipulator Multibody System** (marked as 4 in the Figure 8.1): the inputs to this block are the world frame and the magnitudes of reference for the joints. The outputs are the joints' actual position values in the simulator multibody system and the open chain's end toward the tool block. The robot body goes from the base to the gripper, coupling in an available chain the joints and links. We build all the kinematic chain that constitutes the robot by adding joints and links. It is also possible to import a graphic 3D file for the rigid solid corresponding to each link, which defines its geometric and mechanical characteristics. The robot is an open chain formed by the base, the couples of joint-link, and the gripper. Each link model has one inertia module. It is helpful to define the mass and moments of inertia for the link. A rigid transformation determines the link Center of Mass (CM) and the coordinate frame orientation associated with that CM. For the screw theory dynamics formulations, on many occasions (particularly for ID simulations) it is convenient to define the orientation of this frame associate with the CM of the link, as the orientation of the stationary spatial frame. This simulation is concerned with FK, and the actual dynamics characteristics do not affect the simulator's performance. Nonetheless, to work with a model that mimics well the real manipulator, it is highly recommended to build the multibody system as accurately as possible to the mechanism's dynamics. Besides, in doing so, this manipulator block will be helpful for the simulations of robot dynamics in successive sections of this chapter. For each joint and link model, there are also two rigid transformation modules necessary to define the orientation of the link coordinate system, according to the Denavit–Hartenberg (DH) classical parameters. Even though we use screw theory for all our maths, the Simscape™ system uses the DH approach to define all joints motion on the "Z" axis. This precedent is critical for understanding the function of these two transformations.
- **Tool Center Point (TCP)** (marked as 5 in the Figure 8.1): this block contains an aspheric object to which a reference system is associated, as well as two displays to indicate the position and orientation (X-Y-Z) of that object concerning the spatial reference system. This TCP object is located precisely at the point of the robot tool that we consider as an end-effect for accomplishing a task. The input of this block proceeds from the block where the serial link manipulator's mechanical structure is constructed and, more specifically, from the output given by the gripper. This block stands out independently to compare its pose in space with the NOAP block target. The dashboard defines the target in two possible ways, either manually or automatically with a predefined trajectory file. The difference between the pose of both objects becomes evident numerically and graphically. First, it is easy to check from the displays on the complete module's screen if the position and orientation of the TCP and NOAP coincide. Second, from the Simscape™ mechanical explorer is also evident the contrast between both objects (i.e., the TCP ball and the NOAP cube) in the 3D visualization.
- **Gripper** (marked as 6 in the Figure 8.1): this block contains the gripper and develops a visual aid during the simulations to make evident when the TCP

Robotics Simulation

pose of the robot reaches the NOAP target pose, both in translation and in rotation. When the pose of both objects matches (or has a negligible difference), a signal closes the gripper so that it is visually recognizable. The mechanics manipulation (Mason, 2001) of these simulations is simple and only consists of closing and opening this gripper.

- **Algorithms and Controller** (marked as 7 in the Figure 8.1): this block is the core of the simulations that we will present in this chapter. It includes the implementation of the FK screw theory geometric algorithms for diverse mechanical problems. Besides, it contains a simple control (e.g., PID) for the simulated robot digital twin. Typically, the block's inputs are the targets (e.g., NOAP), which might come from the desired trajectory for specific tasks. In this FK section, the inputs are the reference position for the joints coming from the dashboard. Therefore, the algorithm to the multibody system is merely pass-through. Another information necessary for the robot's control is the feedback with the current positions from the robot's digital twin joints. The block's outputs are the desired values for the magnitudes of the joints, which are positions for the FK simulators.

Simscape™ has a tool to visualize the designs in a 3D environment, the "Simulation Mechanics Explorer" (see Figure 8.2). The device comprises a visualization pane to view the model, a tree view pane to explore the model hierarchy, and a properties pane to see the individual component parameters. Altogether it constitutes a very convenient environment to check the screw theory algorithms and solutions for the robots. Besides, there is another tool, the "Video Creator and Recording," which allows configuring videos from the simulation to understand the algorithms' performance better.

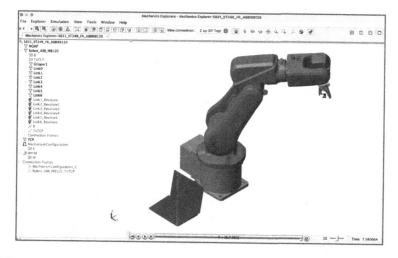

FIGURE 8.2 Simulation mechanics explorer with ABB IRB120 manipulator.

To extend the knowledge regarding FK is recommendable to practice with the simulators of different manipulators. There are several exercises with typical robot architectures in the following sections. They will permit us to consolidate the skills to apply the screw theory formulations. These examples have different joints and configurations. The solved robot architectures are Puma (e.g., ABB IRB120), Bending Backwards (ABB IRB1600), Gantry (e.g., ABB 6620LX), Scara (e.g., ABB IRB910SC), Collaborative (e.g., UNIVERSAL UR16e), and a Redundant manipulator (e.g., KUKA IIWA).

It is advisable to review the Screw Theory Toolbox for Robotics ST24R (Pardos-Gotor, 2021a) to understand better the concepts working in the examples of the following sections. Besides, there are also some videos available with the results of the simulations (Pardos-Gotor, 2021b), which help to better understand the exercises.

8.3.2 Puma Robots (e.g., ABB IRB120)

There is a complete simulator for this Puma type manipulator (Figure 8.1). The complete model operates as described in the previous section for the general solution to FK simulation. This example has the conventional stationary spatial frame (i.e., on the base) with the "Z" axis up and the gravity potential in that direction.

We include a recording for better watching this robot's simulation performing FK (Video 8.3.2)[2]. The recording lasts for 10 seconds and has both automatic and manual command periods. On the left-hand side of the screen, we can check the commands on the simulator dashboard. On the right-hand side, we see the evolution of the manipulator on the mechanics' explorer. The simulation gets started with predefined trajectories applied to the robot joints. At a specific moment, the "safety" button activates the freezing of the manipulator. Once the button is released, the robot retakes the latest target of the FK trajectory still evolving. Finally, for the second simulation period, the manual control from the dashboard activates the joint knobs to move the manipulator.

All files necessary to complete this FK system, including the screw theory functions, are inside a compressed folder (Simulator 8.3.2)[3] in the internet hosting for the software of this book.

8.3.3 Puma Robots (e.g., ABB IRB120) "Tool-Up"

This simulation for the Puma robot has the tool oriented upward in the reference robot pose. The complete model works as explained in Section 8.3.1 for the general solution to FK simulation. The only difference here is for the reference home position of the robot, chosen to demonstrate the adaptability of the screw theory design to different requirements. This example illustrates how we can with the POE flexibility select the most convenient reference (home) pose for the manipulator' application.

In this exercise, the tool in the home pose is pointing up. But this is not the only difference with the previous simulation, and the stationary spatial frame (i.e., the one situated on the base) has the "Y" axis up. This design goes out of the DH convention but shows more flexibility for the screw theory approach. This election affects the axis to apply the gravity potential in the simulation. Altogether the implementation is

Robotics Simulation

equally straightforward. However, realize this screw theory elasticity is very convenient for some applications.

We include a video for better watching this robot's simulation performing FK (Video 8.3.3)[4]. The recording lasts for 10 seconds and has both automatic and manual command periods.

All files necessary to complete this FK system, including the screw theory functions, are inside a compressed folder (Simulator 8.3.3)[5] in the internet hosting for the software of this book.

8.3.4 BENDING BACKWARDS ROBOTS (E.G., ABB IRB1600)

These simulations see a slightly different robot architecture, a Bending Backwards manipulator. This design does not need to rotate the robot to reach for things behind it, swinging the arm backward to extend the robot's working range. The complete model works as explained in Section 8.3.1 for the general solution to FK simulation. This example has the conventional stationary spatial frame (i.e., the one situated on the base) with the "Z" axis up and the gravity potential in that direction.

We include a video for better watching this robot's simulation performing FK (Video 8.3.4)[6]. The recording lasts for 10 seconds and has both automatic and manual command periods. On the left-hand side of the screen, we can check the commands on the simulator dashboard. On the right-hand side, we see the evolution of the manipulator on the mechanics' explorer. The simulation gets started with predefined trajectories applied to the robot joints. At a specific moment, the "safety" button activates the freezing of the manipulator. Once the button is released, the robot retakes the latest target of the FK trajectory still evolving. Finally, for the second simulation period, the manual control from the dashboard activates the joint knobs to move the manipulator.

All files necessary to complete this FK system, including the screw theory functions, are inside a compressed folder (Simulator 8.3.4)[7] in the internet hosting for the software of this book.

8.3.5 GANTRY ROBOTS (E.G., ABB IRB6620LX)

These simulations deal with this Gantry robotic architecture, which combines the advantages of both a long linear axis for the first joint plus five revolute joints. We know how the treatment of the prismatic joint is the same with screw theory. The complete model works as explained in Section 8.3.1 for the general solution to FK simulation. Nonetheless, we must pay attention to the model as it has some novelties and differences from the earlier examples. The main difference comes from the fact that this robot has a first prismatic joint. This fact affects several things in the simulator, like the "Gain" module in the dashboard, the standard type in the multibody tree, or the "Simulink-PS" converter blocks, which must use "m" instead of "rad." The translation of this prismatic joint develops on a "Z" axis, following the DH convention demanded by Simulink®. This example shows the design flexibility provided by the screw theory. For instance, the stationary spatial frame has the "Y" axis up. This election affects the axis to apply the gravity potential in the simulation.

We include a video for better watching this robot's simulation performing FK (Video 8.3.5)[8]. The recording has two periods, one for automatic control and another with a manual command to the joints.

All files necessary to complete this FK system, including the screw theory functions, are inside a compressed folder (Simulator 8.3.5)[9] in the internet hosting for the software of this book.

8.3.6 SCARA ROBOTS (E.G., ABB IRB910SC)

These simulations have a two parallel joint axes mechanism, advantageous for many assembly operations. For the sake of showing the freedom given by the screw theory in the selection of the coordinate systems, we have chosen the typical virtual reality spatial frame orientation. This example has only four joints, but the general approach applies. The complete model works as explained in Section 8.3.1 for the general solution to FK simulation. Nonetheless, this model is quite different from previous examples. This robot has only four joints, and the third is prismatic. These characteristics impact the design of the simulator. For instance, the dashboard has only four knobs, and for the third joint, the type in the multibody tree must be prismatic with the "Simulink-PS" converter blocks defined in "m" instead of "rad." The translation of this prismatic joint moves on a "Z" axis, following the DH convention. This fact obliges to define the coordinate frames correctly in the design of the mechanical tree.

We include a video to watch this robot FK simulation (Video 8.3.6)[10]. The recording has two periods, with automatic and manual commands to the joints. There are four knobs, the safety button, and the selector on the dashboard on the screen's left-hand side. On the right-hand side, we see the evolution of the manipulator on the mechanics' explorer.

All files necessary to complete this FK system, including the screw theory functions, are inside a compressed folder (Simulator 8.3.6)[11] in the internet hosting for the software of this book.

8.3.7 COLLABORATIVE ROBOTS (E.G., UNIVERSAL UR16E)

These simulations deal with a typical collaborative manipulator made for human-robot cooperation in the workspace. The complete model works as explained in Section 8.3.1. This collaborative robot has six revolute joints, and in that sense, the model is quite like the built for the Puma and Bending Backward robots of some previous simulations. This exercise has the conventional stationary spatial frame on the manipulator base, with the "Z" axis up and the gravity potential in that direction.

We include a video for better watching this robot's simulation performing FK (Video 8.3.7)[12]. This example has two periods for both automatic and manual commands. On the left-hand side of the screen, we can check the operations on the simulator dashboard. On the right-hand side, we see the evolution of the robot on the mechanics' explorer. The simulation gets started with predefined trajectories. At a specific moment, the "safety" button makes the manipulator freeze. Then, for the second period of the simulation, we use manual control from the dashboard knobs. During the whole video, check how the TCP (i.e., the ball with a coordinate system) of the robot follows the NOAP (i.e., the cube with a coordinate frame) target.

All files necessary to complete this FK system, including the screw theory functions, are inside a compressed folder (Simulator 8.3.7)[13] in the internet hosting for the software of this book.

8.3.8 REDUNDANT ROBOTS (E.G., KUKA IIWA)

These simulations have the novelty of solving a robot with seven DoF. The performance of the simulator is good working for any number of joints. The complete model works as explained in Section 8.3.1. This example is different from others in several ways. This robot has seven joints of revolute type, which makes this manipulator a redundant mechanism. This fact affects this simulator's design and makes the dashboard have seven knobs for the seven joints of this robot. We must pay attention because, in the automatic mode, the file with the input trajectories must have seven signals. The same case happens for the output files with the joint position and velocity information, which are vectors with seven signals. The model must be redesigned as well for the feedback and command lines to the control module.

We include a video to watch this robot FK simulation (Video 8.3.8)[14]. This example has two periods for both automatic and manual commands. On the screen's left-hand side, see the dashboard, hosting seven knobs for the seven joints, the safety button, and the selector for manual or automatic control. On the right-hand side, watch the manipulator on the mechanics' explorer. The FK commands move the NOAP object, and the TCP object follows in both position and orientation.

All files necessary to complete this FK system, including the screw theory functions, are inside a compressed folder (Simulator 8.3.8)[15] in the internet hosting for the software of this book.

8.4 INVERSE KINEMATICS SIMULATIONS

In this section, we present simulations for all the IK exercises of Chapter 4. The simulation experiments with the actual geometry and dynamics of the manipulators, even though this section only simulates IK formulations. Chapter 4 acknowledged how alluring it is to obtain geometric closed-form solutions for the IK problems with the canonical subproblems and the POE of the screw theory. In the simulations, we apply the IK algorithms to the 3D mechanical system's kinematics, an excellent digital twin of the real robot. Some dynamics for the used manipulators are approximations to the actual commercial robots, but this does not affect the validation of the IK theory. The tools for these simulations are Simulink® environment, with Simscape™, and the ST24R library.

For these simulations, there are two possibilities to generate paths for the robot tool end-effector in the task-space, either in real time with the dashboard's commands or offline with some input file. However, path planning is always online in the joint space. In any case, only a point-to-point trajectory is considered, with the joint magnitude limit respected. The target is to make the end-effector reach the tool path planning in the task-space as strictly as possible. The joint trajectory timing inherits from the task-space path time scaling.

8.4.1 GENERAL SOLUTION TO INVERSE KINEMATICS SIMULATION

There are complete models for all IK simulators of the following exercises. The structure of each simulator model has seven blocks at the first level (see an example in Figure 8.3). All the necessary files to make any simulator running are in a compressed folder in the internet repository[16].

For anyone with a minimum experience working with are MATLAB®, Simulink®, Simscape™, the understanding of the simulator architecture will be easy to grasp with the descriptions for the model blocks already presented for the general solution to FK simulation (Section 8.3.1). We detail hereafter only the differences to the available FK simulator model.

- **Simulation Dashboard** (marked as 2 in Figure 8.3): it serves to manually define the tool target that we want the robot to reach, both in position and orientation. Three sliders permit to determine the Cartesian translations, and three knobs the Euler rotations (in X-Y-Z). A selector also indicates which of all possible IK solutions we decide to apply for the robot to follow the trajectory. In this same selector, we have one more position (i.e., "home") to send the robot to the initial reference pose. One switch (i.e., "freeze") is the mechanism and leaves the robot fixed in its current pose. Another button permits to transfer from the manual definition of targets, by using three sliders and three knobs (i.e., "manual"), to the offline definition of a target with a predefined trajectory file (i.e., "auto"). We can see the controls of this dashboard in Figure 8.4.
- **Robot Target (NOAP)** (marked as 3 in the in Figure 8.3): this block contains a cubic object to which a reference system is associated and two displays to indicate the position and orientation (X-Y-Z) of that object concerning the spatial reference system. It serves to define the targets that the robot tool should

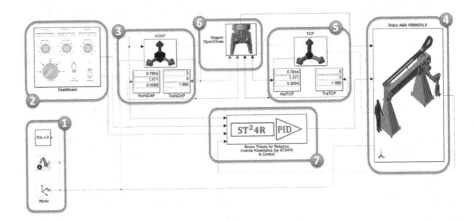

FIGURE 8.3 ABB IRB6620LX IK simulator complete model.

Robotics Simulation

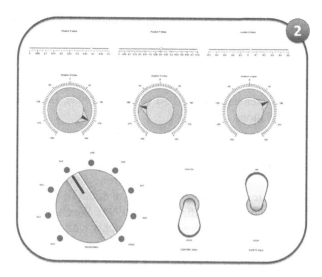

FIGURE 8.4 IK simulator dashboard for a six DoF manipulator.

reach. The block's inputs come out of the dashboard block and are the desired positions and orientations for the TCP. The outputs of this block correspond to the cube's position and orientation, which are then employed to define the homogenous transformation matrix used as input to the algorithm of solving the IK. In building the NOAP object, we use several modules: cartesian joint, gimbal joint, rigid transform, solid, or transform sensor. For more details, see Simscape™ reference documents.

- **Algorithms and Controller** (marked as 7 in the in Figure 8.3): this module is the core of the simulation for implementing geometric solutions to the manipulator IK problem, suitable for real-time applications. Three inputs to this block come from the dashboard: the target NOAP (in fact, this input has six signals for position and orientation (X-Y-Z), the value of the selector to indicate which of the available IK solutions we choose, and the signal from the dead-man safety button (to freeze the robot). Another input for the robot's control is feedback with the actual joint positions, which come from the manipulator module with the mechanical open-chain tree (i.e., the robot's digital twin). The control block has the command values for the positions of the joints. This command signal is compared with the feedback actual joint positions and applied to a PID control block. Finally, the output goes out to the simulator's robot joints as a command reference with position signals. To follow this simulation and precisely the IK function inside this module is a good idea to review Chapter 4 and the robot exercises. It proves the flexibility of the screw theory different algorithms to solve the IK of the manipulators. All of them give geometric closed-form solutions based on the application of other canonical subproblems. It is possible to test any algorithms presented in Chapter 4 by changing the function included in this controller module for the proper manipulator.

For the manual control, we define the reference for the target with the sliders (i.e., position) and knobs (i.e., orientation) from the dashboard. We generate offline a reference trajectory for the robot's tool in 3D (i.e., position and orientation X-Y-Z) for the automatic control. This offline tool trajectory has a kind of sinusoidal, random shape, which acts as reference entering as input to the IK algorithm. All trajectories avoid having closed kinematic chains (Park and Kim, 1999), as this aspect is out of the scope of the simulations.

The manual and automatic controls pass the targets to the controller one by one for the simulator to respond in real time. There is no resolution for the complete joint trajectory offline. When dealing with a single target inside the robot's dexterous workspace, the screw theory algorithm responds with several exact geometric solutions for the joint positions. The closed-form nature of the IK screw theory solutions makes the online simulation very efficient (Park 1994). We choose one out of the set of solutions to define the robot configuration to comply with the task. If the tool target is out of the workspace, we recall that the algorithms developed with ST24R always give at least one approximation. We apply the solutions to the robot joints to make the TCP reach the tool target in position and orientation. We repeat the process for all target points of the whole trajectory. The TCP follows the targets quite well, even with a simple controller. The joints must follow their trajectories to develop the task defined for the tool. It is feasible to check how the joint magnitudes remain inside the physical limits of the commercial manipulators.

To extend the knowledge regarding IK is appropriate to practice with the simulators of different manipulators. There are several exercises with typical architectures in the following sections. The robots are Puma (e.g., ABB IRB120), Bending Backwards (ABB IRB1600), Gantry (e.g., ABB 6620LX), Scara (e.g., ABB IRB910SC), Collaborative (e.g., UNIVERSAL UR16e), and a Redundant manipulator (e.g., KUKA IIWA).

It is advisable to review the Screw Theory Toolbox for Robotics ST24R (Pardos-Gotor, 2021a) to understand better the concepts working in the examples of the following sections. Besides, there are also some videos available with the results of the simulations (Pardos-Gotor, 2021b), which help to better understand the exercises.

8.4.2 Puma Robots (e.g., ABB IRB120)

There is a complete simulator for this Puma type manipulator. The model operates as described in the previous section for the general solution to IK simulation. This example has the conventional stationary spatial frame (i.e., on the base) with the "Z" axis up and the gravity potential in that direction. The path planning is solved using the screw theory IK Paden-Kahan (PK) (Paden, 1986 and Kahan, 1983) and Pardos-Gotor (PG) subproblems. In this case the subproblems are "PG7+PK2+PK1" to build this algorithm presented in Chapter 4. Afterward, it is easy to extend the same model, creating new functions for all algorithms valid also for solving the IK of this robot, which we can change inside the controller block.

This simulation can also debunk some misunderstandings regarding the fact that the POE is a relative map that requires a particular reference position to solve the IK.

Robotics Simulation 229

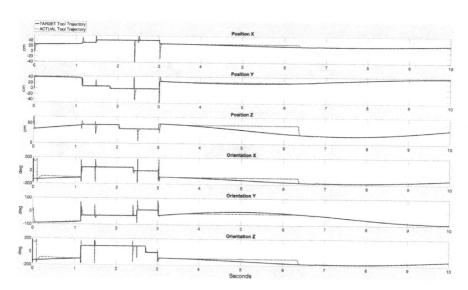

FIGURE 8.5 ABB IRB120 IK Simulation end-effector target and actual trajectories.

It suffices to have only one unique reference initial pose to solve any IK configuration. The robot can freeze itself at any moment or position during a trajectory, and afterward, the movement is resumed. The only reference position needed is the initial one.

We add a plot with the result of a simulation, contrasting the evolution of the end-effector target (Figure 8.5 solid line) and actual trajectories (Figure 8.5 dotted line). The represented simulation starts with the tracking of a trajectory in automatic mode. Then, the control passes to the manual, and we change the reference for positions and orientations. For some time, we freeze the robot to keep the tool at its pose. Afterward, we reactivate the automatic control for leaving the manipulator to track the evolving trajectory again. All the exercise works in real time, and even with a simple controller, the result is quite good because of the effectiveness of the screw theory geometric algorithms (see Figure 8.5).

We include two videos to watch the simulation performing IK, with a general view of the robot (Video 8.4.2a)[17] and focused on the tool behavior (Video 8.4.2b)[18]. Both recordings demonstrate the good real-time response of the implementations.

There is a different exercise for the same robot. The innovation here uses a SpaceMouse to move the target in a manual mode for translation and rotation. This device offers an intuitive, effortless, and precise 3D navigation in applications where one cannot experiment with a standard mouse and keyboard. This new device substitutes the sliders and knobs in the dashboard block model (Figure 8.6).

The SpaceMouse has six DoF and can manipulate digital content inside 3D applications. In this simulation model, the device moves (both translation and rotation) the cubic object with the attached frame, which we use to define the target "NOAP." With the SpaceMouse is easier to make the robot follow different trajectories in real time.

This simulation performance is equivalent to the previous one to follow the automatic input trajectory or targets defined manually (in this case, with the SpaceMouse).

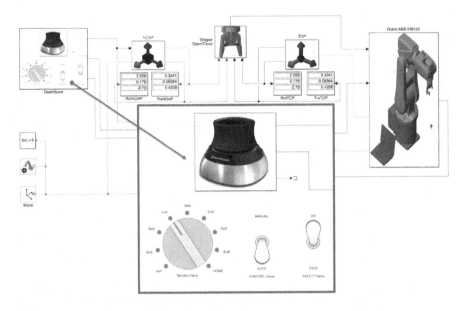

FIGURE 8.6 ABB IRB120 IK simulator complete model with SpaceMouse.

This model's change only affects the instrument that moves the target and not the screw theory IK algorithm (i.e., PG7+PK2+PK1) developed in the previous example.

All files necessary to complete this IK system, including the screw theory functions, are inside two compressed folders (Simulator 8.4.2a[19] and Simulator 8.4.2b[20]) in the internet hosting for the software of this book.

8.4.3 PUMA ROBOTS (E.G., ABB IRB120) "TOOL-UP"

Another simulation for the Puma robot is here, but in this case, the tool is oriented upward in the reference robot pose. We understand the model better with all detailed explanations of Section 8.4.1. for the general solution to IK simulation.

The only difference here is for the reference home position of the robot. This example illustrates how with the screw theory it is possible choosing the most convenient reference (home) pose for the robot according to the application. In this exercise, the tool in the home pose is pointing up. Besides, the stationary spatial frame has the "Y" axis up. This election affects the axis to apply the gravity potential in the simulation. The joint path planning is solved using the screw theory IK "PG7+PK6+PK1" algorithm presented in Chapter 4. Afterward, it is easy to extend the same model, creating new functions for all algorithms valid for solving the IK of this robot, which we can change inside the controller block.

We add a plot with the result of a simulation, contrasting the evolution of the end-effector target (Figure 8.7 solid line) and actual trajectories (Figure 8.7 dotted line). The represented simulation shows a trajectory tracking testing the automatic and manual control. All the exercise works in real time, and the result is quite good because of the effectiveness of the screw theory geometric algorithms (Figure 8.7).

Robotics Simulation

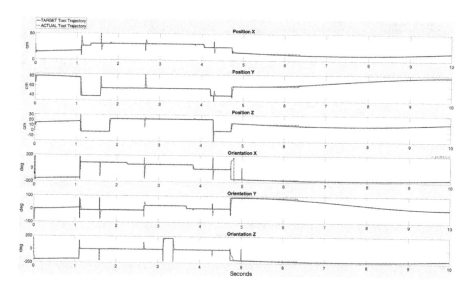

FIGURE 8.7 ABB IRB120 "Tool-Up" IK Simulation end-effector target and actual trajectories.

We include two videos for better watching the simulation performing IK, with a general view of the robot (Video 8.4.3a)[21] and focused on the tool behavior (Video 8.4.3b)[22]. Both recordings demonstrate the good real-time response of the implementations.

All files necessary to complete this IK system, including the screw theory functions, are inside a compressed folder (Simulator 8.4.3)[23] in the internet hosting for the software of this book.

8.4.4 BENDING BACKWARDS ROBOTS (E.G., ABB IRB1600)

This IK simulator sees a slightly different robot architecture, a Bending Backwards manipulator. This design does not need to rotate the robot to reach for things behind it. Swinging the arm backward extends the working range of the robot. We can understand the complete model in detail with all explanations of Section 8.4.1. for the general solution to IK simulation.

The joint path planning is solved using the screw theory IK "PG7+PK6+PK1" algorithm presented in Chapter 4. Afterward, it is easy to extend the same model, creating new functions for all algorithms valid for solving the IK of this robot, which we can change inside the controller block.

We add a plot with the result of a simulation, contrasting the evolution of the end-effector target (Figure 8.8 solid line) and actual trajectories (Figure 8.8 dotted line). The represented simulation shows a trajectory tracking testing the automatic and manual control. All the exercise works in real time, and the result is quite good because of the effectiveness of the screw theory geometric algorithms (Figure 8.8).

We include two videos for better watching the simulation performing IK, with a general view of the robot (Video 8.4.4a)[24] and focused on the tool behavior

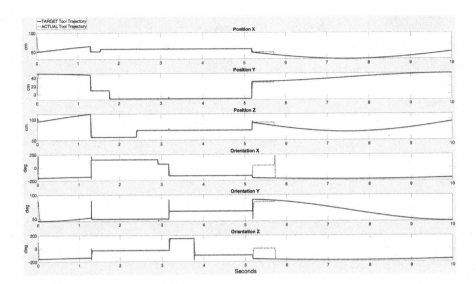

FIGURE 8.8 ABB IRB1600 IK Simulation end-effector target and actual trajectories.

(Video 8.4.4b)[25]. Both recordings demonstrate the good real-time response of the implementations.

All files necessary to complete this IK system, including the screw theory functions, are inside a compressed folder (Simulator 8.4.4)[26] in the internet hosting for the software of this book.

8.4.5 Gantry Robots (e.g., ABB IRB6620LX)

This IK simulator deals with this Gantry robotic architecture, which combines the advantages of both a long linear axis for the first joint plus five revolute joints. We know how the treatment of the prismatic joint is the same with screw theory. We can understand the complete model in detail with all explanations of Section 8.4.1. for the general solution to IK simulation.

This model has some differences from the previous examples. The first joint is prismatic, affecting the type in the multibody tree, and the "Simulink-PS" converter block, which must use "m" instead of "rad." With the screw theory, the rules to determine the stationary spatial frame are flexible, and here it has the "Y" axis up. This election affects the application of the gravity potential in the simulation. The joint path planning is solved using the screw theory IK "PG1+PG4+PG6+PK1" (Pardos-Gotor, 2018) algorithm presented in Chapter 4. Afterward, it is easy to extend the same model, creating new functions for all algorithms valid for solving the IK of this robot, which we can change inside the controller block.

We add a plot with the result of a simulation, contrasting the evolution of the end-effector target (Figure 8.9 solid line) and actual trajectories (Figure 8.9 dotted line). The represented simulation shows a trajectory tracking testing the automatic and manual control. All the exercise works in real time, and the result is quite good because of the effectiveness of the screw theory geometric algorithms (Figure 8.9).

Robotics Simulation

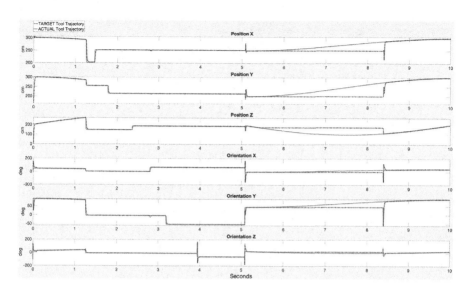

FIGURE 8.9 ABB IRB6620LX IK Simulation end-effector target and actual trajectories.

We include two videos for better watching the simulation performing IK, with a general view of the robot (Video 8.4.5a)[27] and focused on the tool behavior (Video 8.4.5b)[28]. Both recordings demonstrate the good real-time response of the implementations.

All files necessary to complete this IK system, including the screw theory functions, are inside a compressed folder (Simulator 8.4.5)[29] in the internet hosting for the software of this book.

8.4.6 SCARA ROBOTS (E.G., ABB IRB910SC)

This IK simulator has a two parallel joint axes mechanism, advantageous for many assembly operations. For the sake of showing the freedom given by the screw theory in the selection of the coordinate systems, we have chosen the typical virtual reality spatial frame orientation. This example has only four joints, but the general approach applies. We can understand the complete model in detail with all explanations of Section 8.4.1. for the general solution to IK simulation.

This model has only four joints, and the third is prismatic, affecting the standard joint type in the multibody tree and the "Simulink-PS" converter block, which must use "m" instead of "rad" for the third joint. With the screw theory, it is possible to choose the stationary spatial frame with the "Y" axis in the vertical. This election affects the axis to apply gravity. The joint path planning is solved using the screw theory IK "PG1+PG4+PK1" algorithm presented in Chapter 4. Afterward, it is easy to extend the same model, creating new functions for all algorithms valid for solving the IK of this robot, which we can change inside the controller.

We add a plot with the result of a simulation, contrasting the evolution of the end-effector target (Figure 8.10 solid line) and actual trajectories (Figure 8.10 dotted line). The represented simulation shows a trajectory tracking testing the automatic

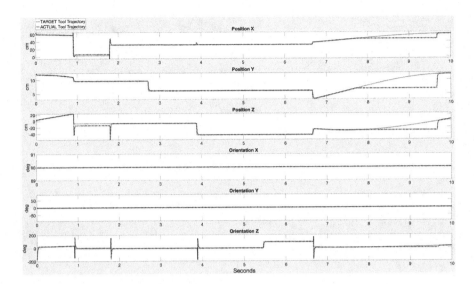

FIGURE 8.10 ABB IRB910SC IK Simulation end-effector target and actual trajectories.

and manual control. All the exercise works in real time, and the result is quite good because of the effectiveness of the screw theory geometric algorithms (Figure 8.10).

We include two videos for better watching the simulation performing IK, with a general view of the robot (Video 8.4.6a)[30] and focused on the tool behavior (Video 8.4.6b)[31]. Both recordings demonstrate the good real-time response of the implementations.

All files necessary to complete this IK system, including the screw theory functions, are inside a compressed folder (Simulator 8.4.6)[32] in the internet hosting for the software of this book.

8.4.7 COLLABORATIVE ROBOTS (E.G., UNIVERSAL UR16E)

This IK simulator deals with a typical collaborative manipulator made for human-robot cooperation in the workspace. We can understand the complete model in detail with all explanations of Section 8.4.1. for the general solution to IK simulation.

The joint path planning is solved using the IK "PG5+PG3+PK1+PG8" algorithm presented in Chapter 4. Afterward, it is easy to extend the same model, creating new functions for all algorithms valid for solving the IK of this robot, which we can change inside the controller.

We add a plot with the result of a simulation, contrasting the evolution of the end-effector target (Figure 8.11 solid line) and actual trajectories (Figure 8.11 dotted line). The represented simulation shows a trajectory tracking testing the automatic and manual control. All the exercise works in real time, and the result is quite good because of the effectiveness of the screw theory geometric algorithms (Figure 8.11).

We include two videos for better watching the simulation performing IK, with a general view of the robot (Video 8.4.7a)[33] and focused on the tool behavior

Robotics Simulation 235

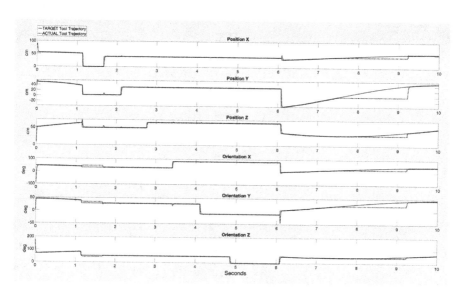

FIGURE 8.11 UNIVERSAL UR16e IK Simulation end-effector target and actual trajectories.

(Video 8.4.7b)[34]. Both recordings demonstrate the good real-time response of the implementations.

All files necessary to complete this IK system, including the screw theory functions, are inside a compressed folder (Simulator 8.4.7)[35] in the internet hosting for the software of this book.

8.4.8 REDUNDANT ROBOTS (E.G., KUKA IIWA)

This IK simulator has the novelty of solving a robot with seven DoF. The performance of the formulations is good working for any number of joints. A redundant robot has more than the minimal number of DoF required to complete a task. The robot can have an infinite number of joint configurations to hit a specific tool target. The extra DoF in redundant manipulators are helpful to develop some motion strategies. We can understand the complete model in detail with all explanations of Section 8.4.1. for the general solution to IK simulation.

The joint path planning is solved using the screw theory IK algorithm "PK1+PK3+PK2+PK2+PK2+PK1" presented in Chapter 4. Afterward, it is easy to extend the same model, creating new functions for all algorithms valid for solving the IK of this robot, which we can change inside the controller. For instance, the "PK1+PK3+PG6+PG6+PG6+PK1" is also an interesting algorithm to test.

We add a plot with the result of a simulation, contrasting the evolution of the end-effector target (Figure 8.12 solid line) and actual trajectories (Figure 8.12 dotted line). The represented simulation shows a trajectory tracking testing the automatic and manual control. All the exercise works in real time, and the result is quite good because of the effectiveness of the screw theory geometric algorithms (Figure 8.12).

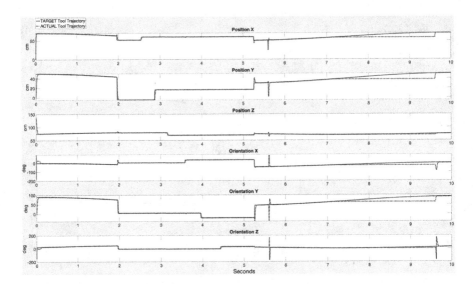

FIGURE 8.12 KUKA IIWA IK Simulation end-effector target and actual trajectories.

We include two videos for better watching the simulation performing IK, with a general view of the robot (Video 8.4.8a)[36] and focused on the tool behavior (Video 8.4.8b)[37]. Both recordings show the good real-time response of the implementations.

All files necessary to complete this IK system, including the screw theory functions, are inside a compressed folder (Simulator 8.4.8)[38] in the internet hosting for the software of this book.

8.5 DIFFERENTIAL KINEMATICS SIMULATIONS

This section presents simulations for all the exercises presented in Chapter 5 of this book. With the explanations of Chapter 7 for Trajectory Generation is possible to understand those algorithms better.

In Chapter 5, we have realized how practical it is to obtain closed-form solutions from geometric Jacobian. It permits to work differential kinematics (DK) problems with excellent performance. The screw theory provides a very natural description of the robot geometric Jacobian, highlighting the robot mechanics and has none of a local analytic representation's weaknesses. We can calculate this geometric Jacobian without any differentiation, which is an enormous advantage for computer implementations. Actually, for the DK simulations we need the inverse geometric Jacobian matrix (Penrose, 1955).

The simulations of this section are for the trajectory generation based on DK. Remember that this approach is an approximative alternative to IK. The fundamental strategy is the integration of the joint velocities to generate the joint trajectories.

The tools to test these simulations are MATLAB®, Simulink®, Simscape™ & ST24R.

DK is a method to move the robot from one end-effector pose to another without calculating the IK. We get the new joint magnitudes with the joint coordinates for the initial tool pose and the joint velocities' integration.

We provide a reference trajectory for the robot tool in the practical application, discretized according to specific step size. We start the algorithm by knowing the robot configuration with its initial joint positions, which imply the actual tool pose by FK. We get the tool velocities from the difference between the configurations of the point of interest in the tool reference trajectory and the actual tool pose. By the integration of the geometric DK, we obtain the joint coordinates increments. There can be many formulations to compute the integration, but even the Euler explicit integration method performs well. The process goes on for the following points of the expected reference tool trajectory.

Pay attention to the fact that this formulation is moving along the tangent instead of the actual curve. It is only one approximation to the exact trajectory, but when the discretization of targets is small enough, the results are more than acceptable. It is possible even to check with the simulators' manual control that this approach works "surprisingly well" even for huge jumps between two successive targets.

There are limitations for this DK approach, which may emerge sometimes. For instance, moving along a trajectory with a particular robot configuration may leave unreachable targets inside the robot's workspace. The DK algorithm tries to follow a linear velocity between targets, triggering the problem for some trajectories. Still, if some joints' position limits reached their maximum magnitude, this makes some motion impossible. The manipulator gets stacked and does not reach the final target. The solution for those events is to choose a different robot configuration or define a feasible alternative trajectory (e.g., turnaround) between the two targets. One way to select a different robot configuration can be to use the IK algorithms explained in Chapter 4. They provide several solutions (exact or approximate) for any tool target, which correspond to different robot configurations. For an application that does not accept approximative solutions, it is possible to reprogram it to adapt the IK algorithms.

As a summary, we can say that there are two possibilities to generate paths for the robot tool in the task-space, either offline with some predesigned file or in real time with the dashboard commands. However, path planning in joint space is always online. Then, in any case, only a joint point-to-point trajectory is considered, with trapezoidal velocity interpolation in real time. The joint positions and velocities respect their constraints for keeping the safety limits of the robot actuators. The trajectory planning's core goal is to make the robot joints pass through the target series (i.e., start, via, goal) with the position's accuracy. The goal is to make the tool end-effector reach as strictly as possible the shape and timing of the task-space trajectory. In other words, the joint path time scaling does not maximize the speed of joint motion. The joint trajectory planning timing inherits from the task-space path time scaling.

We have learned some lessons throughout the simulations presented so far, always with the underlying screw theory. For robots with available IK closed-form algorithms, we have exact geometric solutions for the different manipulator configurations when the target is inside the workspace (also, the ST24R implementation gives approximative solutions when the targets are out). Nevertheless, sometimes there is no geometric solution for the IK of the mechanism. In other cases, the only available answer is numeric, or the target is not in the robot's dexterous workspace. For all those situations, a distinctive approach and probably the best is to solve the trajectory generation using the screw theory DK. It provides an excellent geometric approximation without the need for differentiation.

8.5.1 GENERAL SOLUTION TO DIFFERENTIAL KINEMATICS SIMULATION

There are complete models for all DK simulators of the following exercises. The structure of each simulator model has seven blocks at the first level (see an example in Figure 8.13). All the necessary files to make any simulator running are in a compressed folder in the internet repository[39].

For anyone with a minimum experience working with are MATLAB®, Simulink®, Simscape™, the understanding of the simulator architecture will be easy to grasp with the descriptions for the model blocks already presented for the general solution to FK simulation (Section 8.3.1). We detail hereafter the differences from the general simulation architecture.

- **Simulation Dashboard** (marked as 2 in the in Figure 8.13): this dashboard for DK is new in the simulation model (Figure 8.14). A central switch permits to transfer from the manual online (i.e., "manual") definition of targets to the offline description of a predefined trajectory (i.e., "auto"). We use three sliders for the translations and three knobs for rotations in X-Y-Z. Another switch selects the application of the DK algorithm to follow the target (i.e., "tracking"). The other function of that switch (i.e., "home") sends the robot to the initial reference pose. One more button is for freezing the mechanism and leaves the robot fixed in its current posture (i.e., "stop"). The alternative (i.e., "on" position) allows the algorithms and simulator to command the manipulator.

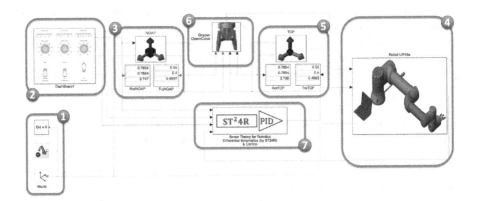

FIGURE 8.13 UNIVERSAL UR16e DK simulator complete model.

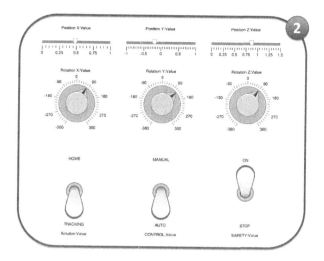

FIGURE 8.14 DK simulator dashboard.

- **Algorithms and Controller** (marked as 7 in the in Figure 8.13): this module is the core of the simulation for implementing the manipulator DK, suitable for real-time applications. This block's inputs and outputs are precisely the same as those used in the simulator of the previous section for IK (see Section 8.4.1 for more explanations). It is an excellent idea to review Chapter 5 with the exercises for better following the algorithm called from the DK function. We introduced the advantages of the geometric Jacobian and twist velocity concepts provided by the screw theory.

To extend the knowledge regarding DK is desirable to practice with the simulators of different manipulators. They will permit us to consolidate the skills to apply the screw theory formulations. These examples have different configurations. The solved robot architectures are Puma (e.g., ABB IRB120), Bending Backwards (ABB IRB1600), Gantry (e.g., ABB 6620LX), Scara (e.g., ABB IRB910SC), Collaborative (e.g., UNIVERSAL UR16e), and a Redundant manipulator (e.g., KUKA IIWA).

It is advisable to review the Screw Theory Toolbox for Robotics ST24R (Pardos-Gotor, 2021a) to understand better the concepts working in the examples of the following sections. Besides, there are also some videos available with the results of the simulations (Pardos-Gotor, 2021b), which help to better understand the exercises.

8.5.2 Puma Robots (e.g., ABB IRB120)

There is a complete DK simulator for this Puma type manipulator. The model is more understandable with all explanations of Section 8.5.1. We generate a reference trajectory for the robot's tool in 3D (i.e., position and orientation X-Y-Z). This reference enters as input to the DK algorithm. For each point of the trajectory, the tool's velocity and the geometric Jacobian permit to obtain the velocity of the joints.

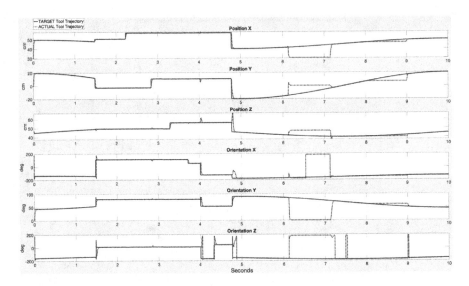

FIGURE 8.15 ABB IRB120 DK Simulation end-effector target and actual trajectories.

The integration of each joint velocity gives the incremental change in the joint trajectory magnitude. Applying the resulting magnitudes to the robot joints, the TCP follows the target in terms of both position and orientation.

We add a plot with the result of a simulation, contrasting the evolution of the end-effector target (Figure 8.15 solid line) and actual trajectories (Figure 8.15 dotted line). The represented simulation starts with the tracking of a trajectory in automatic mode. Then, the control passes to the manual, and we change the reference for positions and orientations. For some time, we send the robot to the home configuration, and subsequently, the control goes to manual and freeze the robot to keep the tool at its pose. Afterward, we reactivate the automatic control for leaving the manipulator to track the evolving trajectory planning again. All the exercise works in real time, and even though there is a simple controller, the result is quite good for the trajectory tracking (see Figure 8.15).

We include a video for better watching the simulation performing of DK (Video 8.5.2)[40]. On the left-hand side of the screen, we have the dashboard. On the right-hand side, we see the evolution of the robot in the mechanics' explorer. In the simulation recorded we test all the features available in the dashboard, such as manual and automatic control or the possibility to freeze the robot or sending it home. The recording demonstrates the good real-time response of this implementation.

All files necessary to complete this DK system, including the screw theory functions, are inside a compressed folder (Simulator 8.5.2)[41] in the internet hosting for the software of this book.

8.5.3 PUMA ROBOTS (E.G., ABB IRB120) "TOOL-UP"

Another simulation for the Puma robot is here for DK, but in this case, the tool is oriented upward in the reference robot pose. We understand the model better with all explanations of Section 8.5.1. for the general solution.

Robotics Simulation

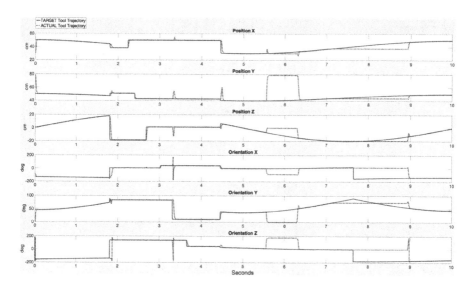

FIGURE 8.16 ABB IRB120 "Tool-Up" DK Simulation end-effector target and actual trajectories.

We generate an end-effector trajectory planning either with a complete task or with the dashboard for a single target. This reference enters as input to the DK algorithm. For each point of the trajectory with the geometric Jacobian and the tool's velocity, it is possible to get in real time the velocity of the joints. The integration of each joint velocity gives the increment in the joint magnitude. Applying the resulting magnitudes to the robot joints, the TCP follows the target in position and orientation.

We add a plot with the result of a simulation, contrasting the evolution of the end-effector target (Figure 8.16 solid line) and actual trajectories (Figure 8.16 dotted line). We test all the features available in the dashboard, such as manual and automatic control or the possibility to freeze the robot or sending it home. Despite the approximative nature of the DK geometric algorithms, the result of the trajectory tracking is good (see Figure 8.16).

We include a video for better watching the simulation performing DK (Video 8.5.3)[42]. In the recorded simulation, we test all the features available in the dashboard, such as manual and automatic control or the possibility to freeze the robot or sending it home. The recording demonstrates the good real-time response of the implementation.

All files necessary to complete this DK system, including the screw theory functions, are inside a compressed folder (Simulator 8.5.3)[43] in the internet hosting for the software of this book.

8.5.4 BENDING BACKWARDS ROBOTS (E.G., ABB IRB1600)

This DK simulator sees a slightly different robot architecture, a Bending Backwards manipulator. This design does not need to rotate the robot to reach for things behind

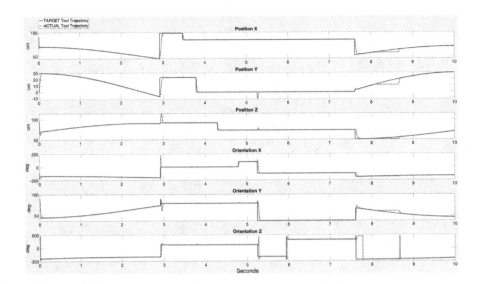

FIGURE 8.17   ABB IRB1600 DK Simulation end-effector target and actual trajectories.

it. Swinging the arm backward extends the working range of the robot. We understand the model better with all explanations of Section 8.5.1 for the general solution.

We generate an end-effector trajectory planning either with a complete task or with the dashboard for a single target. This reference enters as input to the DK algorithm. For each point of the trajectory, the tool's velocity and the geometric Jacobian permit to obtain the velocity of the joints in real time. The integration of each joint velocity gives the increment in the joint trajectory magnitude. Applying the resulting magnitudes to the robot joints, the TCP follows the target in position and orientation.

We add a plot with the result of a simulation, contrasting the evolution of the end-effector target (Figure 8.17 solid line) and actual trajectories (Figure 8.17 dotted line). We test all the features available in the dashboard, such as manual and automatic control or the possibility to freeze the robot or sending it home. Despite the approximative nature of the DK geometric algorithms, the result of the trajectory tracking is good (see Figure 8.17).

We include a video for better watching the simulation performing DK (Video 8.5.4)[44]. In the recorded simulation, we test all the features available in the dashboard, such as manual and automatic control or the possibility to freeze the robot or sending it home. The recording demonstrates the good real-time response of this implementation.

All files necessary to complete this DK system, including the screw theory functions, are inside a compressed folder (Simulator 8.5.4)[45] in the internet hosting for the software of this book.

8.5.5   GANTRY ROBOTS (E.G., ABB IRB6620LX)

This DK simulator deals with this Gantry robotic architecture, which combines the advantages of both a long linear axis for the first joint plus five revolute joints. We know

Robotics Simulation 243

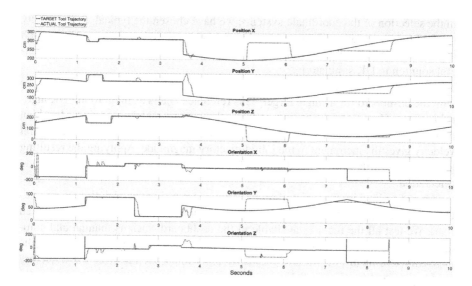

FIGURE 8.18 ABB IRB6620LX DK Simulation end-effector target and actual trajectories.

how the treatment of the prismatic joint is the same with screw theory. We understand the model better with all explanations of Section 8.5.1 for the general solution.

We generate an end-effector trajectory planning either with a complete task or with the dashboard for a single target. This reference enters as input to the DK algorithm. For each point of the trajectory, the tool's velocity and the geometric Jacobian permit to obtain the velocity of the joints in real time. The integration of each joint velocity gives the increment in the joint trajectory magnitude. Applying the resulting magnitudes to the robot joints, the TCP follows the target in position and orientation.

We add a plot with the result of a simulation, contrasting the evolution of the end-effector target (Figure 8.18 solid line) and actual trajectories (Figure 8.18 dotted line). We test all the features available in the dashboard, such as manual and automatic control or the possibility to freeze the robot or sending it home. Despite the approximative nature of the DK geometric algorithms, the result of the trajectory tracking is good (see Figure 8.18).

We include a video for better watching the simulation performing DK (Video 8.5.5)[46]. In the recorded simulation, we test all the features available in the dashboard, such as manual and automatic control or the possibility to freeze the robot or sending it home. The recording demonstrates the good real-time response of the implementation.

All files necessary to complete this DK system, including the screw theory functions, are inside a compressed folder (Simulator 8.5.5)[47] in the internet hosting for the software of this book.

8.5.6 SCARA ROBOTS (E.G., ABB IRB910SC)

This DK simulator has a two parallel joint axes mechanism, advantageous for many assembly operations. For the sake of showing the freedom given by the screw theory in the selection of the coordinate systems, we have chosen the typical virtual reality spatial frame orientation. This example has only four joints, but the general approach applies. We understand the model better with all detailed Section 8.5.1 for the general solution to DK simulation.

We generate an end-effector trajectory planning either with a complete task or with the dashboard for a single target. This reference enters as input to the DK algorithm. For each point of the trajectory, the tool's velocity and the geometric Jacobian permit to obtain the velocity of the joints in real time. The integration of each joint velocity gives the increment in the joint trajectory magnitude. Applying the resulting magnitudes to the robot joints, the TCP follows the target in position and orientation.

We add a plot with the result of a simulation, contrasting the evolution of the end-effector target (Figure 8.19 solid line) and actual trajectories (Figure 8.19 dotted line). We test all the features available in the dashboard, such as manual and automatic control or the possibility to freeze the robot or sending it home. Despite the approximative nature of the DK geometric algorithms, the result of the trajectory tracking is good (see Figure 8.19).

We include a video for better watching the simulation performing DK (Video 8.5.6)[48]. In the recorded simulation, we test all the features available in the dashboard, such as manual and automatic control or the possibility to freeze the robot or sending it home. The recording demonstrates the good real-time response of this implementation.

All files necessary to complete this DK system, including the screw theory functions, are inside a compressed folder (Simulator 8.5.6)[49] in the internet hosting for the software of this book.

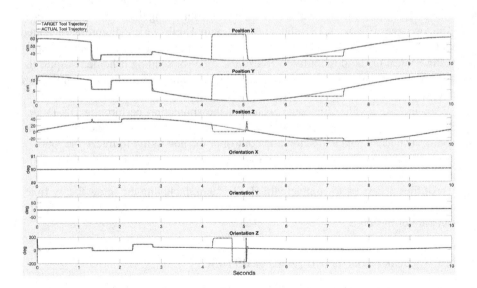

FIGURE 8.19   ABB IRB910SC DK Simulation end-effector target and actual trajectories.

8.5.7 COLLABORATIVE ROBOTS (E.G., UNIVERSAL UR16E)

This DK simulator deals with a typical collaborative manipulator made for human-robot cooperation in the workspace. We understand the model better with all detailed explanations of Section 8.5.1 for the general solution.

We generate an end-effector trajectory planning either with a complete task or with the dashboard for a single target. This reference enters as input to the DK algorithm. For each point of the trajectory, the tool's velocity and the geometric Jacobian permit to obtain the velocity of the joints in real time. The integration of each joint velocity gives the increment in the joint trajectory magnitude. Applying the resulting magnitudes to the robot joints, the TCP follows the target in position and orientation.

We add a plot with the result of a simulation, contrasting the evolution of the end-effector target (Figure 8.20 solid line) and actual trajectories (Figure 8.20 dotted line). We test all the features available in the dashboard, such as manual and automatic control or the possibility to freeze the robot or sending it home. Despite the approximative nature of the DK geometric algorithms, the result of the trajectory tracking is good (see Figure 8.20).

We include a video for better watching the simulation performing DK (Video 8.5.7)[50]. In the recorded simulation, we test all the features available in the dashboard, such as manual and automatic control or the possibility to freeze the robot or sending it home. The recording demonstrates the good real-time response of this implementation.

All files necessary to complete this DK system, including the screw theory functions, are inside a compressed folder (Simulator 8.5.7)[51] in the internet hosting for the software of this book.

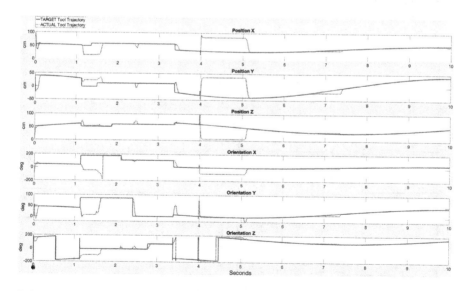

FIGURE 8.20 UNIVERSAL UR16e DK Simulation end-effector target and actual trajectories.

8.5.8 Redundant Robots (e.g., KUKA IIWA)

This DK simulator has the novelty of solving a robot with seven DoF. The performance of the formulations is good working for any number of joints. A redundant robot has more than the minimal number of DoF required to complete a task. Therefore, the robot can have an infinite number of joint configurations to reach a specific tool target. The extra DoF in redundant manipulators are helpful to develop some motion strategies. We understand the model better with all detailed explanations of Section 8.5.1 for the general solution.

We generate an end-effector trajectory planning either with a complete task or with the dashboard for a single target. This reference enters as input to the DK algorithm. For each point of the trajectory, the tool's velocity and the geometric Jacobian permit to obtain the velocity of the joints in real time. The integration of each joint velocity gives the increment in the joint trajectory magnitude. Applying the resulting magnitudes to the robot joints, the TCP follows the target in position and orientation.

We add a plot with the result of a simulation, contrasting the evolution of the end-effector target (Figure 8.21 solid line) and actual trajectories (Figure 8.21 dotted line). We test all the features available in the dashboard, such as manual and automatic control or the possibility to freeze the robot or sending it home. Despite the approximative nature of the DK geometric algorithms, the result of the trajectory tracking is good (see Figure 8.21).

We include a video for better watching the simulation performing DK (Video 8.5.8)[52]. In the recorded simulation, we test all the features available in the dashboard, such as manual and automatic control or the possibility to freeze the robot or sending it home. The recording demonstrates the good real-time response of this implementation.

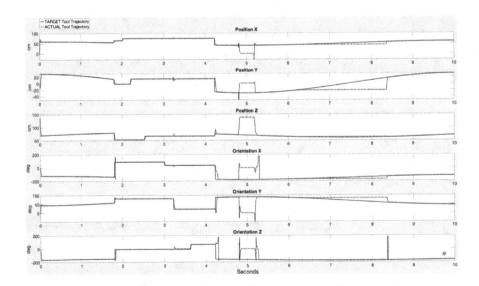

FIGURE 8.21 KUKA IIWA DK Simulation end-effector target and actual trajectories.

All files necessary to complete this DK system, including the screw theory functions, are inside a compressed folder (Simulator 8.5.8)[53] in the internet hosting for the software of this book.

8.6 INVERSE DYNAMICS SIMULATIONS

In this last section of simulations, we present exercises for all the ID algorithms introduced in Chapter 6 of this book (Stokes and Brockett, 1996). It is also convenient to read the explanations of Chapter 7 for trajectory generation. The goal is to test the complete geometric formulation for solving the ID problem of a manipulative robot with screw theory and its extensions. The simulators can implement the functions presented for the ID non-recursive Lagrange and recursive Newton–Euler formulations (Brockett et al., 1993). For the classical screw theory closed-form Lagrange algorithm, we showed that screw theory offers the fantastic possibility of solving robot dynamics utilizing linear algebra techniques (Strang, 2009). Even more, the tools of the se(3) Lie algebra allow expressing this dynamics problem practically by geometry. Employing a set of POE operations, we can represent the Mass, Coriolis, and Potential matrices. The Recursive Newton-Euler Algorithm (RNEA) (Featherstone, 2016) uses the spatial vector algebra jointly with the POE, and the efficiency is higher (Sipser, 2021).

The simulators used are like those employed in the previous section for DK. Nonetheless, there are some crucial differences. First, torques and forces apply to the manipulator digital twin joints instead of positions (Raibert and Craig, 1981). Second, the control law, which is implemented by software instead of by a Simulink® specific block. And third, the joint velocities feedback, which, besides the joint position feedback, is necessary to implement the computed torque control software algorithm (Sastry, 1999). The tools to test these simulations are MATLAB®, Simulink®, Simscape™, and the ST24R library.

In the implementation of ID, it is necessary to know the robot joints' position for each target of the trajectory planning. We apply the DK algorithm presented in Chapter 5 to solve these joint positions corresponding to the required tool trajectory. We must remember that this approach is an approximative alternative to IK, based on integrating the joint velocities to generate the joint trajectories. There are other alternatives to solve the joint trajectory planning, such as the IK algorithms of Chapter 4, but we do not use them for this ID simulations section.

We provide a reference trajectory for the robot tool, discretized according to specific step size. The algorithm starts by knowing the joints' initial positions. We get the tool velocities from the difference between the point of interest in the reference target end-effector trajectory planning and the actual tool pose. The integration of the DK joint velocities gives the joint position increments.

There are two possibilities to generate paths for the robot tool end-effector in the task-space, either in real time with the dashboard's commands or offline with some input file. However, path planning in joint space is always online. Consequently, in any case, only a point-to-point joint trajectory is considered, with trapezoidal motion interpolation in real time. The joint motion respects its position and velocity constraints of the robot actuators. The accelerations have a limit for keeping the safety limits of the robot joints torques and forces. The core goal is to make the robot joints pass through the series of points (i.e., start, via, goal) as accurately as position. The target is to make the tool end-effector reach as strictly as possible the shape of the task-space path in time. In other words, we do not define the joint path time scaling to minimize the time of joint motion, neither to optimize the robot energy or productivity. The joint trajectory planning timing inherits from the task-space path time scaling.

The DK application for getting the joint positions is an approximation solution that makes the motion tangent to the actual trajectory. Nonetheless, when the discretization of targets is small enough, the results are promising. However, the limitations of the DK approach may emerge with some robot configurations. For instance, moving along a particular trajectory may leave unreachable targets, even though they are inside the robot's workspace. The reason is that the DK algorithm tries to follow linear velocity between consecutive targets. Still, if some joint position already reached its limits, the motion becomes impossible, and the manipulator might get stacked. The solution for those events is to choose a different robot configuration or define an alternative trajectory (e.g., turnaround) between the two targets. Nevertheless, this DK is an excellent methodology when we do not have a closed-form solution for the IK, the only available answer is numeric, or the target is out of the robot's dexterous workspace.

With the robot twists and dynamics parameters (i.e., mass and moments of inertia for each link), plus the calculated joint positions and the defined velocities and accelerations for the joints, there is enough information to implement both ID algorithms. For the first, it is possible to compute the Mass, Coriolis, and Potential matrices of the Lagrange formulation. For the second, it is feasible to calculate the Newton–Euler link forces. Either way, the result gives the necessary joint torques or forces to move the robot. The process repeats itself for the following points of the expected tool reference trajectory.

We need to apply a control law to correct many kinds of errors following an accurate trajectory (Khalil, 2014). In previous sections, we implemented some Simulink® blocks to perform simple PID controls. Differently, these ID simulators build control with software inside the algorithm of the controller. We are interested in geometric methods to control (Jurdjevic, 1997). The only constraints considered are joint physical limits (Liu and Li, 2002). The code implements the "Computed Torque Control Law" presented in Chapter 6 (Ploen, 1997). This regulator adds state feedback to obtain the control equation, including two terms to compensate for the velocity and position errors. It requires adjusting two constant gain matrices, one for the velocity error "K_v" and another for the position error "K_p." Nonetheless, these simulators can be great sandboxes to develop and test more advanced control techniques for many types of controllers.

These exercises' goal is more to prove the theoretical approach rather than to get excellent computational performance. With the second algorithm, the RNEA plus POE, the simulation speed is good even with all the code in interpreted MATLAB®, demonstrating the much better performance of this approach.

All added, we suggest studying the following simulators for experimentation. Even with all the explained practical limitations, the ID formulation's performance is remarkable, as it is evident when using the manual control from the dashboard.

Robotics Simulation

For instance, for two consecutive targets very far away in terms of position and orientation, the robot develops an excellent jumping trajectory to reach the new target with great effectiveness and efficiency.

8.6.1 GENERAL SOLUTION TO ID SIMULATION

There are complete models for all ID simulators of the following exercises. The structure of each simulator model has seven blocks at the first level (see an example in Figure 8.22). All the necessary files to make any simulator running are in a compressed folder in the internet repository[54].

For anyone with a minimum experience working with are MATLAB®, Simulink®, Simscape™, the understanding of the simulator architecture will be easy to grasp with the descriptions for the model blocks already presented for the general solution to FK simulation (Section 8.3.1). We detail hereafter the differences from the general FK simulator architecture.

- **Manipulator Multibody System** (marked as 4 in the Figure 8.22): the inputs to this block are the reference world coordinate frame and the magnitudes of reference for the joints (i.e., results from the ID control algorithm). However, in this simulator, these reference values must be the joint torques or forces, impacting the definition of the "Simulink-PS" converters and the revolute joint action parameters (i.e., "Nm" and "N"). The modules' outputs are the position and velocity values from the joints in the simulator multibody system, which are the feedback for the computed torque control law. Besides, another output is the end of the open chain toward the TCP.
- **Algorithms and Controller** (marked as 7 in the Figure 8.22): this module is the core of the ID simulation for the robot. This module keeps some inputs, like the tool target definition in terms of position and orientation, the signal for sending the robot to home pose or following the target, stopping the robot, or releasing the motion, and the joints position feedback from the manipulator

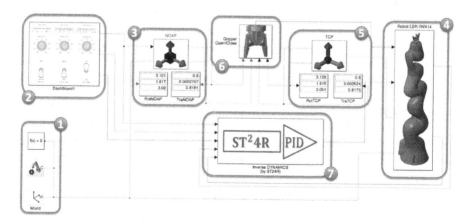

FIGURE 8.22 KUKA IIWA ID simulator complete model.

model. Besides, this block adds new inputs, with the joint velocities feedback, also coming from the robot model. There are two functions available for both ID approaches, the classical of Lagrange and the RNEA with POE in the repository. The control law is implemented by software instead of by a Simulink® block. For this simulator, the reference command signals are the joint torques (Nm) or forces (N).

It is possible to test manual and automatic controls to generate simulation tool targets in position and orientation. To show the performance of ID automatic tracking, we create a reference trajectory for the robot's tool in 3D with some sinusoidal form, with a random shape. This reference enters the ID control algorithm, whose outcomes are the torques and forces to the robot joints. In the figures of the exercises, it is possible to check how the TCP (dotted line) follows the trajectory targets NOAP (solid line). The algorithms keep the joint trajectories inside the actual physical limits of the robot for both position and velocity.

The joint magnitudes information (i.e., position, velocity, torques, and forces) is recorded in files by the simulators for each running of the system, and this data is available for plotting and study. The same happens for the information regarding the end-effector trajectories, either target or actual.

The simulations produce a video for a better understanding of the ID robot's performance. The typical registration tests all dashboard functionalities. For example, it gets started with the predefined trajectory and the automatic control mode in operation. Then, it switches for freezing the robot. When the button is released, the robot pursues the trajectory again in automatic control.

Another functionality shows how to send the robot to its home pose. Afterward, the control commutes to manual. In this control mode, the three sliders permit defining the cartesian position for the tool target. With the three knobs, it is possible to determine the Euler orientation for the same target. Finally, the control goes back to automatic, and the robot moves to follow the trajectory planning at its current point of evolution.

To extend the knowledge regarding ID is very convenient to practice with the simulators of different manipulators. They will permit us to consolidate the skills to apply the screw theory formulations. These examples have different configurations. The solved robot architectures are Puma (e.g., ABB IRB120), Bending Backwards (ABB IRB1600), Gantry (e.g., ABB 6620LX), Scara (e.g., ABB IRB910SC), Collaborative (e.g., UNIVERSAL UR16e), and a Redundant manipulator (e.g., KUKA IIWA).

It is advisable to review the Screw Theory Toolbox for Robotics ST24R (Pardos-Gotor, 2021a) to understand better the concepts working in the examples of the following sections. Besides, there are also some videos available with the results of the simulations (Pardos-Gotor, 2021b), which help to better understand the exercises.

8.6.2 Puma Robots (e.g., ABB IRB120)

There is a complete simulator for ID of this Puma type manipulator. We understand the model better with all detailed explanations of Section 8.6.1 for the general model.

Robotics Simulation 251

FIGURE 8.23 ABB IRB120 ID Simulation TCP target and actual 3D trajectories.

We generate an end-effector trajectory planning either with a complete task or with the dashboard for a single target. This reference enters the chosen ID algorithms, which can be either Lagrange or RNEA. Applying the resulting magnitudes to the robot joints, the TCP follows the target in position and orientation.

We add a 3D plot with the result of a simulation for the automatic control over a sinusoidal input trajectory. It is possible to contrast the evolution of the TCP position target (Figure 8.23 solid line) and actual trajectories (Figure 8.23 dotted line). The result of the trajectory tracking is good.

We include a recording for better watching another simulation performing ID (Video 8.6.2)[55]. In the recorded simulation we test all the features available in the dashboard, such as manual and automatic control or the possibility to freeze the robot or sending it home. The recording demonstrates the good real-time response of the implementation.

All files necessary to complete this ID system, including the screw theory functions, are inside a compressed folder (Simulator 8.6.2)[56] in the internet hosting for the software of this book.

8.6.3 PUMA ROBOTS (E.G., ABB IRB120) "TOOL-UP"

This simulation for the ID of the Puma robot has the tool oriented upward in the home pose. The model is the same as the previous simulation, except for the initial reference. We understand the model better with all detailed explanations of Section 8.6.1 for the general model.

We generate an end-effector trajectory planning either with a complete task or with the dashboard for a single target. This reference enters the chosen ID algorithm, the non-recursive of Lagrange or the RNEA. Applying the resulting magnitudes to the robot joints, the TCP follows the target in position and orientation.

We add a 3D plot with the result of a simulation for the automatic control over a sinusoidal input trajectory. It is possible to contrast the evolution of the TCP position

FIGURE 8.24 ABB IRB120 "Tool-Up" ID Simulation TCP target and actual 3D trajectories.

target (Figure 8.24 solid line) and actual trajectories (Figure 8.24 dotted line). The result of the trajectory tracking is good.

We include a recording for better watching another simulation performing ID (Video 8.6.3)[57]. In the recorded simulation we test all the features available in the dashboard, such as manual and automatic control or the possibility to freeze the robot or sending it home. The recording demonstrates the good real-time response of this implementation.

All files necessary to complete this ID system, including the screw theory functions, are inside a compressed folder (Simulator 8.6.3)[58] in the internet hosting for the software of this book.

8.6.4 BENDING BACKWARDS ROBOTS (E.G., ABB IRB1600)

This simulator for ID sees a slightly different robot architecture, a Bending Backwards manipulator. This design does not need to rotate the robot to reach for things behind it. Swinging the arm backward extends the working range of the robot. We understand the model better with all detailed explanations of Section 8.6.1 for the general model.

We generate an end-effector trajectory planning either with a complete task or with the dashboard for a single target. This reference enters the chosen ID algorithm, the non-recursive of Lagrange or the RNEA. Applying the resulting magnitudes to the robot joints, the TCP follows the target in position and orientation.

We add a 3D plot with the result of a simulation for the automatic control. In this exercise for the sake of introducing a variation, the automatic trajectory it is a kind of rectangle with several stretches of straight lines. It is possible to contrast the evolution of the TCP position target (Figure 8.25 solid line) and actual trajectories (Figure 8.25 dotted line). The result of the trajectory tracking is good enough.

We include a recording for better watching another simulation performing ID (Video 8.6.4)[59]. In the recorded simulation we test all the features available in the

Robotics Simulation

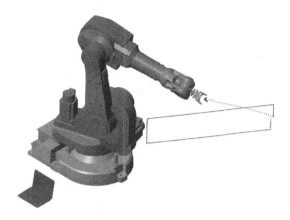

FIGURE 8.25 ABB IRB1600 ID Simulation TCP target and actual 3D trajectories.

dashboard, such as manual and automatic control or the possibility to freeze the robot or sending it home. The recording demonstrates the good real-time response of the implementation.

All files necessary to complete this ID system, including the screw theory functions, are inside a compressed folder (Simulator 8.6.4)[60] in the internet hosting for the software of this book.

8.6.5 GANTRY ROBOTS (E.G., ABB IRB6620LX)

This simulator for ID deals with this Gantry robotic architecture, which combines the advantages of both a long linear axis for the first joint plus five revolute joints. We know how the treatment of the prismatic joint is the same with screw theory. We understand the model better with all detailed explanations of Section 8.6.1. for the general model.

We generate an end-effector trajectory planning either with a complete task or with the dashboard for a single target. This reference enters the ID algorithm, which can be Lagrange or RNEA. Applying the resulting magnitudes to the robot joints, the TCP follows the target in position and orientation.

We add a 3D plot with the result of a simulation for the automatic control over a sinusoidal input trajectory. It is possible to contrast the evolution of the TCP position target (Figure 8.26 solid line) and actual trajectories (Figure 8.26 dotted line). The result of the trajectory tracking is good.

We include a recording for better watching another simulation performing ID (Video 8.6.5)[61]. In the recorded simulation we test all the features available in the dashboard, such as manual and automatic control or the possibility to freeze the robot or sending it home. The recording demonstrates the good real-time response of this implementation.

All files necessary to complete this ID system, including the screw theory functions, are inside a compressed folder (Simulator 8.6.5)[62] in the internet hosting for the software of this book.

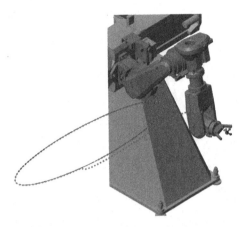

FIGURE 8.26 ABB IRB6620LX ID Simulation TCP target and actual 3D trajectories.

8.6.6 SCARA ROBOTS (E.G., ABB IRB910SC)

This simulator for ID has a two parallel joint axes mechanism, advantageous for many assembly operations. For the sake of showing the freedom given by the screw theory in the selection of the coordinate systems, we have chosen the typical virtual reality spatial frame orientation. This example has only four joints, but the general approach applies. We understand the model better with all detailed explanations of Section 8.6.1 for the general model.

We generate an end-effector trajectory planning either with a complete task or with the dashboard for a single target. This reference enters the ID algorithm, which can be Lagrange or RNEA. Applying the resulting magnitudes to the robot joints, the TCP follows the target in position and orientation.

We add a 3D plot with the result of a simulation for the automatic control over a sinusoidal input trajectory. It is possible to contrast the evolution of the TCP position target (Figure 8.27 solid line) and actual trajectories (Figure 8.27 dotted line). The result of the trajectory tracking is good.

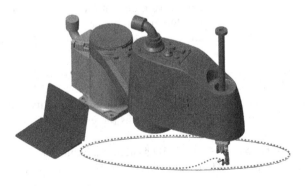

FIGURE 8.27 ABB IRB910SC ID Simulation TCP target and actual 3D trajectories.

Robotics Simulation

We include a recording for better watching another simulation performing ID (Video 8.6.6)[63]. In the recorded simulation we test all the features available in the dashboard, such as manual and automatic control or the possibility to freeze the robot or sending it home. The recording demonstrates the good real-time response of the implementation.

All files necessary to complete this ID system, including the screw theory functions, are inside a compressed folder (Simulator 8.6.6)[64] in the internet hosting for the software of this book.

8.6.7 COLLABORATIVE ROBOTS (E.G., UNIVERSAL UR16E)

This simulator for ID deals with a typical collaborative manipulator made for human-robot cooperation in the workspace. We understand the model better with all detailed explanations of Section 8.6.1 for the general model.

We generate an end-effector trajectory planning either with a complete task or with the dashboard for a single target. This reference enters the chosen ID algorithm, which can be Lagrange or RNEA. Applying the resulting magnitudes to the robot joints, the TCP follows the target in position and orientation.

We add a 3D plot with the result of a simulation for the automatic control over a sinusoidal input trajectory. It is possible to contrast the evolution of the TCP position target (Figure 8.28 solid line) and actual trajectories (Figure 8.28 dotted line). The result of the trajectory tracking is good.

We include a recording for better watching another simulation performing ID (Video 8.6.7)[65]. In the recorded simulation we test all the features available in the dashboard,

FIGURE 8.28 UNIVERSAL UR16e ID Simulation TCP target and actual 3D trajectories.

such as manual and automatic control or the possibility to freeze the robot or sending it home. The recording demonstrates the good real-time response of this implementation.

All files necessary to complete this ID system, including the screw theory functions, are inside a compressed folder (Simulator 8.6.7)[66] in the internet hosting for the software of this book.

8.6.8 Redundant Robots (e.g., KUKA IIWA)

This simulator for ID has the novelty of solving a robot with seven DoF. The performance of the formulations is good working for any number of joints. A redundant robot has more than the minimal number of DoF required to complete a task. Therefore, the robot can have an infinite number of joint configurations to reach a specific tool target. The extra DoF in redundant manipulators are helpful to develop some motion strategies. We understand the model better with all detailed explanations of Section 8.6.1 for the general model.

We generate an end-effector trajectory planning either with a complete task or with the dashboard for a single target. This reference enters the chosen ID algorithm, which can be Lagrange or RNEA. Applying the resulting magnitudes to the robot joints, the TCP follows the target in position and orientation.

We add a 3D plot with the result of a simulation for the automatic control over a sinusoidal input trajectory. It is possible to contrast the evolution of the TCP position target (Figure 8.29 solid line) and actual trajectories (Figure 8.29 dotted line). The result of the trajectory tracking is good.

We include a recording for better watching another simulation performing ID (Video 8.6.8)[67]. In the recorded simulation, we test all the features available in the dashboard, such as manual and automatic control or the possibility to freeze the robot or sending it home. The recording demonstrates the good real-time response of this implementation.

All files necessary to complete this ID system, including the screw theory functions, are inside a compressed folder (Simulator 8.6.8)[68] in the internet hosting for the software of this book.

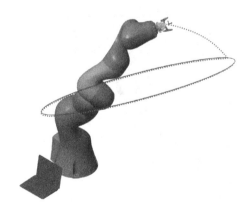

FIGURE 8.29 KUKA IIWA ID Simulation TCP target and actual 3D trajectories.

8.7 SUMMARY

This chapter includes simulations that allow putting into practice all the screw theory knowledge, functions, and algorithms presented throughout this book. Additionally, the employed simulators can be sandboxes were to test new algorithms, developments, and applications with the simulation tools put at our disposal.

The robotics simulators create applications for a robot without relying on the actual machine, thus saving time and costs. The simulator usually allows us to create objects in a virtual world so our robot can interact with them, allowing the testing of different mechanism's behaviors. All modern robot simulators have the capability for 3D environment modeling.

We choose the set of tools provided with MATLAB® because it is feasible and easy to design hardware platforms and analyze 3D rigid-body mechanics. It is possible to work directly with existing CAD and URDF files. It is viable to add constraints, such as friction, and model multi-domain systems with electrical, hydraulic, pneumatic, and other components. Once in operation, it is also handy to reuse design models as a robot "Digital Twin" to optimize real robots' performance. MATLAB® provides a software environment convenient for robotics researchers and engineers because it works intrinsically integrated with Simulink® and Simscape® packages to design and tune algorithms and generate code. This software environment is precious because it provides well-recognized instruments to support our developments well.

The complete screw theory for robotics theoretical background presented in this book is coded in the ST24R library. This toolbox encompasses a collection of functions to analyze the robot kinematics and dynamics through formulas of Lie groups' theory and the geometry of screws. Naturally, the toolbox covers all the necessary screw mathematics to follow the examples and simulations. The ST24R toolbox has functions for POE, conversion between data types, position and rotation representation, homogeneous transformations, twists, screws, wrenches, geometric Jacobian, adjoint transformations, gravity wrench matrix, or spatial vector algebra.

The software for ST24R is developed mainly for teaching purposes. Therefore, it is not optimized for industrial use. Our code's main aim is to understand better the screw theory concepts rather than optimize the software's performance. Nevertheless, it is possible to appreciate how this new mathematical approach's advantages are so significant that they will become the cornerstone of compelling applications.

The ST24R functions implementing the canonical IK subproblems always give a result (exact or approximate). Therefore, the algorithms will provide an inaccurate but practical result in having a problem without an exact solution. We program these functions in this fashion because they are beneficial for the simulations and lead to many robots' natural motion solutions. If there is a need for a different approach, it is easy to modify the functions' code for new aims and utility.

The simulations included in this chapter allow us to explore interactively all robot exercises presented throughout this book. The fundamental architectures and manipulators used to put to the test the theoretical screw theory formulations are:

- Puma robot (e.g., ABB IRB120), with two different home positions.
- Bending Backwards robot (e.g., ABB IRB1600).

- Gantry robot (e.g., ABB IRB6620LX).
- Scara robot (e.g., ABB IRB910SC).
- Collaborative robot (e.g., UR16e).
- Redundant robot (e.g., KUKA IIWA).

For all these robots, exercises are corresponding to the main screw theory concepts presented in this book in Chapters 3, 4, 5, and 6, which are: FK, IK, DK, and ID.

In each section, the first exercise presents excellent detail for all simulator components: Complete Model, Configuration, Dashboard, Target, Multibody System, TCP, Gripper, and Control Module. Also, other functionalities are explained, such as Mechanics Explorer or Video Creator. The presentation of the rest exercises in the section is brief, as the simulators are practically the same, and the main change resides in the Multibody System, which obviously must correspond to the robot of that example. Nonetheless, when some particularity arises for a robot, it is adequately explained.

There are links to Github and YouTube repositories for all exercises, where it is possible to find respectively the necessary files to build and run the simulators and some videos. There are several examples for the same robot, with the corresponding files. When it is illustrative to present different algorithms for the same robot, there are several exercises in the same section. Such is the case with canonical subproblems to solve IK problems.

The performance of these simulations is remarkable even when using DK, which is an approximation solution that makes the motion to be tangent to the exact trajectory. Something similar happens in the section with simulations for ID. Even when the discretization of targets is not tiny, the results are reassuring. It is advisable to check with the simulators' manual control that this approach works surprisingly well for massive jumps between two successive targets. These results demonstrate the quality of the applied underlying screw theory for robotics, which is this text's goal.

Having arrived here, we cover the scope of this book along with these eight chapters. Now it is time to summarize and highlight the conclusions of this work.

NOTES

1 Pardos-Gotor, J.M. (2021). *Screw Theory in Robotics*. Github. https://github.com/DrPardos Gotor/Screw-Theory-in-Robotics/tree/master/Simulations
2 Pardos-Gotor, J.M. (2021). *Screw Theory in Robotics*. Youtube. https://youtu.be/BRpQykx A1kg
3 Pardos-Gotor, J.M. (2021). *Screw Theory in Robotics*. Github. https://github.com/DrPardosGotor/Screw-Theory-in-Robotics/blob/master/Simulations/Simulator_8.3.2.zip
4 Pardos-Gotor, J.M. (2021). *Screw Theory in Robotics*. Youtube. https://youtu.be/y0WyJ_govsg
5 Pardos-Gotor, J.M. (2021). *Screw Theory in Robotics*. Github. https://github.com/DrPardos Gotor/Screw-Theory-in-Robotics/blob/master/Simulations/Simulator_8.3.3.zip
6 Pardos-Gotor, J.M. (2021). *Screw Theory in Robotics*. Youtube. https://youtu.be/kDRpTL 9gKug
7 Pardos-Gotor, J.M. (2021). *Screw Theory in Robotics*. Github. https://github.com/DrPardos Gotor/Screw-Theory-in-Robotics/blob/master/Simulations/Simulator_8.3.4.zip

Robotics Simulation

8 Pardos-Gotor, J.M. (2021). *Screw Theory in Robotics*. Youtube. https://youtu.be/WZ8Qs xHLgYo
9 Pardos-Gotor, J.M. (2021). *Screw Theory in Robotics*. Github. https://github.com/DrPardos Gotor/Screw-Theory-in-Robotics/blob/master/Simulations/Simulator_8.3.5.zip
10 Pardos-Gotor, J.M. (2021). *Screw Theory in Robotics*. Youtube. https://youtu.be/lm6iDM 47q_E
11 Pardos-Gotor, J.M. (2021). *Screw Theory in Robotics*. Github. https://github.com/DrPardos Gotor/Screw-Theory-in-Robotics/blob/master/Simulations/Simulator_8.3.6.zip
12 Pardos-Gotor, J.M. (2021). *Screw Theory in Robotics*. Youtube. https://youtu.be/lm6iDM 47q_E
13 Pardos-Gotor, J.M. (2021). *Screw Theory in Robotics*. Github. https://github.com/DrPardos Gotor/Screw-Theory-in-Robotics/blob/master/Simulations/Simulator_8.3.7.zip
14 Pardos-Gotor, J.M. (2021). *Screw Theory in Robotics*. Youtube. https://youtu.be/bWWuP stq0B4
15 Pardos-Gotor, J.M. (2021). *Screw Theory in Robotics*. Github. https://github.com/DrPardos Gotor/Screw-Theory-in-Robotics/blob/master/Simulations/Simulator_8.3.8.zip
16 Pardos-Gotor, J.M. (2021). *Screw Theory in Robotics*. Github. https://github.com/DrPardos Gotor/Screw-Theory-in-Robotics/tree/master/Simulations
17 Pardos-Gotor, J.M. (2021). *Screw Theory in Robotics*. Youtube. https://youtu.be/9NYtn Gs8oPY
18 Pardos-Gotor, J.M. (2021). *Screw Theory in Robotics*. Youtube. https://youtu.be/JrDcrL8y CXM
19 Pardos-Gotor, J.M. (2021). *Screw Theory in Robotics*. Github. https://github.com/DrPardos Gotor/Screw-Theory-in-Robotics/blob/master/Simulations/Simulator_8.4.2a.zip
20 Pardos-Gotor, J.M. (2021). *Screw Theory in Robotics*. Github. https://github.com/DrPardos Gotor/Screw-Theory-in-Robotics/blob/master/Simulations/Simulator_8.4.2b.zip
21 Pardos-Gotor, J.M. (2021). *Screw Theory in Robotics*. Youtube. https://youtu.be/CyXXnd WMb0c
22 Pardos-Gotor, J.M. (2021). *Screw Theory in Robotics*. Youtube. https://youtu.be/wtzLO rgcYeo
23 Pardos-Gotor, J.M. (2021). *Screw Theory in Robotics*. Github. https://github.com/DrPardos Gotor/Screw-Theory-in-Robotics/blob/master/Simulations/Simulator_8.4.3.zip
24 Pardos-Gotor, J.M. (2021). *Screw Theory in Robotics*. Youtube. https://youtu.be/GmkO uIXnEsE
25 Pardos-Gotor, J.M. (2021). *Screw Theory in Robotics*. Youtube. https://youtu.be/OeAoA rtecp4
26 Pardos-Gotor, J.M. (2021). *Screw Theory in Robotics*. Github. https://github.com/DrPardos Gotor/Screw-Theory-in-Robotics/blob/master/Simulations/Simulator_8.4.4.zip
27 Pardos-Gotor, J.M. (2021). *Screw Theory in Robotics*. Youtube. https://youtu.be/hNrXM 134m54
28 Pardos-Gotor, J.M. (2021). *Screw Theory in Robotics*. Youtube. https://youtu.be/WgqNK 2u3AIs
29 Pardos-Gotor, J.M. (2021). *Screw Theory in Robotics*. Github. https://github.com/DrPardos Gotor/Screw-Theory-in-Robotics/blob/master/Simulations/Simulator_8.4.5.zip
30 Pardos-Gotor, J.M. (2021). *Screw Theory in Robotics*. Youtube. https://youtu.be/diwycx 11IZs
31 Pardos-Gotor, J.M. (2021). *Screw Theory in Robotics*. Youtube. https://youtu.be/XxLT16 ZxCjo
32 Pardos-Gotor, J.M. (2021). *Screw Theory in Robotics*. Github. https://github.com/DrPardos Gotor/Screw-Theory-in-Robotics/blob/master/Simulations/Simulator_8.4.6.zip
33 Pardos-Gotor, J.M. (2021). *Screw Theory in Robotics*. Youtube. https://youtu.be/ZvhaZ s3rs5A

34 Pardos-Gotor, J.M. (2021). *Screw Theory in Robotics*. Youtube. https://youtu.be/cf8JG5kTliE
35 Pardos-Gotor, J.M. (2021). *Screw Theory in Robotics*. Github. https://github.com/DrPardosGotor/Screw-Theory-in-Robotics/blob/master/Simulations/Simulator_8.4.7.zip
36 Pardos-Gotor, J.M. (2021). *Screw Theory in Robotics*. Youtube. https://youtu.be/ilmSE7yN3t8
37 Pardos-Gotor, J.M. (2021). *Screw Theory in Robotics*. Youtube. https://youtu.be/5wePEFRhAu0
38 Pardos-Gotor, J.M. (2021). *Screw Theory in Robotics*. Github. https://github.com/DrPardosGotor/Screw-Theory-in-Robotics/blob/master/Simulations/Simulator_8.4.8.zip
39 Pardos-Gotor, J.M. (2021). *Screw Theory in Robotics*. Github. https://github.com/DrPardosGotor/Screw-Theory-in-Robotics/tree/master/Simulations
40 Pardos-Gotor, J.M. (2021). *Screw Theory in Robotics*. Youtube. https://youtu.be/xjyEdzYGWOQ
41 Pardos-Gotor, J.M. (2021). *Screw Theory in Robotics*. Github. https://github.com/DrPardosGotor/Screw-Theory-in-Robotics/blob/master/Simulations/Simulator_8.5.2.zip
42 Pardos-Gotor, J.M. (2021). *Screw Theory in Robotics*. Youtube. https://youtu.be/X7sb5C6qqPI
43 Pardos-Gotor, J.M. (2021). *Screw Theory in Robotics*. Github. https://github.com/DrPardosGotor/Screw-Theory-in-Robotics/blob/master/Simulations/Simulator_8.5.3.zip
44 Pardos-Gotor, J.M. (2021). *Screw Theory in Robotics*. Youtube. https://youtu.be/rfen2AQe70w
45 Pardos-Gotor, J.M. (2021). *Screw Theory in Robotics*. Github. https://github.com/DrPardosGotor/Screw-Theory-in-Robotics/blob/master/Simulations/Simulator_8.5.4.zip
46 Pardos-Gotor, J.M. (2021). *Screw Theory in Robotics*. Youtube. https://youtu.be/wlfmBqWUXbU
47 Pardos-Gotor, J.M. (2021). *Screw Theory in Robotics*. Github. https://github.com/DrPardosGotor/Screw-Theory-in-Robotics/blob/master/Simulations/Simulator_8.5.5.zip
48 Pardos-Gotor, J.M. (2021). *Screw Theory in Robotics*. Youtube. https://youtu.be/zre89PvZq9I
49 Pardos-Gotor, J.M. (2021). *Screw Theory in Robotics*. Github. https://github.com/DrPardosGotor/Screw-Theory-in-Robotics/blob/master/Simulations/Simulator_8.5.6.zip
50 Pardos-Gotor, J.M. (2021). *Screw Theory in Robotics*. Youtube. https://youtu.be/1bJ09mql0i8
51 Pardos-Gotor, J.M. (2021). *Screw Theory in Robotics*. Github. https://github.com/DrPardosGotor/Screw-Theory-in-Robotics/blob/master/Simulations/Simulator_8.5.7.zip
52 Pardos-Gotor, J.M. (2021). *Screw Theory in Robotics*. Youtube. https://youtu.be/n16zeLkBrL4
53 Pardos-Gotor, J.M. (2021). *Screw Theory in Robotics*. Github. https://github.com/DrPardosGotor/Screw-Theory-in-Robotics/blob/master/Simulations/Simulator_8.5.8.zip
54 Pardos-Gotor, J.M. (2021). *Screw Theory in Robotics*. Github. https://github.com/DrPardosGotor/Screw-Theory-in-Robotics/tree/master/Simulations
55 Pardos-Gotor, J.M. (2021). *Screw Theory in Robotics*. Youtube. https://youtu.be/-Rsa-PGFCbE
56 Pardos-Gotor, J.M. (2021). *Screw Theory in Robotics*. Github. https://github.com/DrPardosGotor/Screw-Theory-in-Robotics/blob/master/Simulations/Simulator_8.6.2.zip
57 Pardos-Gotor, J.M. (2021). *Screw Theory in Robotics*. Youtube https://youtu.be/UDevZRV0bIE
58 Pardos-Gotor, J.M. (2021). *Screw Theory in Robotics*. Github. https://github.com/DrPardosGotor/Screw-Theory-in-Robotics/blob/master/Simulations/Simulator_8.6.3.zip
59 Pardos-Gotor, J.M. (2021). *Screw Theory in Robotics*. Youtube. https://youtu.be/QjqOCSwmtss

60 Pardos-Gotor, J.M. (2021). *Screw Theory in Robotics*. Github. https://github.com/DrPardos Gotor/Screw-Theory-in-Robotics/blob/master/Simulations/Simulator_8.6.4.zip
61 Pardos-Gotor, J.M. (2021). *Screw Theory in Robotics*. Youtube. https://youtu.be/OMs_6WWUZr0
62 Pardos-Gotor, J.M. (2021). *Screw Theory in Robotics*. Github. https://github.com/DrPardos Gotor/Screw-Theory-in-Robotics/blob/master/Simulations/Simulator_8.6.5.zip
63 Pardos-Gotor, J.M. (2021). *Screw Theory in Robotics*. Youtube. https://youtu.be/GRh2iz DpWaQ
64 Pardos-Gotor, J.M. (2021). *Screw Theory in Robotics*. Github. https://github.com/DrPardos Gotor/Screw-Theory-in-Robotics/blob/master/Simulations/Simulator_8.6.6.zip
65 Pardos-Gotor, J.M. (2021). *Screw Theory in Robotics*. Youtube. https://youtu.be/BXT2Rq MiIo4
66 Pardos-Gotor, J.M. (2021). *Screw Theory in Robotics*. Github. https://github.com/DrPardos Gotor/Screw-Theory-in-Robotics/blob/master/Simulations/Simulator_8.6.7.zip
67 Pardos-Gotor, J.M. (2021). *Screw Theory in Robotics*. Youtube. https://youtu.be/rRjgErf tvNk
68 Pardos-Gotor, J.M. (2021). *Screw Theory in Robotics*. Github. https://github.com/DrPardos Gotor/Screw-Theory-in-Robotics/blob/master/Simulations/Simulator_8.6.8.zip

9 Conclusions

"There's nothing more practical than a good theory."
—Gilbert K. Chesterton

9.1 SUMMARY

Research and development with robots continue to be a tremendous technological challenge because they are complex systems with many degrees of freedom (DoF) and restrictions. This text aims to introduce effective and efficient algebraic and geometric solutions to solve mechanics problems in real-time robotics.

The best designs, solutions, and applications are conceived with the elegance of thought, and abstraction saves time in the long run. For this reason, nonstandard mathematical formulations are presented, based on the screw theory, and Lie algebras. The new algorithms have a differential and computational geometry approach for solving problems in a closed-form way. Their performance is tested with several real manipulators.

The importance of screw theory is widely recognized in robotics applications, as it allows better capture of the physical characteristics of a robot through a geometric description. However, these tools have remained relatively inaccessible for many researchers because they require a new language (e.g., screws, twists, wrenches, adjoint transformation, spatial vector algebra) and minimal dedication to studying these mathematical foundations. This book has shown an exact route to overcome these difficulties, with an accessible and visual approach to these methodologies.

Many algorithms for robots have the basis of standard algebraic alternatives, which end in confusing implementations due to the complexity of handling the calculation details. Conversely, at the heart of screw theory is a simple, concrete, high-level, and genuinely significant geometric interpretation from the point of view of mechanics and geometry.

This book will encourage more profound research about screw theory for robotics in the future. It will reveal that it is not difficult to develop better algorithms and solutions for some new robotics challenges and applications by applying these screw theory techniques.

At the beginning of this book, we aimed for some targets and goals already significantly reached. These **ACHIEVEMENTS** are as follows:

- Provide a highly visual approach for screw theory mechanics.
- Give new ideas to solve archetypical robotics' mechanical problems, which can be a source of inspiration for further applications.
- Hand over a solid base of understanding and learning screw theory basics.
- Formulate abstract screw theory for robotics concepts tangibly and practically.

- Exemplify how complex robot mechanical problems are tackled better with the powerful tools of the screw theory.
- Provide gratification by solving complex mechanics problems with an elegant approach and minimal code.
- Restrict the number of equations to those necessary for screw theory fundamentals.
- Encourage the research for good theories and not so much for implementations as the best way to speed up great results.
- Give rise to inspiration for developing new and better solutions using these screw theory methodologies.

There are great **BENEFITS** and advantages in the use of screw theory for robotics projects, available for all roboticists:

- It gives a complete and geometric representation of the mechanics, which dramatically simplifies the analysis for robotics.
- The robot's resulting equations are very efficient and convenient because the primary mathematical expression is the matrix exponential.
- It supplies a precise geometric, global representation of mechanics, greatly simplifying analysis for robotics through twists and wrenches.
- It offers the possibility to develop new geometric algorithms using Product of Exponentials (POE) to solve Inverse Kinematics (IK) problems.
- The natural and explicit description of the geometric Jacobian (without differentiation) has none of the drawbacks of local analytic representation when working with robot velocities.
- There are exact geometric solutions for the kinematics and dynamics of robots with many DoF instead of turning to numerical solutions.
- The availability of systematic, elegant, and geometrically meaningful solutions for the mechanics of many robot architectures (e.g., Stanford, Bending Backwards, Puma, Gantry, Scara) can be used as a foundation and learning for other developments.
- There are several specialized software libraries available (e.g., "ST24R" in this book, "RBT" by P. Corke, or "ModernRobotics" by K. Lynch and F. Park).
- The screw theory is an excellent opportunity for expanding the lessons learned to develop more efficient and practical algorithms for a wide variety of robotic mechanisms (e.g., industrial manipulators, humanoids, autonomous vehicles, drones, soft robots).

The use of the mathematical tools from Lie groups and algebras and the canonical formalism of the POE permits a practical approach for creating algorithms suitable for the mechanical control of complex robots with many DoF with significant advantages.

The fundamentals and developments are attractive because they bring forth a unified mathematical framework of geometric algorithms for robotics.

The independence of the formulations for the choice of reference systems is another advantage. The equations for mechanical parameters are straightforward

Conclusions

since matrices' exponential is the basic primitive for the differential geometric formulation.

The screw theory algorithms give exact solutions that do not have the most popular numerical solutions' convergence problems. As an example of these geometric algorithms' advantages, we have seen how IK's resolution with the POE provides multiple exact solutions. That is impossible with a numerical approach, which points only to one solution, which, besides, is many times only an approximation.

When working with velocities, the screw theory provides the great advantage of solving Differential Kinematics (DK) without differentiation with the geometric Jacobian.

The classical screw theory and the extending constructs of Featherstone, with the introduction of the Spatial Vector Algebra, showed how to formulate the dynamics of rigid multibody systems with algorithms of remarkable efficiency. Besides, we introduced an innovation giving support to the kinematics of the Recursive Newton–Euler Algorithm (RNEA) with the POE.

In short, the screw theory algorithms described throughout the chapters of this text are a handy tool for real-time robotics applications. Hereafter, we review some of the main messages of the different chapters of this book. For those who have followed the book exercises, these will be good reminders.

9.1.1 Introduction

The study of mechanical systems has sparked interest in the scientific world for millennia. There are examples of this from the ancient Greek world, through medieval times and the Renaissance (e.g., see Figure 9.1 with a bioengineering analysis by Leonardo), up to the commercial era of industrial robotics in the 20th century.

During the last decades, numerous researchers have applied differential geometric methods to approach the study of rigid body chains, the primary mathematical tool being the Theory of Screws and Lie algebras. Some mathematicians presented an excellent introduction to Lie special Euclidean group and its algebra. Furthermore, they showed the geometric meaning of these theories by relating them to theory of

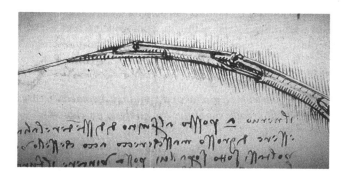

FIGURE 9.1 A bioengineering study from a codex by Leonardo da Vinci. (Photograph by the author.)

screws. A crucial step in the contemporary revival of these theories came with applying Lie algebras to robot kinematics by introducing the POE.

The complexity of robot mechanics makes some elegant mathematical expressions a paramount issue to build up efficient solutions (Siciliano and Khatib, 2016). Many developments employ the Newton–Euler equations with recursive algorithms to deal with the mechanics of rigid body open chain. However, geometric formulations based on the Lagrange equations are also effective. Anyway, almost all state-of-the-art techniques use the essential screw theory representation of six-dimensional vectors for kinematics and dynamics or some expansions of these concepts.

These screw theory formulations and some extensions lead to numerically stable solutions and unambiguous geometrical interpretation. Therefore, this approach is the most suitable for real-time applications.

A truly advanced humanoid robot would possibly be the best system to serve humans and society well.

9.1.2 Mathematical Tools

We introduce the robot's mathematical description as a rigid multibody mechanism and shortly review the standard mathematical homogeneous representation.

There is a brief historical itinerary for the development of the screw theory: first in kinematics, with an introduction to the screw concept of Chasles's theorem, to demonstrate that a translation along a line (screw axis) followed (or preceded) by a rotation about that line can produce the most general rigid body motion; and second in dynamics, with Poinsot's theorem, to demonstrate that any system of forces acting on a rigid body can be replaced by a force along a line (screw axis) followed (or preceded) by a torque about that line (Murray et al., 2017).

There is an introduction to some new screw theory concepts: "Screw," "Twist - ξ," and "Wrench - $\mathcal{F}$." As a critical tool, we present the screw exponential, which equates to the homogeneous representation. The natural extension for the robotic multibody systems mechanics is the POE fundamental formulation (Brockett, 1983). These screw theory concepts apply to the most characteristic articulations used in robotics, pure translation prismatic joints (Figure 9.2a) and pure rotation revolute joints (Figure 9.2b).

There are two exercises solved with both homogeneous and screw theory representations. It illustrates the use of the two mathematical tools introduced in this chapter.

The Screw Theory provides a global and genuinely geometric representation of mechanics, which greatly simplifies the analysis for robotics.

9.1.3 Forward Kinematics

The concept of Forward Kinematics (FK) describes the relationship between the joints' motion and the motion of the rigid body chain that forms the robot. More

Conclusions

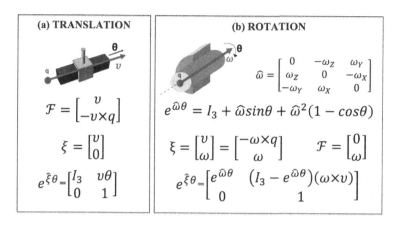

FIGURE 9.2 (a) Basic screw theory formulations for prismatic joint. (b) Basic screw theory formulations for revolute joint.

precisely, the FK problem's interest is to obtain the configuration or pose (i.e., position and orientation) for the tool or end-effector, given the motion of the robot joints (i.e., magnitudes $\theta_1...\theta_n$). The screw theory POE formalizes the robot joints' FK (see example in Figure 9.3) and has many advantages and benefits.

- There is no fixed rule to define the spatial and tool coordinate systems.
- There is no need to define a coordinate system for each robot link.
- To define the kinematics of the robot, the only necessary information to identify is for each joint, the axis, and any point on the rotational axis.
- The twists have an exact geometrical meaning. In any case, with identifying the joint axis and any point on it, the twists' definition is immediate.
- It is very convenient to solve the FK mapping, with the product of all joint screw exponentials and the tool configuration at home position.

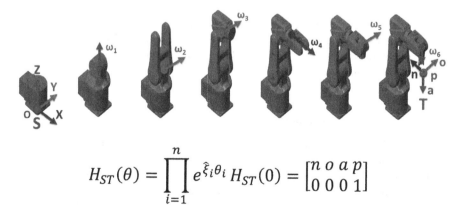

$$H_{ST}(\theta) = \prod_{i=1}^{n} e^{\hat{\xi}_i \theta_i} H_{ST}(0) = \begin{bmatrix} n & o & a & p \\ 0 & 0 & 0 & 1 \end{bmatrix}$$

FIGURE 9.3 The Forward Kinematics manipulator analysis with Product of Exponentials (POE).

There are some FK examples with typical robot architectures applied with commercial manipulators: ABB IRB120, ABB IRB1600, ABB IRB6620LX, ABB IRB910SC, UNIVERSAL UR16e, and KUKA IIWA.

The Forward Kinematics always has a unique solution, and the essential Products of Exponential formulation makes easy the treatment of the resulting robot motion equations.

9.1.4 Inverse Kinematics

The Inverse Kinematics (IK) problem must obtain the joint magnitudes, which, once applied, make the robot tool achieve the desired configuration (i.e., position and orientation). The analytical difficulty of IK is quite significant, as the problem can have none, one, or multiple solutions out of a system of nonlinear coupled equations.

It is possible to develop geometric algorithms using the POE to solve IK for complex mechanisms with many DoF. The method's cornerstone is the availability of several canonical subproblems numerically stable and geometrically meaningful. To solve the IK complexity, we must reduce the entire problem into appropriate canonical subproblems (Paden, 1986). There are three already classical Paden–Kahan (PK) canonical subproblems (Kahan, 1983). In addition, we present eight innovative Pardos-Gotor (PG) canonical subproblems for different geometries (see examples for PG7 and PG8 in Figure 9.4). Besides extending this approach, it is possible to create new subproblems (Pardos-Gotor, 2018). Moreover, the screw theory formulations provide the complete set of possible exact solutions, opening alternatives to choose the better outcome according to the application.

We showed some systematic, elegant, and geometrically meaningful solutions for the IK of some well-known robotics architectures: ABB IRB120, ABB IRB1600, ABB IRB6620LX, ABB IRB910SC, UNIVERSAL UR16e, and KUKA IIWA. For some robots, we have several equivalent algorithms, as the approach is not limited.

There are some systematic, elegant, and geometrically meaningful solutions based on the POE for solving the Inverse Kinematics with efficient and effective algorithms based on Canonical Subproblems giving the complete set of exact closed-form solutions.

9.1.5 Differential Kinematics

Differential Kinematics (DK) defines the relationship between the end-effector's velocities (linear and angular) and joint velocities. Central to DK is the Jacobian concept as the operator relating those velocities for both the forward and inverse DK problems.

A critical idea is a difference between the concepts of analytic and geometric Jacobian. The geometric screw theory Jacobian is obtained directly without any differentiation, using the joints' twist (Figure 9.5a) and the adjoint transformation (Figure 9.5b). We introduce more screw theory terminology as the "Velocity Twist"

Conclusions

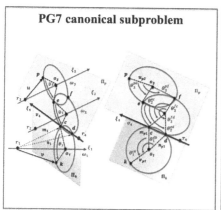
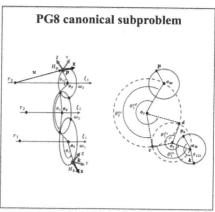

FIGURE 9.4 Two examples of Pardos-Gotor Canonical subproblems (i.e., PG7 and PG8).

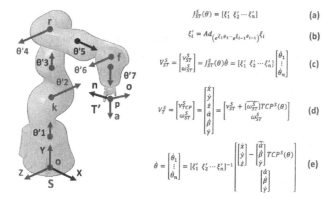

FIGURE 9.5 KUKA IIWA DK schematic. (a) Geometric Jacobian. (b) Mobile twist definition. (c) Twist velocity. (d) Forward Differential Kinematics. (e) Inverse Differential Kinematics.

(Figure 9.5c). With all these concepts, it is possible to solve the forward DK (Figure 9.5d) and inverse DK (Figure 9.5e) with algebraic formulations and a precise geometric meaning.

We introduce the concept of robot singularity, which corresponds to those configurations at which the Jacobian matrix drops rank. Singular configurations can demand unacceptable velocities for some joints, eventually dangerous for the mechanism. Therefore, the robot motion design must avoid any singular configuration.

There are some DK examples with typical robot architectures applied with commercial manipulators: ABB IRB120, ABB IRB1600, ABB IRB6620LX, ABB IRB910SC, UNIVERSAL UR16e, and KUKA IIWA.

We can also use the Jacobian to solve IK when we do not have and closed-form solution by integrating joint velocities. This feature will be handy also for trajectory generation as shown in Chapter 7.

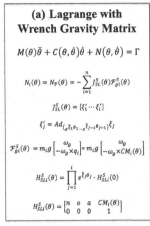 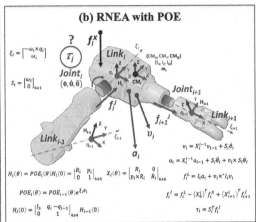

FIGURE 9.6 Inverse dynamics with Gravity Wrench matrix for Lagrange formulation and RNEA with POE schematic.

The Differential Kinematics treatment of velocities is more operative with the very natural and explicit description of the Geometric Jacobian without differentiation and has none of the drawbacks of the local analytic representation.

9.1.6 Inverse Dynamics

The subject of robot dynamics is still an open area of research because of the complex mechanical structures. Our goal is to provide robot Inverse Dynamics (ID) algorithms that rely on methods from the screw theory and some of its extensions. The approach is geometric, with many advantages to capture the robot's physical features.

The first formulation presented is the classical of Lagrange with a closed-form solution based on screw theory. This approach counts with three novel expressions for the Potential or Gravity matrix. Two of which are complete geometric formulas (e.g., see Gravity Wrench matrix definition in Figure 9.6a). They allow completing the screw theory ID with a total geometric and algebraic expression of guaranteed convergence.

Featherstone introduced the spatial vector algebra extending the classical screw theory (Featherstone, 2016). He presented the RNEA to solve the ID problem. We introduce an innovation that uses the POE for the kinematics analysis of the RNEA (Figure 9.6b). This idea provides flexibility and clarity. The efficiency of this recursive algorithm is much better than the classical approach by some orders of magnitude (Sipser, 2021).

Both perspectives support the ID exercises for some typical robotics architectures applied to commercial manipulators: ABB IRB120, ABB IRB1600, ABB IRB6620LX, ABB IRB910SC, UNIVERSAL UR16e, and KUKA IIWA.

Conclusions

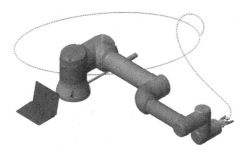

FIGURE 9.7 Manipulator target end-effector trajectory generation and TCP tracking in a 3D model.

The Inverse Dynamics problem is solved with two effective screw theory algorithms, the non-recursive Lagrange with the new Gravity Wrench Matrix and the RNEA with the Product of Exponentials.

9.1.7 Trajectory Generation

A tool trajectory specifies as a function of time the robot end-effector path (i.e., set of pose configurations). We can see the trajectory as a combination of a path, a purely geometric description of the sequence of poses, and the timing between configurations. Then, the joint trajectory planning and generation consist of constructing the path plus time scaling so that once applied to the robot actuators, the tool reaches the desired configurations.

The joint trajectory planning has a clear relationship with the solution for the IK and DK problem (Lynch and Park, 2017). An IK geometric solution for any tool target can generate a joint trajectory planning formed by a series of points and the time scaling to develop the necessary position, velocities, and accelerations. Another way to create the joint trajectory is to use the velocity of the robot joints at a specific configuration and integrate the inverse DK solution to give the joints path increments. Afterward, the method repeats this recursive process for all trajectory points.

The joint trajectory generation consists of creating a time scaling for following such joint path planning, introducing constraints such as position, velocity, and acceleration. For doing that, it is possible to apply several interpolation algorithms (e.g., trapezoidal, polynomial).

There are several exercises for trajectory generation with some typical robotics architectures applied to commercial manipulators: ABB IRB120, ABB IRB1600, ABB IRB6620LX, ABB IRB910SC, UNIVERSAL UR16e (see an example of trajectory generation and tracking in Figure 9.7), and KUKA IIWA.

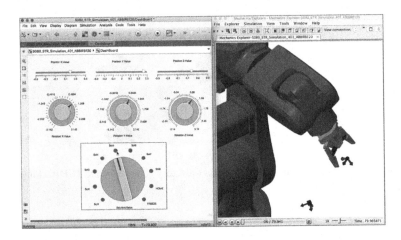

FIGURE 9.8 Simulator with a dashboard and the mechanical explorer with the 3D robot model.

For Trajectory Generation, the screw theory inverse kinematics joint trajectory planning is the right approach for significant changes in the path magnitudes, whereas differential kinematics is suitable for small changes in the path dimensions while keeping the robot configuration.

9.1.8 Robotics Simulation

The simulations included allow to explore interactively all robot exercises presented throughout this book. There are manipulators used to put to the test the theoretical screw theory formulations, such as a Puma robot (e.g., ABB IRB120) with two different home positions, a Bending Backwards robot (e.g., ABB IRB1600), a Gantry robot (e.g., ABB IRB6620LX), a Scara robot (e.g., ABB IRB910SC), a Collaborative robot (e.g., UR16e), and a Redundant robot (e.g., KUKA IIWA).

Simulators correspond to the main covered topics: Forward Kinematics, Inverse Kinematics, Differential Kinematics, and Inverse Dynamics. The general configuration of the simulators presents a command dashboard and a mechanics explorer with the three-dimensional (3D) virtual reality digital model for the robotics manipulators (see example in Figure 9.8).

The "ST24R" (Screw Theory Toolbox for Robotics) library gives software support to all theoretical backgrounds. ST24R contains the underlying code with all the functions to run the simulations.

There are links to internet GitHub (Pardos-Gotor, 2021a) and YouTube (Pardos-Gotor, 2021b) repositories for all simulations. It is possible to find, respectively, the necessary files to build and run the simulations and some videos. We choose MATLAB® to create the models because it provides a software environment practical for robotics researchers and engineers (Corke, 2017). It works intrinsically integrated with Simulink® and Simscape® packages to design and tune algorithms and generate code (MathWorks, 2021b).

The employed simulators can be sandboxes to test new algorithms, developments, and applications with the tools at our disposal

> *The Simulators are robust and work well even for big steps in the discretization of trajectories or for huge jumps between two successive targets, as can be checked with the differential kinematics and inverse dynamics simulations.*

9.2 FUTURE PROSPECTS

There is a lot of work still on finding robots that are extensively useful to our society. There are not yet commercial robotics platforms with truly mass distribution. As a modest contribution to improving this situation, we hope the elegance of the screw theory in robotics underlying this text's works can encourage further progress in this line of research in defense of geometric approaches to solving the mechanical problems of robotics.

In the field of robot dynamics, we propose to include other essential concepts, such as dynamics in the workspace, interaction with the environment, integration of constraints in dynamics, or dynamics control using screw theory and Lie algebras tools. Sometimes, control and regulation need to be very sophisticated because the robot's dynamic model is never perfect and is affected by disturbances of many kinds.

There is a plan for the extension of the "ST24R" software library, including new functions, classes, and algorithms, and the migration to other programming languages that may be of general interest, such as C ++ or Python. Additionally, integrate everything into ROS.

We feel that the technological progress in the years to come will include the incursion of screw theory in some other fields beyond industrial manipulators. Some of them are Virtual Reality, Dexterous Manipulation, Robotics in Hazardous Environments, Underwater and Construction Robots, Agriculture Robotics, Drones, Micro-robotics, Soft Robots, or Medical applications. The growth potential of these areas is enormous. For many of these specialized fields, the technology of multi-fingered robot hands and manipulation could be crucial.

An essential work for the future of robotics will deal with Computational Intelligence. It will be paramount to take advantage of the screw theory fundamentals for the leading research investigating the role of artificial intelligence in robotics. Other cutting-edge technologies might be considered, for example, evolutionary computation, morphological analysis, machine learning, and pattern recognition for robotics.

The geometric approach advantages have been recognized for some years by practitioners of screw theory and its extensions. This mathematical point of view is genuinely innovative, and we trust in its potential development for a great future of robotics technology. We intend to transmit our enthusiasm for screw theory to a broader audience.

Most of these progressive ideas were out of this book's scope and are left, with some time and fortune, for future editions of this book.

Epigram

"Intellige ut credas, crede ut intelligas"

—Augustinus Hipponensis

References

Abraham, R.A., & Marsden, J.E. (1999). *Foundations of mechanics*. Perseus Publishing.

Agahi, D., & Kreutz-Delgado, K. (1994). A star topology dynamic model for efficient simulation of multilimbed robotic systems. In *Proceedings IEEE International Conference on Robotics and Automation*, 352–357, San Diego, CA, USA.

Arbulú, M., Pardos-Gotor, J.M., Cabas, L., Staroverov, P., Kaynov, D., Pérez, C., Rodriguez, M., & Balaguer, C. (2005). Rh-0 humanoid full size robot's control strategy based on the Lie logic technique. *Proceedings of 2005 5th IEEE-RAS International Conference on Humanoid Robot*. https://ieeexplore.ieee.org/document/1573579

Ball, R.S. (1900). *A treatise on the theory of screws*. Cambridge University Press.

Barrientos, A., Peñin, L.F., Balaguer, C., & Aracil R. (2007). *Fundamentos de Robótica*. McGraw Hill.

Bloch, A.M. (2003). *Nonholonomic mechanics and control*. New York: Springer.

Brockett, R.W. (1983). Robotic manipulators and the product of exponentials formula. In *Proceedings of International Symposium on Mathematical Theory of Networks and Systems*, Beer Sheba, Israel.

Brockett, R.W., Stokes, A., & Park, F. (1993). A geometrical formulation of the dynamical equations describing kinematic chains. *Proceedings IEEE International Conference on Robotics and Automation*, 2, 637–641, Atlanta, GA, USA.

Bullo, F., & Lewis, A.D. (2004). *Geometric control of mechanical systems*. Springer.

Ceccarelli, M. (2000). Screw axis defined by Giulio Mozzi in 1763 and early studies on helicoidal motion. *Mechanism and Machine Theory*, 35, 761–770.

Chen, Q., Zhu, S., & Zhang, X. (2015). Improved inverse kinematics algorithm using screw theory for a six-DoF robot manipulator. *International Journal of Advanced Robotic Systems*, 12(10):1.

Choset, H., Lynch, K.M. (2005). *Principles of robot motion: Theory, algorithms, and implementations*. Cambridge, MA: MIT Press.

Corke, P. (2017). *Robotics, vision & control: Fundamental algorithms in MATLAB*. Springer.

Craig, J. (2004). *Introduction to robotics: Mechanics and control*. Prentice-Hall.

Davidson, J.K., & Hunt, K.H. (2004). *Robots and screw theory*. Oxford University Press.

Denavit, J., & Hartenberg, R.S. (1955). A kinematic notation for lower-pair mechanisms based on matrices. *Journal of Applied Mechanics*, 23, 215–221.

Dimovski, I., Trompeska, M., Samak, S., Dukovski, V., & Cvetkoska, D. (2018). Algorithmic approach to geometric solution of generalized Paden–Kahan subproblem and its extension. *International Journal of Advanced Robotic Systems*, 15(1), 1729881418755157.

Duffy, J. (1990). The fallacy of modern hybrid control theory that is based on "orthogonal complements" of twist and wrench spaces. *Journal of Robotic Systems*, 7(2): 139–144.

Featherstone, R. (2016). *Robot dynamics algorithms*. Springer.

Greenwood, D.T. (2006). *Advanced dynamics*. Cambridge University Press.

Husty, M.L. (1996). An algorithm for solving the direct kinematics of general Stewart-Gough platforms. *Mechanism and Machine Theory*, 31(4), 365–380.

Jurdjevic, V. (1997). *Geometric control theory*. Cambridge University Press.

Kahan, W. (1983). *Lectures on computational aspects of geometry*. Department of Electrical Engineering and Computer Sciences, University of California, Berkeley. Unpublished.

Khalil, H.K. (2014). *Nonlinear control*. Pearson.

Khatib, O. (1987). A unified approach for motion and force control of robot manipulators: The operational space formulation. *IEEE Transactions on Robotics and Automation*, RA3(1), 43–53.

Lee, S.H., Kim, J., Park, F.C., Kim, M., & Bobrow, J.E. (2005). Newton type algorithms for dynamics-based robot movement optimization. *IEEE Transactions on Robotics, 21*(4), 657–667.

Li, Z. (1990). Geometrical considerations of robot kinematics. *International Journal of Robotics and Automation, 5*(3), 139–145.

Lilly, K.W., & Orin, D.E. (1994). Efficient dynamic simulation of multiple chain robotic mechanisms. *Journal of Dynamic Systems, Measurement, and Control, 116*(2), 223–223.

Liu, G., & Li, Z. (2002). A unified geometric approach to modelling and control of constrained mechanical systems. *IEEE Transactions on Robotics and Automation, 18*(4), 574–587.

Lynch, K.M., & Park, F.C. (2017). *Modern robotics–mechanics, planning & control.* Cambridge University Press.

Marsden, J.E., & Ratiu, T.S. (1999). *Introduction to mechanics and symmetry.* New York: Springer-Verlag.

Martin, B.M., & Bobrow, J.E. (1997). Minimum effort motions for open chain manipulators with task dependent end-effector constraints. In *Proceedings of IEEE International Conference on Robotics and Automation.* Albuquerque, NM: IEEE.

Mason, M.T. (2001). *Mechanics of robotic manipulation.* MIT Press.

MathWorks. (2021a). *Robotics system toolbox.* https://uk.mathworks.com/products/robotics.html

MathWorks. (2021b). *Simscape multibody.* https://uk.mathworks.com/products/simmechanics.html

Millman, R.S., & Parker, G.D. (1997). *Elements of differential geometry.* Upper Saddle River, NJ: Prentice-Hall.

Murray, R.M., Li, Z., & Sastry, S.S. (2017). *A mathematical introduction to robotic manipulation.* CRC Press. (Original work published 1994)

Ohwovoriole, M.S., & Roth, B. (1981). An extension of screw theory. *Journal of Mechanical Design, 103*(4): 725–735.

Paden, B. (1986). *Kinematics and Control of Robot Manipulators.* PhD thesis, Department of Electrical Engineering and Computer Sciences, University of California, Berkeley.

Paden, B., & Sastry, S.S. (1988). Optimal kinematic design of 6R manipulators. *International Journal of Robotics Research, 7*(2), 43–61.

Pardos-Gotor, J.M. (2018). *Screw theory for robotics: A practical approach for modern robot KINEMATICS.* Amazon Fulfilment.

Pardos-Gotor, J.M. (2019). *Algo de Geometría para Humanoides.* Amazon Fulfilment.

Pardos-Gotor, J.M. (2021a). *Screw theory in robotics.* Github. https://github.com/DrPardosGotor/Screw-Theory-in-Robotics

Pardos-Gotor, J.M. (2021b). *Screw theory in robotics.* Youtube. https://www.youtube.com/channel/UC4MFs0PncA6fopjfvrmgR4Q

Pardos-Gotor, J.M., & Balaguer, C. (2005). RH0 humanoid robot bipedal locomotion and navigation using lie groups and geometric algorithms. *IEEE/RSJ International Conference on Intelligent Robots and Systems, IROS.* https://ieeexplore.ieee.org/document/1545288

Park, F.C. (1991). *Optimal kinematic design of mechanisms.* PhD thesis, Division of Applied Sciences, Harvard University.

Park, F.C. (1994). Computational aspects of the product of exponentials formula for robot kinematics. *IEEE Transactions on Automatic Control, 39*(3), 643–647.

Park, F.C., Bobrow, J.E., & Ploen, S.R. (1995). A lie group formulation of robot dynamics. *The International Journal of Robotics Research, 14*(6), 609–618.

Park, F.C., & Kim, J. (1999). Singularity analysis of closed kinematic chains. *ASME Journal of Mechanical Design, 121*(1), 32–38.

Penrose, R. (1955). A generalized inverse for matrices. *Proceedings of the Cambridge Philosophical Society, 51*(3), 406–413.

References

Ploen, S. (1997). *Geometric algorithms for the dynamics and control of multibody systems*. Ph. D. Thesis, University of California, Irvine.

Raibert, M.H., & Craig, J.J. (1981). Hybrid position/force control of manipulators. *ASME Journal of Dynamic Systems, Measurement, and Control, 102*, 126–133.

Rodriguez, G., Jain, A., & Kreutz-Delgado, K. (1991). A spatial operator algebra for manipulator modelling and control. *International Journal of Robotics Research, 10*(4), 371–381.

Sastry, S.S. (1999). *Nonlinear systems: Analysis, stability, and control*. New York: Springer-Verlag.

Selig, J.M. (2005). *Geometric fundamentals of robotics*. Springer-Verlag.

Siciliano, B., & Khatib, O. (2016). *Book of robotics*. Springer.

Siciliano, B., Sciavicco, L., Villain, L., & Oriole, G. (2009). *Robotics: Modelling, planning and control*. Springer.

Sipser, M. (2021). *Introduction to the theory of computation*. Course Technology Inc.

Stokes, A., & Brockett, R. (1996). Dynamics of kinematic chains. *The International Journal of Robotics Research, 15*(4): 393–405.

Strang, G. (2009). *Introduction to linear algebra*. Wellesley-Cambridge Press.

Tsai, L.W. (1999). *Robot analysis*. Wiley-Interscience.

Yue-Sheng, T., & Ai-Ping, X. (2008). *Extension of the second paden-kahan sub-problem and its' application in the inverse kinematics of a manipulator*. IEEE Conference on Robotics, Automation and Mechatronics, Chengdu, China.

Index

A

ABB IRB120, 35–36
  differential kinematics, 126–127, 135–138
  forward kinematics, 38–40
  inverse dynamics Lagrange, 161–165
  inverse dynamics RNEA, 180–185
  inverse kinematics, 82–90
  trajectory generation, 199–203
ABB IRB1600
  differential kinematics, 139–140
  forward kinematics, 41–42
  inverse dynamics Lagrange, 166–167
  inverse dynamics RNEA, 186–187
  inverse kinematics, 94–96
  trajectory generation, 204–207
ABB IRB6620LX
  differential kinematics, 140–141
  forward kinematics, 42–43
  inverse dynamics Lagrange, 166–167
  inverse dynamics RNEA, 187
  inverse kinematics, 96–99
  trajectory generation, 207–208
ABB IRB910SC
  differential kinematics, 142–145
  forward kinematics, 44–45
  inverse dynamics Lagrange, 168
  inverse dynamics RNEA, 188
  inverse kinematics, 99–102
  trajectory generation, 208–209
adjoint transformation, 128, 131
angular velocity, 129, 197
approach vector, 22

B

base frame, 36, 154
bending backwards robots, *see* ABB IRB1600
body frame, 179

C

CAD, 216
Cartesian, 193
center of mass, 154–155, 157–158, 176
Chasles's theorem, 266
Christoffel symbols, 155
closed-form solution, 53–56
collaborative robots, *see* UNIVERSAL UR16e
computational performance, 89–90, 248
computed torque control, 170–171, 249
controller, 170
Coriolis matrix, 155
cross product, 24
  of force spatial vectors, 178, 184
  of motion spatial vectors, 177, 183

D

degrees of freedom, 17, 51
Denavit-Hartenberg, 33–36
dexterous
  manipulation, 120, 273
  workspace, 56, 106–109
differential equations, 3, 151
differentiation, 120–126, 132–134
discretization, 54, 198–199
dynamics, 151–153
  inverse, 158–161, 174–180
  simulation, 247–250

E

effectiveness, 53–56, 178–180
efficiency, 53–56, 178–180
end-effector, 21–22
  configuration, 119
  path, 198–199
  pose, 120
  target, 225–228
  trajectory, 197–199
  velocity, 127–129
Euler, 80, 151–152, 174
  angles, 27
  orientation, 198
  rotation, 197
exponential, 29–30
  coordinates, 26–29
  map, 37–38
  matrix, 80
  representation, 24, 38–40
  screw, 27–29, 80–82

281

F

feedback, 221
  linearization, 170–171
force, 151
  control, 4
  spatial vector, 172–173
friction, 153

G

Gantry robots, *see* ABB IRB6620LX
generalized
  coordinates, 152
  forces, 24, 155
  velocities, 155
geometric solutions, 53–55, 158
global
  characterization, 153
  representation, 80
gravity, 151
  action vector, 159
  matrix, 155–157
  symbolic, 155–156
  twist, 156
  wrench, 156–157

H

homogeneous
  matrix, 18
  representation, 21–22
  transformation, 26–27

I

inertia, 151
  matrix, 154
  moment, 158–159, 176
  product, 159, 176
  tensor, 159, 176
intersecting line, 74–75

J

Jacobian
  analytic, 120–122
  geometric, 127–131
joint
  prismatic, 28–29
  revolute, 27–28
  space, 170–171
  twist, 27–29, 130

K

kinematic chain, 220
kinematics
  differential, 119–122, 131–134
  forward, 33–35
  inverse, 51–53, 80–82
  simulation, 218–219, 225–228, 236–239
KUKA IIWA
  differential kinematics, 146–147
  forward kinematics, 46–47
  inverse dynamics Lagrange, 168–169
  inverse dynamics RNEA, 189–190
  inverse kinematics, 107–113
  trajectory generation, 211–212

L

Lagrange, 152–153
Lie, 26
  algebra, 26, 131
  group, 131
  special Euclidean group, 26
link
  coordinates matrix, 35
  frame, 154

M

manipulator Jacobian, *see* Jacobian geometric
mobile
  system, 20–21
  twist, 128
motion
  spatial vector, 172
  subspace, 175

N

Newton-Euler, 174
non-recursive Lagrange algorithm, 158–161
normal vector, 22

O

open chain manipulator, 33
orientation vector, 22
orthogonal space, 24

Index

P

Paden-Kahan
  canonical subproblems, 57–58
  PK1, 59
  PK2, 60–61
  PK3, 61–63
Pardos-Gotor
  canonical subproblems, 63
  PG1, 63–65
  PG2, 65–66
  PG3, 66–67
  PG4, 67–69
  PG5, 69–70
  PG6, 70–74
  PG7, 74–77
  PG8, 77–79
passivity property, 153
PDI control, 171
Plücker coordinates, 172–173
Poinsot's theorem, 26
potential matrix, 155
product of exponentials, 36–38
Puma robots, *see* ABB IRB120

R

recursive Newton-Euler algorithm, 174–180
redundant robots, *see* KUKA IIWA
reference configuration, 36
relative motion, 27
rigid body motion, 17
RNEA, *see* recursive Newton-Euler algorithm
Rodrigues's formula, 24
rotation around one single axis
  applied to a line, 69–70
  applied to a plane, 69–70
  applied to a point, 59
rotation around three axes
  applied to a point, 74–77
  applied to a pose, 77–79
rotation around two subsequent axes
  crossing, 60–61
  parallel, 67–69
  skewed, 70–74
rotation at a given distance, 61–63
rotation matrix, 18

S

Scara robots, *see* ABB IRB910SC
screw, 24
  axis, 25–28
  exponential, 27–29
  motion, 26
  pitch, 25
  rotation, 24–25
  theory, 26
  translation, 29
SE(3), 26
se(3), 26
simulation, 215
  differential kinematics, 236–239
  forward kinematics, 218–222
  inverse dynamics, 247–250
  inverse kinematics, 225–228
singularity, 125, 145
skew-symmetric matrix, 24–25
SLERP, 197
SO(3), 24
so(3), 24
SpaceMouse, 229–230
spatial
  frame, 19–20
  vector algebra, 172–174
  velocity, 119
special Euclidean group, *see* SE(3)
special orthogonal group, *see* SO(3)
stability, 170
Stanford robot, 37
stationary system, 20–21

T

target, 54, 237, 248
tool
  center point, 21
  frame, 36
  system, 21
trajectory
  generation, 193–197
  planning, 197–199
translation along a single axis, 63–64
translation along two subsequent axes, 65–66
translation at a given distance, 66–67
twist, 24–25
  rotation, 27–28
  translation, 28–29
  velocity, 129

U

UNIVERSAL UR16e
  differential kinematics, 146

forward kinematics, 45–46
inverse dynamics Lagrange, 168–169
inverse dynamics RNEA, 188–189
inverse kinematics, 102–107
trajectory generation, 210–211
URDF, 216

V

vector
  approach, 22
  displacement, 18
  normal, 22
  orientation, 22
  velocity, 119–120, 129
  twist, 129

W

workspace, 170
wrench, 26–29
  gravity, 156–157
  rotation, 27–28
  translation, 28–29

Printed in the United States
by Baker & Taylor Publisher Services